No. 2806
$19.95

HOW TO
PLAN, CONTRACT AND BUILD YOUR OWN HOME

RICHARD M. SCUTELLA AND DAVE HEBERLE
ILLUSTRATIONS BY JAY MARCINOWSKI

TAB TAB BOOKS Inc.
Blue Ridge Summit, PA 17214

FIRST EDITION
FIRST PRINTING

Library of Congress Cataloging in Publication Data

Scutella, Richard M.
How to plan, contract, and build your own home.

Includes index.
1. Dwellings—Design and construction—Handbooks,
manuals, etc. I. Heberle, Dave. II. Title.
TH4811.S39 1987   690'.8   86-30024
ISBN 0-8306-1906-2
ISBN 0-8306-2806-1 (pbk.)

Contents

Part 2: How to Build It   63

Introduction

We live in the age of information. We know more about practically everything. Motion pictures and television bring political unrest, the successes and tragedies of space exploration, the jungle, and treasures up from the ocean depths into our living rooms. We know more about more.

To acquire such a cosmopolitan array of information, we've had to trade off much of the basic knowledge that our fathers and their fathers and grandfathers had once known. Granted, they had learned such knowledge not by choice, but by necessity. A few hundred years ago, for example, people grew their own food, doctored their own sick, and built their own homes—with their own hands. They took care of all their basic needs by themselves.

Thanks to the collective progress realized over the past few centuries, much of that all-around knowledge has become of little use to the average person. How many people must still be able to shoe a plowhorse, dig their own well, deliver their own babies, or even teach their own children how to read and write?

Today, if you're an accountant, you're an accountant. You have to keep up with an avalanche of changing accounting information to service your clients. If you're a farmer, you're a farmer, and you have to sub-scribe to the latest agricultural techniques to be a success. Or, if you're a builder of houses, you're a builder of houses.

If you can't perform in whatever business you're in, and if you don't really know the ins and outs of the trade, you'll be supplanted by other professionals who do. Specialists who keep up to date with the evolving nature of their business will acquire the competitive edge needed to stay ahead of their peers. Consequently, people no longer have the time, need, or inclination to acquire many of the basic skills our forefathers found unavoidable.

Take the subject of houses, for example. Because the public has largely turned the job of housebuilding over to a group of professional builders, not many people really understand exactly what a house consists of, or how all the parts must mesh together to make a satisfactory dwelling.

And that's the reason for this book. The idea is simple enough: by understanding houses, you can better arrive at an understanding house—one that suits you perfectly, one that is constructed to give maximum enjoyment and value with minimum investment.

Everybody has to live somewhere; that's a fact. You can live at someone else's place for free, with parents or benevolent friends. Or you can make do at a place provided by an accommodating employer—government or otherwise. You can reside at someone else's place and pay rent. Then again—you can choose to live in something you own. And at some point in their lives, most people prefer the latter option. This book has been written for people in search of a modern home situated in either a subdivision down the street or on five acres out in the boondocks. Although it focuses on new construction, it also proposes what to look for in existing houses, new or old.

Without knowing better, people considering the purchase of a new house will approach one or more builders with a sketch and floorplan ripped out of a magazine. Then they'll ask the builder how much he'll charge to complete that same house for them. Depending on the amounts and quality of materials used, or rather, depending on what specifications or "specs" are followed, the cost of a 2,000-square-foot two-story house can vary by tens of thousands of dollars. By providing only a simple generic floorplan as a building guide, eager buyers overlook important pre-construction choices that should, in fact, be made by the buyers themselves. Instead of exercising their own wishes in the planning and construction of their new house, these trusting buyers leave consequential decisions completely up to the builder.

Another irony is that, when faced with such a complex and expensive task as the purchase of a new house, most people remain interested merely in the basic floorplan, the amount of square footage of living space, the appearance of the exterior sheathing, and other incidentals such as the color of the flooring, carpeting, and the style and stain of kitchen cabinets. Rarely do they care what thickness the outer walls will be. Or on which side of the house the garage should be located. They leave detail after detail to the builder's discretion.

Naturally, in his competitive business, the builder provides what the buyer wants in a manner most advantageous to the builder himself. He tends to use materials that he's been using on other houses he's built, and he probably gets them in bulk, at a discount. Unless otherwise requested, this usually means he'll provide the absolute minimum specs needed to satisfy local building codes even when, for a few dollars more, substantial long-range savings and additional conveniences can be enjoyed by the buyers.

Most minimum building codes can be satisfied with economy-grade materials. "Economy grade" is a mis-leading term at best, because materials in that class actually end up costing *more* than materials of high quality, due to frequent maintenance and repairs needed, and shorter lifespans. That can mean early replacements, all at extra cost and inconvenience to the homeowner. And marginal products, because they're less durable, can lead to a very annoying and even uncomfortable house. Unfortunately, marginal quality materials are found not only in low-priced houses, but in many high-priced dwellings as well.

High-quality flooring, paneling, wiring, heating, and many other products cost only a little more, by and large, than the same products of marginal quality. Certainly the installation costs are about the same in either case; an identical amount of labor is required to put down a new floor of the best or worst vinyl. All things considered, studies indicate that the initial price of a house built with high-quality materials will run only about 8 to 10 percent more than an identical economy-grade dwelling.

The argument to go with quality materials is a persuasive one. But even before that comes the question: do you want to build a new house or move into an existing one?

It's entirely possible to find an older or recently built house that meets most of your needs and is constructed similar to the guidelines described in the following chapters. Such houses can be difficult to locate, but in many cases, due to circumstances of the present owners, you can get more house for the money, but not without certain tradeoffs: the house might be in a location you don't like, or it might have everything you asked for *except* a basement, a den, or a two-car garage.

With new construction, you have the opportunity to custom design your own house. With new construction you'll end up with more built-in conveniences, with better insulation and more energy-efficient appliances and heating/cooling systems. There are also less repairs to worry about and less time spent maintaining things, especially with many of today's maintenance-free items. New homes generally require lower down payments, with better financing terms available to owners. Kitchens can be loaded with modern appliances and built-ins. There are better roofing materials, flooring, easy-care carpeting, windows, and exterior sheathings. And it's also a nice feeling to move into a house knowing that you and your family will be the first to live there. It has that wonderful sweet new-house smell of sawdust, plaster, and carpeting.

New homes are clean to the eye and touch, and hold their value well when soundly constructed. Everything is under warranty. You know exactly what went into the place, having periodically inspected its erection. You know that it's a structurally sound dwelling, built with the latest high-quality materials.

No matter what you decide to go with—new construction or buying an existing home—buying any house is a major investment. Let's face it; finding or arriving at the right house for you can be a tough (though enjoyable) process. It's not like buying a stereo, for example, where you can walk into a stereo store and listen to different brands and different models, turn them on, one after another. You can pick up a Consumer Reports and read reviews on them. You can shop for the same models all over town.

But a house? The two biggest words in real estate have always been "buyer beware," and for no small reason. There are endless possibilities open to house buyers and endless pitfalls.

A house is a lot more than a "hedge against inflation" or a pleasant alternative to paying rent. And sure, owning property will probably give you the urge to become involved with the community, and will probably encourage you to send out roots. But most of all, a house is a dwelling in which human drama unfolds. Children grow up in them. Marriages and other relationships flourish or flounder in them, in unique environments created within the home's outer shell. In homes we grow up, learn things, spend time together, eat, play, party, laugh, fight, cry, make love, pray, and entertain one another. Homes reflect our personalities and uniqueness, and they deserve to be acquired as the result of careful deliberation instead of happenstance.

Everybody has to live somewhere. If you have a say in the matter, then exercise it. Get involved with selecting what you're going to live in. It only makes sense. Why settle for less? Why put up with a building that doesn't meet your and your family's basic requirements? Especially if within those same means you have the ability to attain a dwelling far superior, with an optimal plan custom tailored to your needs.

Talking new construction, you have to familiarize yourself with building jargon (it's not tough) and prepare yourself for entering into a close and beneficial relationship with a general contractor who builds houses for a living. You must understand the advantages and disadvantages of the various types of homes available in order to make an educated choice: ranch, Cape Cod, two-story, split foyer or split-level. You should know how to arrange the rooms you choose for the greatest convenience of both initial construction and everyday living.

Beyond that, you should figure out if you might want to enlarge the house at a later date. If so, a few relatively minor modifications up front can mean a lot of savings later. You need to know how to match the home you want to a building site. And you need to know additional do's and don'ts along the way. In fact, in many cases, just knowing what *not* to do will lead you to a correct choice. It's a complex mix of qualities and factors that can make a difference between ending up with an extremely pleasant and valuable home, or a disaster.

But don't worry. Again, it's not that hard to acquire a lot more house than you ever thought you could afford—one built with conveniences and quality not usually included with your neighbor's house. And it will be a lot of fun.

To get the most out of this book, first read the table of contents for a general idea of what's covered. Then take it a chapter at a time, section by section, at a leisurely pace.

After you've read the entire book, proceed, if you're ready, full of the confidence that adequate preparation develops. Arrive at the house you want—in your mind and on paper. Use the checklists and summary points throughout the book so you don't ignore any important considerations.

While it's true that most real estate and house-building books include sections about mortgage financing, and are full of examples of such, this book does not. You'll find no references to interest rates or prices of new and existing houses. That's because any volume declaring itself an authority on financing (or even using examples within the financial realm) runs a risk of becoming hopelessly outdated even before the book is out of the printer's door. So to avoid that embarrassment, this book defers financial advice to establishments you'll have to approach anyway: your community banks and savings and loans. Find out what alternatives you have by going directly to the lending officers. That's their job, and you'll find their information to be more current and updated than any you'd read elsewhere.

Determine what construction costs are running by going to open houses and by asking builders and real estate agents you'll meet at those open houses. Realtors can provide valuable and free financial information, perhaps a helpful summary of mortgage rates in your area. And they can be instrumental in putting together packages requiring creative financing (especially when a sale hinges upon you getting the loan).

Together, these people can tell you how much money the bank will provide, and how much money the bank thinks you can afford based on your dependents, income, work history, credit, stability, and shoe size.

Now let's get into it. Let's begin with some style. On to the homework!

Part 1

What to Build

To arrive at the best possible home for yourself, first take a step back—away from houses and house construction in general— and review some basic concepts about what a home can and should contain. You need to think about the pros and cons of the various house styles and types available, and how you can arrive at floorplans that will most favorably suit your present and future needs. You should realize that features can be built into a new house, at minor cost, that will permit efficient and convenient expansion of living space years later. And you should also be mildly familiar with typical house construction plans, prints, and drawings.

It's best to arrive at your "ideal" house on your own, with the help of books or articles, before approaching builders who will naturally, even with good intentions, lead you toward house types they prefer or think that you should prefer. But if you already have a building site picked out, and your brother-in-law happens to be a builder, so be it. Still read the chapters about what to build. If your initial plans (or those of your brother-in-law) are correct, then you'll be able to proceed full of confidence. On the other hand, if you discover that a different type of house would be better for you, don't let anyone tell you otherwise.

Don't be one of the thousands of homebuyers who are just on the receiving end of a new house, passively accepting whatever happens along. Let there be conscious reasons for everything you end up with.

Chapter 1

A House Divided

The modern house consists of at least six main types of areas:

■ Living and Recreation Space. In any one dwelling, this can include a living room, family room, den, library, music room, sunroom, or other rooms and niches planned for entertainment, relaxation, hobbies, or study.

■ Food Preparation Space. The kitchen ranks number one here, followed by additional food storage areas such as nearby walk-in pantries and "satellite" serving platforms, bars, and grills conveniently located in family rooms and other living and recreational spaces.

■ Dining Space. Depending on the floorplan, this can be a separate formal dining room adjacent to the kitchen, or an open dining area having the kitchen on one side and the living room or family room on another, or—when economy is desired—a portion of the actual kitchen itself can be reserved for a dining table or booth.

■ Bathing and Washing Space. This means bathrooms, both full and half, plus toilets, hand sinks, showers, saunas, and indoor hot tubs.

■ Sleeping and Dressing Space. Bedrooms, dressing nooks, and related storage areas.

■ Service and Storage Space. Everything else in a house fits in here: basement, attic, stairs, hallways, laundry rooms, and garages.

LIVING AND RECREATION SPACE

We've come a long way since our primitive ancestors stalked through forests in search of daily sustenance. Indeed, prehistoric life was tough. There was no plumbing, no cable television, no supermarkets, doctors, or economists. Instead you had one continuous struggle against the elements, with little time to do anything but attempt to satisfy the most basic of needs. And even if the typical primitive *had* the time, he still wouldn't have known how to put a formal living room to good use.

Okay, so he did have a den—in the strictest sense of the word (Fig. 1-1). Back then a shelter was just that—a few walls against which to huddle. There was a roof for protection from rain, sleet, and snow, a place early man could hunker down in, relatively safe from predators.

Even though our primitive forerunners had the protection of rough shelters, they were still lucky just to

Fig. 1-1. A primitive den.

make it through their teens in one piece. In fact, as late as the early 1900s, the expected American lifespan was only 46 years for men and 49 for women. That didn't leave much free time for recreation.

Thanks to revolutionary medical and pharmaceutical advances, look how long people are living today. Consider also, especially during the latter half of this century, how work weeks have grown shorter and family sizes smaller. Automobiles, trucks, and aircraft have given us mobility, and sped up the delivery of time-saving conveniences to our doors. New leisure time has created a demand that greater attention be paid to the living and recreational space in modern dwellings.

Simply put, a house's living and recreation rooms should contain enough space to satisfy the needs of its occupants. It means enough space for general and specific leisure activities such as entertaining friends, watching television, listening to music, practicing musical instruments, gaming, reading, studying, writing, bookkeeping, children's play, and plain old relaxing . . . all in comfort, and if need be, privacy.

Living and recreation rooms must also be adequate in size to hold sufficient furniture for comfortable seating. At the same time they should be large enough to permit the rearrangement of major pieces of furniture into a variety of positions while still maintaining good traffic circulation within the room.

Depending on a dwelling's size and floorplan, living and recreational space may be planned as a number of individual rooms, or may be distilled into a single, all-purpose area.

The Living Room

If one such room is all you desire, or all you can afford at the outset, that room should probably be the living room. At one time referred to as a parlor, the living room was originally designed to entertain guests in a formal fashion. There are two main types of living rooms: traditional and open.

A traditional living room is a room placed away from the kitchen and other work or recreational rooms in the house. This out-of-the-way placement allows at least two separate entertainment centers (including the kitchen in a smaller house) so that more than one family member can plan activities with outside friends at the same time, without infringing on each other's privacy. The traditional living room is usually situated at a dead end location to minimize unwanted interruptions and through traffic. Doors can be considered for additional privacy.

As mentioned earlier, if the house you plan will not accommodate other recreation rooms beyond a living room, then consider that your living room will have to be used for whatever leisure time activities you enjoy. When a single room must serve a variety of functions, a more open type of room is often the most practical choice. In fact, due to the overall reduction of square footage in recent years for cost reasons, open planning is becoming increasingly important because it lends the impression of maximum space for the money, and does, because there are less walls to take up space, give you more square footage of usable space. It serves well in the more contemporary and expensive plans, and in households having few or no children, where privacy is not as important as it would be to a large family. However, a danger in open-type rooms is that spaces must be carefully planned so not to appear jumbled and haphazardly thrown together.

The Family Room

A second recreational room that has grown popular in the past few decades is the family room. Instead of wearing out the living room furnishings, informal activities such as children's play, listening to music, and lounging on sofas while watching television can be comfortably pursued in a more casual environment.

A home that contains both a living room and a family room will typically have the main television set placed in the family room. And like it or not, television has become a mainstay in most modern households. On average, the tube is on for over seven hours per day, 365 days per year in every household that has one. Videocassette players are practically as common as toasters.

Because people want the television where groups of family members and guests can watch it comfortably, it has a definite impact on interior decoration. This means a wall or corner is effectively removed from the placement of seating and other furniture. Thus, when the room is originally sized and laid out, the television placement should be taken into account.

If you plan to have a separate family room in addition to a living room, then consider locating it next to the kitchen, where only a few steps will separate food and drink from leisure time activities, and where a parent can still work in the kitchen while supervising children at play.

Another desirable feature for a family room is an outdoors access, commonly provided by sliding glass doors that lead to a patio, terrace, or deck. Although some people add doors to the family room to seclude it from the rest of the house, most active families prefer an open-type plan allowing easy movement to and from the kitchen and fostering efficient communications between the two areas.

Fireplaces

Whether you decide upon one living room/family room combination, or multiple living and recreation rooms, one feature to consider—even in the warmer climates—is a fireplace.

When planned from the beginning, one or more fireplaces can be integrated in the house in such a manner as to save space, materials, and money. For instance, if the back of a fireplace is located inside the garage, you can reduce the number of expensive finishing bricks or stones normally needed to construct an outside wall chimney. Not only that, but instead of dissipating heat outside through an outer chimney wall, some heat will radiate into the garage, where it will do some good. The ash cleanout door can be built so it opens at a convenient height within the garage, where messy ashes and grits can be removed and disposed of in one efficient step. No more trudging across carpets, or climbing stairs.

If you plan to put a fireplace in a first floor living or family room, consider two other points: First, if your house will have a basement, do you plan on doing much socializing there? If so, you might want to include a second fireplace. Then an economical way would be to align both fireplaces one right over another.

Secondly, due to their very nature, fireplaces demand comfortable seating around them, and need freedom from internal traffic and other interference. Thus a door next to a fireplace is poor design. Anyone entering or exiting the room through such an access becomes an immediate intruder, an interruption to the conversation group. A door near a fireplace also prevents the placement of furniture on that side of the hearth, creating wasted space that in turn will effectively shrink the amount of usable space in a large living or family room down to that of a much smaller area.

Windows and Glass Doors

An important feature common to all living and recreation rooms are windows and glass sliding doors. While you should make sure that living and recreation rooms are bright and cheerful from natural light, and plenty of ventilation is provided, too much glass—especially sliding glass doors and floor-to-ceiling windows—can pose a number of irritating problems:

1. If too much wall space consists of glass there might be no place to arrange furniture unless you decide to block off some of the glass with a piece of furniture such as a sofa or plush chair.

2. If you deliberately decide to place furniture in front of glass, consider how the furniture will look from outside, too.

3. Remember that too many glass walls will severely restrict possibilities for hanging pictures and other decorative works of art.

4. When you entertain in glass-lined living or recreation rooms, large panes of glass can be distracting. During the day, people find themselves gazing out the windows instead of paying attention to the conversation, and at night huge panes of blackened glass make certain individuals uncomfortable: they feel that they're being watched from the outside.

The inclination to overuse glass is especially strong on sites having dramatic views, when the owners are naturally moved toward taking full advantage of those views. In this situation you must be careful not to get carried away. Instead, strike a happy balance between beautiful views and functional rooms.

Along the same lines, in most settings (except in rural locations where a house is tucked back from the main road), avoid an oversized picture window in the front of the house. Such a window invariably gets covered with drapes or blinds anyway—for privacy. When it's not covered, the residents feel like goldfish in a bowl. And typically, all that can be seen through the front window is passing traffic. The larger picture windows are far more productive when placed at the side or back of the house, facing a private patio, terrace, yard, or other more intimate and less "public" views.

Built-Ins

To make for a more aesthetically pleasing appearance, it's wise to include sufficient built-ins for living and recreation rooms, especially to accommodate the items you want to store there. Books, records, momentos, knick-knacks, card tables, and even fireplace wood are a few common objects to keep in mind.

Built-in bookcases and shelves are installed most efficiently when the carpenters and woodwork stainers are putting up the rest of the house trim.

Patios

Another recreation space to consider when planning a new house is a patio. Because it can easily become an important part of your home living experience, a patio deserves the same careful thought that goes into the arrangement of your interior living and recreation areas.

When pinching dollars, settle first for the foundation and concrete slab, and plan to add a roof, privacy screen, and other conveniences later. Location is a major factor for enjoyment of an open or screened-in patio or deck. Remember that an open concrete patio can get as hot as a city street when it bakes in the sun.

Other Living and Recreation Rooms

Rooms sometimes built into a house are a library, den or study, and a music or other hobby room. There are darkrooms for amateur photographers, billiard rooms for would-be Mosconis, sewing rooms and trophy rooms and rooms designed specifically for personal computers and video games. These are special areas in a house which, depending on the interests of yourself and your family, can greatly enhance the total living experience.

FOOD PREPARATION SPACE

For the bachelor who scratches his head in bewilderment while attempting to boil water for instant coffee, a kitchen complete with the appliances of his dreams might consist of a frost-free refrigerator, a microwave oven, a double-slotted toaster, and the plainest of sinks. Others, individuals who fancy themselves a step or two below award-winning French chefs, need wide expanses of counter space, double ovens, microwaves, electric grills and barbecues, three-tubbed stainless steel sinks, boxes of hand appliances and piles of pots, pans, and multi-purpose utensils.

Food is always being highlighted by the media. Sales of cookbooks and culinary magazines have reached best-selling proportions. Talk shows serve up celebrity cooks. Medical studies stress how a healthy diet can ward off heart disease, stress, and even cancer. And there's no denying that everyone has to eat. It's not something we can elect to pass up.

In any household, the primary food preparation area is the kitchen.

Kitchen Size

The overall size of your kitchen should depend on the following points:

1. The size of your family and the number of individuals in your family who like to cook. Usually, the bigger the family, the bigger the kitchen. And don't exclude the children. Psychologists say that youngsters, especially teenagers, should be encouraged to learn how to cook, and that culinary creativity helps a child's overall development.

2. Do family members and many friends and guests tend to congregate in the kitchen? Then make the kitchen large enough to accommodate plenty of seating space.

3. Do you approve of or insist upon having meals other than breakfast in the kitchen? If you prefer nightly suppers in the kitchen, then you'd better plan an eat-in kitchen arrangement, with space for a table and chairs. If only breakfast will be served there, then a bar at which three or four people can comfortably sit is likely to be all the eating space you'll need.

4. What are your shopping habits? If you prefer to go long periods of time between shopping, you'll need ample storage space for canned or packaged goods, as well as a roomy refrigerator and probably a separate freezer. Additional base and wall cabinets might be necessary, and an extra-large food pantry is a must. On the other hand, if your total food and beverage inventory at any one time is likely to consist of a six-pack of beer, a quart of buttermilk, and a few frozen TV dinners, you can get by with a lot less kitchen.

5. You might want to install a small built-in desk in the kitchen, for making out shopping lists, menus, recipes, phone call messages, and financial records. A broom closet keeps long and bulky brooms, mops, sweepers, and ironing boards out of sight in case your house doesn't include a first-floor laundry or utility room.

6. While it's true that a family's kitchen should be a direct reflection of how much that family likes to cook (Why have a big kitchen if you spend most of your time in fast food restaurants?), it should also be an indication of what kinds of cooking are preferred. A lot of baking encourages the installation of double ovens. In fact, if you have the room, consider leaving space in the form of cupboards for a second oven just in case a future potential buyer of your house finds the ability to have a double oven an attractive feature. If you do a lot of entertaining, plan for an indoor grill or barbeque. If fancy presentations are important to you, select any of the other truly marvelous food preparation aids available. Even simple items such as brackets and shelves for condiments and spices, and bookshelves for cookbooks should be carefully planned in advance so enough space is allowed. These built-ins might seem minor, but if you just ignore their placement until everything else is completed, you'll be hard pressed to neatly accommodate them. Finishing touches are often what separates the attractive, efficient kitchen from one that's awkward to work in and always appears cluttered.

Kitchen Functions

No matter what overall size your kitchen is, it still should:

1. Provide adequate working space. Any kitchen can be thought of as a combination of three work areas (Fig. 1-2). The first is food preparation, which includes counter space, utensil storage drawers and cabinets, places to store cutters, knives, food processors, chopping boards, glassware, cups, and plates. The second is food cooking, which includes range, oven, counter space, and storage areas for pots, pans, cooking utensils, seasonings, ingredients and other cooking supplies. The third is food cleaning, which includes single or dual basin sinks, counter space, dishwasher, trash facilities, and perhaps a garbage disposal.

Even the smallest kitchens should be set up so that the items stored at each work area are used for corresponding activities. Near the refrigerator is the place to keep groceries so the sandwich maker has easy access to the peanut butter, jelly, bread, and milk without walking all over the kitchen. Likewise, cooking utensils and aids are best kept by the stove. Cleaning supplies and pot scrubbers should be stored within reach of the sink.

2. Provide sufficient counter space. Counter space at both sides of the sink is crucial. At least three linear feet of counter should be installed between the sink and refrigerator. This is to be able to remove food from storage and put it away with ease, to provide space for cutting and chopping foods or for rolling out doughs. The counter space between the sink and refrigerator is often called the mix center. Near the sink, you need one place to stage dirty cookware and dishes before you rinse and wash them, and another to let them dry off after washing. Sink counter space is still a requirement even if you plan to have an automatic dishwasher.

Fig. 1-2. A kitchen work area.

The range or cooking center should also have counter space on both sides so you can place prepared foods in one place before cooking and afterwards in another while cooked foods are cooling. People frequently don't allow enough counter space here, and family cooks are sorry later on.

Beyond these areas, there should be enough additional counter space to accommodate all your favorite items and appliances such as a mixer, blender, food processor, toaster, crockpot, microwave oven, electric sharpener, can opener, bread box, and even a telephone.

3. Provide ample storage space. In the kitchen, enough storage space can mean the difference between a food preparation area that's easy to organize, easy to work in, and easy to keep clean—or an area that's difficult to work in and always a mess. Neatness and cleanliness count heavily toward a cook's efficiency and enjoyment of his or her work. And it's a proven fact that substantial psychological stress occurs to people who occupy cluttered, disheveled areas.

Here are a few points to keep in mind when planning your kitchen storage:

1. All kitchens should contain cabinets beneath the sink to hold items such as soaps, cleaning utensils, washcloths, and drying towels. If you desire, a garbage disposal can make short work of most food waste and scraps, and an automatic dishwasher can be installed under one side of the sink counter to take care of dirty dishes, pots, pans, glasses, and utensils.

2. There should be a good supply of cabinets and drawers around the food preparation area for utensils, cutters, chopping boards, and glassware. The

range and cooking center also requires cabinet space on both sides for pots, pans, dishes, trays, casserole dishes, strainers, and dozens of miscellaneous objects.

3. A pantry is a helpful addition for storing food, beverages, liquors, a stool, and even a small sweeper for quick cleanup (Fig. 1-3).

4. After making sure you have sufficient base and wall cabinets, drawers, and pantry space, give careful consideration to the size of your refrigerator and freezer units. Analyze your shopping habits again, and plan to purchase large enough refrigerators and freezers so you won't find yourself short of storage space for cold and frozen foods. Do you like to hunt and fish? Is one of your hobbies picking farm-fresh fruits and vegetables? Do you raise your own bumper crops, or prefer to buy meats in bulk?

5. Have an efficient layout. There are four widely accepted arrangements of the three kitchen work centers: the U-shape, the L-shape, the Parallel Wall, and the One-Wall.

The U-shape. With this plan the sink is usually placed in the center leg of a U-shaped counter, between the food storage and cooking centers. The work triangle consists of three relatively short and equal-length distances. This, plus the fact that no through traffic interferes with the triangle, is what makes the U-shape plan the most efficient and desirable arrangement for many kitchens. It's compact, step-saving, and keeps the cooks out of the limelight (Fig. 1-4).

The L-shape. This arrangement fits well on two adjacent walls and provides a good location for dining and/or laundry space on the opposite side of the room. It's not as convenient as the U-shape, but it's the next best thing. This plan can be converted into a U-shape

Fig. 1-3. A pantry.

Fig. 1-4. A U-shaped kitchen.

by the addition of an island or peninsula section of counter and cabinet to work with (Fig. 1-5).

The Parallel Wall. This arrangement has one work center on one wall and the others along an opposite wall. If your house seems to demand a parallel wall or "corridor" style of kitchen, take precautions to prevent

Fig. 1-5. An L-shaped kitchen.

Fig. 1-6. A parallel kitchen.

kitchen traffic from interfering with the work triangle. Try to locate doors so people won't naturally cut through the kitchen when entering or departing through a back or side door. For the sake of whatever traffic you end up with, this corridor between the two walls should be a minimum of 4 1/2 feet wide between facing appliances and equipment. This lets two people easily pass each other while working. Avoid placing the refrigerator or oven where their open doors will block off a frequently used passageway. Otherwise, a work triangle arrangement almost as efficient as that of a U-shape can be constructed using this plan (Fig. 1-6).

The One-Wall. One-wall kitchens are best suited to small houses where space is extremely tight. At best, cabinet and counter space is minimal, and you have no choice but to live with relatively long kitchen traffic patterns. However, if the distance from one end of the cabinetry to the other is close to 10 feet, some degree of efficiency can be realized through carefully laid out appliances, even when a true work triangle is lacking (Fig. 1-7).

The same traffic precautions that apply to the Parallel Wall kitchen also pertain to the One-wall plan.

Any kitchen arrangement can be further improved upon or detracted from by placement of windows and doors. Effective lighting over the sink and main work surfaces is essential. Whenever possible, place a window that opens easily over the sink. It goes there for

light, for ventilation, and even to provide a view to make washing dishes more palatable. Naturally, an electric light is still needed for night work at the sink.

Doors should encourage traffic to go around the kitchen work area instead of through it. This minimizes interruption of the cooks, and the possibility of spilling hot foods on innocent bystanders. The work area should also be out of the way to individuals who enter and exit the house from the rear or side, and should not be directly adjacent to kitchen tables and chairs.

Other points to think about when planning the kitchen include:

1. Someone will be spending a lot of time in the kitchen. Try to arrange the nicest views available through the windows or sliding doors.

2. There should be an exhaust fan or range hood with a built-in fan directly over the range. You need a way to expel cooking smoke, fumes, and odors to keep your kitchen fresh.

3. Plan for plenty of electrical outlets along the kitchen walls and counters. It's frustrating not to have enough outlets for the standard complement of kitchen appliances.

4. Be aware that because a kitchen is considered the heart of any household, it's best situated in a central location, close to dining areas and family entrances near the garage so groceries can be easily carried into

Fig. 1-7. A sidewall kitchen.

the house. If you plan a family room, consider having it adjoin the kitchen along one wall to facilitate traffic, communications, and even the ability to spread out if ever you throw a sizeable party.

DINING SPACE

People have all kinds of theories on what dining should be. Some individuals prefer to dine on the run, and aren't particular about what they eat, where they eat, or even how the food tastes. To them, eating is merely a necessary fact of life, a biological requirement.

Others enjoy taking the time and effort either to prepare or seek out gourmet-style meals. Good meals to them are to be slowly savored in the company of others, in carefully structured atmospheres at home or in fine restaurants.

A typical household leans toward a happy median between the "fast food" meal and candlelit dinner. That's why there are often two dining areas in a typical home—one for quick breakfasts, lunches, and children's meals, and another for more formal dining, which, though pressed into service less frequently, still plays an important role in holiday celebrations and special family events.

In many households, a third dining area consists of an outdoor patio or deck having a gas grill. All three areas, however, should be located near the kitchen food preparation site for greatest convenience.

You have several basic choices to make when deciding upon informal and/or formal dining space: Quick, easy meals and snacks can be served either at an attractive utilitarian bar, which is simply an extension of the kitchen work counter, an overhanging portion of counter that can accommodate three or four bar stools (Fig. 1-8), or a table/chairs or booth/bench arrangement included as part of an extension of the kitchen—often referred to as a breakfast room or nook.

If you don't want a formal dining room then you should probably opt for a table/chairs set that's placed right into the heart of the kitchen, yet out of the cook's way. This arrangement is referred to as an eat-in kitchen. Formal meals can still be served in an eat-in kitchen when special attention is given to items such as appropriate lighting, ventilation, and mood music.

The kind of kitchen dining space you prefer also depends on if you plan to use the table, booth, or bar for other purposes. Certainly, you can hardly play cards at a breakfast bar.

If you want more than a breakfast bar and eat-in kitchen, and you have the space and resources, then go with a formal dining room too. This can be footage borrowed from a living or family room, often delineated by special interior decorations or furnishings such as vinyl flooring, wallcoverings, or a chair rail. Or it could be a separate room of its own.

For ultimate privacy, doors can be installed to completely close off the dining room from the kitchen and living or family room. This can come in handy for special events such as birthday or graduation celebrations.

Fig. 1-8. An eating bar, or breakfast bar.

A dining room implies both enclosed space and service at the table. For this you'll need plenty of table space, chair space, access and serving space, plus room for any china hutch or side table server you want. A chandelier or other suspended lighting fixture, preferably one controlled by a dimmer switch, makes a lot of sense. The ability to lower the lighting level, at little cost, will provide a relaxing and classy atmosphere.

While a separate formal dining room is more traditional, an open-style dining room can fit nicely into an active family's lifestyle. Without the walls of a formal dining room, communications between the kitchen and living or family room are greatly improved. And due to the additional cost of a separate dining room, plus the overall reduction of square footage in many of today's modern houses, the open-style dining room situated directly between the kitchen and living or family room is becoming increasingly popular.

BATHING AND WASHING SPACE

Three questions must be addressed when you consider the bathing, washing, and comfort facilities you want in your new house: Are there enough bathrooms planned? Are the bathrooms large enough? And are the bathrooms conveniently located?

Are There Enough Bathrooms?

We've come a long way since the days of an outhouse behind every barn. Today even a house having one

13

complete bathroom is considered old fashioned.

It's tempting to trade away the relatively high cost of bathroom construction and the space bathrooms require for more of the higher profile living and work areas. But as it is, to ensure a good market value for your house, you should seriously consider at least two bathrooms, and possibly more if you have a large family.

Bathrooms come in both full and half sizes. A full bath includes a toilet, hand sink, bathtub and/or shower (Fig. 1-9). However, most modern premolded units combine a bathtub and shower in the same piece. A half

bath consists of a toilet and hand sink (Fig. 1-10).

To begin with, one full bathroom should be designated for general use, close to the bedrooms. It should be accessible from most areas of the house and should not be reached by traveling through other rooms. A second bath, often directly adjacent to or back-to-back with the main bath so plumbing fixtures can be shared, is usually located in a master bedroom. A third bath, or at least a half bath, can be a great convenience, especially when positioned near the living room, family room, and kitchen.

With the exception of single-level ranch houses, it

Fig. 1-9. A full bath.

Fig. 1-10. A half bath.

is definitely a plus to locate a bathroom on each living level. This results in time and energy savings over the short and long run. If there's one thing that can make a house feel too small, it's standing in line in the morning, waiting to use the only bathroom.

Are The Bathrooms Large Enough?

All bathrooms should be large enough for ease of movement, proper traffic flow, and plenty of storage. Individuals who frequently get dirty at work or at play need roomy bathrooms with easy-to-clean surfaces. So do families with lots of children. People who depend on their appearances, such as models, airline attendants, politicians, salespeople, businessmen and others, might also prefer more spacious personal care areas with generous vanities, wide mirrors, and plusher appointments.

No matter what size full and half bathrooms you decide upon, all the required fixtures and accessories should be provided and sensibly located:

■ Plan built-in storage for towels, soap, toilet paper and tissues, shampoo, and other personal aids.

■ A roomy vanity and medicine cabinet rounds out the basic storage.

■ Many clever and attractive shelving arrangements consisting of materials from woven reeds to glass and chrome can be custom installed into otherwise unusable space.

■ A laundry hamper or built-in laundry chute will save many steps and will help keep bathrooms neat and uncluttered.

■ When no window space is available, a ventilation fan unit can be installed to ensure a turnover of fresh air and prevent stuffiness. If a window is used, though, the bathtub should not be placed beneath it. Sufficient lighting is important, and so is a heat vent.

Are The Bathrooms Conveniently Located?

In addition to being located near the bedrooms, bathrooms should also be planned near the main floor living and working areas. If your family often engages in outdoor activities, at least one half bathroom should be placed near the outside access. That same half bath can then be used by guests and family members who congregate in the living or family room, and by individuals coming from the kitchen and dining areas. This half bath is a real step saver, especially in a sprawling single-story ranch home. And remember, people—especially guests—should be able to use the bathrooms without being seen by everyone else.

The nearer to other plumbing lines you can locate the bathrooms, the better. Placing two baths back-to-back saves on the installation labor and material costs and takes up a minimum of space. So does situating rooms containing plumbing fixtures as close as possible and practical to where the sewer and water lines enter the house—this effectively reduces the length of indoor service piping and allows for fixtures to be drained and vented with a single stack.

SLEEPING AND DRESSING SPACE

We spend practically a third of our lives in bedrooms. When examining how bedrooms will fit into your new house, six factors should be given careful attention: size, layout, windows, closets, noise, and the importance of having a master bedroom.

Size

There should be a rhyme and reason for bedroom sizes. Look at each one separately. Will it be for children or adults? Do you want bunk beds? Or queen and king-size versions? Will a spare bedroom also serve as a sewing room? Are you planning a master bedroom complete with its own bath? In addition to providing space for beds, bedrooms should be able to comfortably handle a small desk, a dresser, and other clothes storage. There should also be enough room for dressing and personal care, and ample window space to provide sufficient air and natural lighting.

Layout

There are no tricks to designing functional bedrooms, but a frequent mistake is to have a poor bed location in relation to the bedroom's traffic pattern. Because bedrooms usually have at least several doors—the entrance door from a hallway, closet doors in various arrangements, and possibly a door to a private bathroom, dressing room, or even to an adjoining child's nursery—these entranceways make continuous wall space for furniture, including the bed, hard to find.

Typical bedroom traffic patterns are from the main bedroom door to the bathroom, clothes closet, or dresser, in that order of frequency. If you have to walk around the end of the bed to reach any of these places, you'll have an awkward traffic pattern. When laying out each bedroom make sure there's a place to locate the bed or beds that won't result in the creation of obstacle courses.

Windows

Bedroom windows, like most other windows, have two primary functions: to provide light and ventilation. Windows are helpful both in cool climates where they let the sun in to illuminate and heat bedrooms (although windows can also be mediums of rapid heat loss when the sun is absent), and in warm climates, where windows on opposite walls provide refreshing cross ventilation breezes. At the same time, heavy-duty window shades or coverings can be used during the day to keep out hot sunrays. In unusual circumstances, when only inside walls enclose a bedroom, the need for cross ventilation can be negated by the installation of central air-conditioning.

When considering the style, efficiency, and placement of bedroom windows, think about the possibility of drafts during cold weather, especially at the head of a bed placed near a window. On the other hand, such drafts can become cooling breezes welcome on hot summer evenings. Weight the pros and cons and remember that certain types of windows positioned over the head of a bed might be difficult to open or even reach.

After deciding on the types of windows you want, make sure they'll provide adequate safety, including an escape route in case of a fire. In a child's bedroom you won't want windows so low that you'll have to worry about children falling out of them. But if the windows are too high off the floor, or too small, then the kids can't reach or escape from them during an emergency. All bedrooms should have at least one easy-to-open window with an opening of not less than 5 square feet, having a minimum clear width of 22 inches and a sill height not more than 48 inches above the floor.

Closets

Closet space is vital to any bedroom. Without closets, orderly clothes storage becomes impossible. In fact, even individuals living in houses with an abundance of storage space always seem to want more closet room as the years go by and possessions keep accumulating.

To head off what could be an eventual problem, plan generous closets that make the most of their space. Consider going with a split-design version: half of it open all the way up so dresses, raincoats, and other long garments can be hung, and the other half consisting of two double-decked pole and shelf arrangements, for shorter garments. Small items can be placed on shelves installed in what otherwise would have been wasted space within conventional closets. Many people are surprised when they learn how much a scientifically designed closet can neatly hold. (Figs. 1-11 and 1-12).

For sheer convenience, there should be an inside light in every closet; not one operated by a pull string. Install an electrical switch either inside or outside the closet doors. Pull strings and chains have a way of getting tucked up on a shelf as something is being put away, then they become difficult to find in the dark.

Don't use a bare light bulb without some kind of glass cover. There's a danger of fire if the bulb is not at least 18 inches from the edge of the nearest shelf or from the closest item of clothing. A better alternative is to install recessed lighting in all closets.

Lastly, if you choose not to store extra replacement linens in a bedroom closet, see that storage for them is available in a different closet located close to the bedrooms, perhaps in a hallway.

Noise

As a rule, bedrooms should be placed together in a part of the house that's protected from outside vehicle noise from nearby streets. As much as possible they should also be secluded from living, entertainment, and working space noise inside the house. Some dwellings lend themselves to a clear-cut separation of sleeping space: 2-stories, for example, usually reserve the second floor for bedrooms. This clustering of bedrooms, generally near bathroom facilities, makes parental supervision easier and simplifies the ease with which a quiet sleeping area is maintained.

The Master Bedroom

It sounds like a throwback to medieval times, when the master and mistress of the house lived out their lives in luxury, catered to hand and foot by indentured servants. Today a master bedroom is one common luxury item that most homeowners still enjoy. A master bedroom offers the utmost in convenience and privacy, and should at the very least include a half bath (a full bath is much preferred), and should be large enough to permit several alternative furniture arrangements.

As mentioned before, the clothes closet should be roomy. Two separate closets—"hers" and "his"—are ideal. It's also a good idea to soundproof the wall between the master bath and master bedroom for the late sleeping/early rising couple.

SERVICE AND STORAGE SPACE

There's nothing glamorous about service and storage space. You can have spectacular living rooms for talk-of-the-town entertaining, modern kitchens that grace the covers of architectural magazines, luxurious master bedrooms, and spacious whirlpool and sauna bathrooms for pampering yourself. But who can get excited about hallways or stairs? Or laundry and utility rooms? Usually the homeowners who have ill-planned ones.

Fig. 1-11. An example of an efficiently laid out closet.

Certainly these behind-the-scene features cannot be excluded from the typical house. Storage areas are also necessary accompaniments to the more popular parts of every dwelling. They include closets, small storage nooks and crannies, plus garages, basements, and attics.

Hallways

There's something disconcerting about a main entrance that opens directly into your living room. Instead, look for a center hallway plan that offers access to any part of the house without leading you conspicuously through living and entertainment areas. On the other hand, excessive hallway footage, along with its special walls, means costly wasted space. To limit the space used by an entrance hall, make the entry a part of a corner of your living room, but still keep it "separate" by installing a different flooring and by breaking up the ceiling line directly above the change of flooring.

Instead of completely writing off necessary hallway space as practically useless except as a pathway between rooms, consider the wall surfaces as potential showcases for artwork and photos. Strategically located lighting fixtures will also help support an individual decorating effort.

Stairways

This kind of house space has had its ups and downs

Fig. 1-12. Another example of an efficiently laid out closet.

over the years. Wrongly positioned, stairways—like hallways—can rob otherwise useful living and working space from any floorplan. To avoid this wasted effort, stairways should be constructed one on top of another whenever possible.

Even though you've probably heard a lot of nice things about the spiral stairway, don't believe all of them. Although this setup, which takes approximately 4 by 4 feet of floor space, is the most compact arrangement you can have, it's also the most expensive, inconvenient, and dangerous. The novelty quickly wears off. Ask anyone who has had to raise children around a spiral.

The most economical and convenient choice is a standard straight stairway taking up about 8 by 3 1/2 feet of floor space. An altered version of the straight stairway that can be very acceptable is the landing stairway, consisting of a half flight of stairs that leads to a rectangular landing and then another half flight to the next floor level. The landing stairway takes up about 7 by 7 feet of floor space. It adds an extra touch to a room at a cost higher than that of a straight stairway, but without the hazards of the spiral.

When planning a basement, consider the practical advantages gained by an outside stairway. Bulky items can be carried from the back or side yard to the basement, and vice versa. This also increases the likelihood that you'll use your basement for storage instead of just dropping everything in the garage.

Laundry and Utility Rooms

Mention "laundry room" as a topic of conversation to anyone and you'll likely receive a polite nod and a yawn. Most homebuyers tend to overlook the importance of both laundry and utility rooms because these areas simply don't have the pizzazz potential of other parts of a house. That's unfortunate. Laundry and utility rooms haphazardly planned are usually tucked somewhere out of the way, with little consideration to convenience.

There are two kinds of utility rooms: those with laundry facilities and those without. Utility rooms without laundry facilities can be located practically anywhere toward the outer reaches of the central living and work areas of the house. Utility rooms having laundry facilities must be positioned more carefully and can be located successfully in a number of places within a house.

A basement is one locale. This can be an economical place, although over the years you'll have to contend with constant stair climbing. If you decide on the basement as your laundry/utility area, give serious thought to a direct outside basement access (Fig. 1-13). A door to the yard lets you hang clothes outside during nice weather. It also lets you enter from the outdoors in wet or dirty clothes that can be changed right in the laundry area. Then you simply wash up at the laundry tub and proceed through the house. A clothes chute from the upstairs levels to the basement laundry basket can be another stepsaver.

A garage or carport is another locale. In warm-climate locations people often choose to put their utility room either in the garage/carport or in an adjoining space called a "mud" room—so named because of its outdoors accessibility. There are no steps to climb (or very few) and with a laundry tub you can enjoy all of the conveniences found in a basement utility room having direct access to the outside. If there's a second story in the house, a laundry chute can be arranged to drop clothes straight into a garage or carport or mud room hamper.

The first floor of the house is yet another alternative. There are two types of in-house laundry rooms that can be positioned on the first floor of any house. First, there's a laundry room that's a full-size room a little smaller than the average bedroom but large enough to hold a washer, dryer, stationary tub, ironing board, soap, bleach, cleansers, a sewing machine and supplies, and if enough space is available, a working surface for sorting clothes before and after washing, drying, and ironing (Fig. 1-14). There should also be a clothes closet with floor space for storing dirty laundry out of sight. The closet should have shelves for laundry sprays, softeners, measuring cups, and scrub brushes. Consider having the clothes rod go only three-quarters across the closet, so one-quarter of the vertical space can be reserved for shelves affixed from top to bottom.

The second type of first floor in-house laundry room is a laundry center that's installed entirely within a closet (Fig. 1-15). This is a more economical route than the full-size room. Although you don't have as much space or versatility, you do have the basic necessities for washing clothes in a first-floor location. The closet in question need be only as deep and as wide as required to accommodate a washer, dryer, and stationary tub. A shelf across the top of the appliances can hold laundry baskets, soap, and other items. You might have room to stand up an ironing board, too. If not, just place it in a kitchen broom closet. A screen or folding louvered door should be installed to conceal the laundry center when no one is using it. This setup is clean, simple, and convenient.

Fig. 1-13. Outside basement stairs.

Fig. 1-14. A full laundry room.

Fig. 1-15. A laundry center closet.

When possible, try to have one of the first floor laundry arrangements. They're usually the best alternatives, and have the most conveniences.

Once you decide on the size and type of first floor laundry you want, a number of other things must be considered when you're planning its location:

1. If it's situated near an outside door, you'll save wear and tear on carpeting. Someone walking into the house after playing football in the rain, jogging, or fishing can easily change clothes and clean up before trooping through the rest of the house. An outside access also makes it easier to hang clothes in the yard.

2. The nearer the kitchen, the better. Clothes washing can be made much more palatable when done in between other tasks. If the laundry is close to the kitchen you can move from one area to another with ease, and accomplish more work in a shorter timespan.

3. If you plan your house so the main bath and bedrooms are on the same floor as the kitchen, consider locating the laundry between the main bath/bedrooms and kitchen. Again, being near the kitchen will allow you to move easily between those work areas. Being near the main bathroom will save steps when collecting soiled clothing. And being near the bedrooms will make it simpler to put away clean clothes. If you plan a 2-story house with the main bath and bedrooms on the second floor, try to include a clothes chute that will usher soiled clothes straight into the laundry room collection hamper.

4. It's nice to have a window in the laundry for the daytime sun. If you can't arrange one due to your

layout (for instance, if you have room for only a hallway closet-type laundry center) then make sure you plan for adequate lighting. Working in the dark causes eyestrain and general fatigue.

As you can see, a variety of arrangements are available to accommodate your laundry center. Some are more convenient than others. No matter where you locate it, here are a few additional considerations:

Make sure there is an outside dryer vent. Otherwise moisture from the clothes dryer will cause an uncomfortable environment that's tough on laundry machines and other appliances, tough on the house (moisture causes mold and mildew, will soften plasterboard, and will rust metal), and even tough on people (it's hard on the lungs).

If you're concerned with the possibility of overflowing wash or rinse water, have a drain installed beneath the washer. With a first-floor laundry not set up over concrete, consider vinyl flooring so spills can be efficiently mopped up without leaving a mess.

Garages

People who have never had a garage don't realize what they're missing. They say that their cars have always been out in the weather, and if not for a little inconvenience during winter, who needs a garage anyway? But once they move to a place that has a garage—even a small detached garage, they'll never go without one again. It's nice to get into a dry car after an all-evening snow storm. It's nice to be able to pull out of the garage in the morning and drive past the neighbors who are stamping their feet and swearing while they scrape sheets of ice from their windshields. And by the same token, in the South or Southwest, it's nice to climb into a car that hasn't been roasting all day in the sun. Persistent sunlight fades paint, rots fabric, and cracks vinyl.

It's also nice to have storage space outside the house for seasonal items that can go six months of the year out of sight secured on garage rafters. It's nice to be able to lock up your cars, bicycles, golf clubs, tools, and spare tires in a garage.

When determining the size of your garage, consider how many vehicles you have (or plan to acquire), as well as the quantity of other items you want to store there. A roomy garage can serve many purposes other than being a car barn. Will you need additional room for workshop or hobby areas? You'll find a garage useful for storing canned foods, holding garage sales in,

and even as a handy play area for children when it's raining. When designing your garage:

1. Insist on at least one floor drain. If the garage will be connected to the house there should be a drain under each car space. Contour the concrete floor to the drains so the surface can be easily cleaned. Without the drains, dirt, mud, and slush will inevitably get tracked into the house.

2. If the garage is built under the same roof line as the rest of the house it will simplify construction efforts and costs. The property assessment will also likely be lower with that kind of arrangement, which will in turn minimize real estate taxes.

3. A garage should conform to the slope of the lot. If necessary, side-sloping lots can be accommodated by lowering the garage to meet the natural lot line, but this increases the number of steps needed from garage to living area, and by doing so takes away from some of the garage's usable space. On the other hand, if the garage is located higher on the same side-sloping lot, the driveway and garage floor will have to be built up substantially. This will be discussed further in Chapter 15, on garages.

4. In warm-climate locations the garage is sometimes used to house the furnace, air conditioner, water heater, and laundry equipment. Whatever items you plan on installing, make sure your garage will be large enough to carry out your designs. If you place any equipment having natural gas pilot lights in a garage, precautions must be taken, especially if flammable liquids are stored there.

5. If the back of your fireplace will protrude into the garage, that must be taken into account. As mentioned earlier, such a setup will save on chimney finishing bricks, and excess heat will be radiated into the garage instead of being lost directly outside.

Basements

The decision to have or not have a basement is often influenced by the regional custom of the area you're building in. Most houses in cold-climate locations will have basements. But in areas having exceptionally high groundwater tables (swampy places, for instance) or not experiencing freezing temperatures during winter, houses do not need and sometimes should not have basements. In general, though, there are a number of advantages to a basement:

1. A basement will provide handy storage space for household materials and outdoors equipment, es-

pecially when an outside access door is installed.

2. It's an ideal out-of-the-way place to put your water heater and furnace. Both of these appliances are simple to service and repair in a basement.

3. A house with a basement is usually easier to protect from wood-destroying insects such as carpenter ants and termites.

4. In a basement, water pipes are less likely to freeze, and wiring and all piping installed beneath the house are easy to get at for repairs and modifications.

5. A basement offers economical potential for future living expansion. A family room, bedroom, sauna, bathroom, workshop, darkroom, or other hobby and game room can be neatly situated within a basement.

Some people, even though they live in areas where a basement is traditionally included with the typical house, have strong feelings that such a feature is a waste of both money and space. They consider basements as dark, dank areas suitable for merely storage and the housing of a furnace and water heater. Here are some frequently mentioned drawbacks to having a basement:

1. The necessary stairway encroaches on usable space both in the basement and on the floor above.

2. There's no doubt that basements *can* be dark, gloomy, wet, and clammy.

3. There's an expense for waterproofing and establishing proper draining around the foundation.

4. There's also the cost of basement flooring, finished walls if desired—plus heat, wiring, and lighting.

5. If the money spent on the basement could be used elsewhere, you could substantially add on to your upstairs living levels.

6. Unlike the rest of the house, the typical basement has little natural light or ventilation.

Attics

Here's a house feature that appears to be nearing extinction, going the same way as the covered sit-down front porch. Years ago, when two- and three-story houses were crammed together along big city streets like upright dominoes, attics were included with every dwelling.

They held (and still do) old chests and cardboard boxes loaded with Christmas decorations, clothes, toys, school papers, books, antiques, and other momentos. Attics have always been cluttered, dusty repositories of family memorabilia—mostly because there was little

other space built anywhere else into the house for storage. If you happen to decide on an attic, remember:

1. Although the attic opening should be located in a concealed, out-of-the-way area, it should be easily accessible when you have to use it. A good place to put a pull-down stairway unit is inside a utility room or spare bedroom closet.

2. The attic opening should not be smaller than 22 inches by 22 inches; preferably it should be larger. Don't settle for anything less than a drop-down staircase or ladder arrangement, especially when the attic opening is situated in the garage.

3. If possible, allow for ample attic height and headroom to enable you to move around without constantly stooping over.

4. Some individuals believe they can store items across the attic floor joints without laying down flooring. Don't fall for that trap. Never settle for a completely unfloored attic. The more you lay down, the better.

5. Provide lighting and an electric outlet in the attic. Attics are generally dark inside. It's an inconvenience having to always carry a flashlight up there, or to string an extension cord every time you want to use a vacuum cleaner, light, or power tool.

6. Provide ventilation in the attic to alleviate harmful heat and moisture buildup during summer.

Storage Space

A sufficient amount of storage space will make a home a much more pleasant place to live in. While some individuals catalog everything they own, and are able to find the most obscure item at a moment's notice, others are so disorganized that they routinely lose anything from their Christmas decorations to last year's swimsuits.

The relatively recent reduction in the overall square footage of houses has wreaked havoc with storage space, and has thus increased the importance properly designed storage space can mean to you. Efficient storage in bedrooms and kitchens is especially critical. Here are a few points to consider when planning your overall storage space:

1. Place a clothes closet near the main entrance.

2. Have a linen closet near the bedrooms and main bathroom to hold sheets, pillow cases, towels, washcloths, comforters and other bulky whites.

3. Try to place a clothes closet near the garage or side entrance.

4. Locate a pantry closet in the kitchen for hold-

ing canned foods, beverages, liquor, and lots of other kitchen items.

5. Make sure that your "live" storage—for items used day-in and day-out, is very accessible. Live storage requires drawers, shelves, closets, and at times, chests. Each storage area should be thought out in advance for particular needs and sized accordingly. And each should be located in the proper place.

6. Your "dead" storage—for things you use only infrequently during the year, such as lawn furniture and snow tires, can be put in out-of-the-way locations. Find dead storage in the most inaccessible spots, in places such as attics, basements, and garages. Use boxes and chests to store smaller items in.

It only stands to reason that as far as possible, your house should provide you with whatever you need for safety, comfort, enjoyment, and privacy. For individuals having particular interests requiring special adaptations, facilities such as the following may be desired: a sauna, steam bath, hot tub, greenhouse, elaborate garden, fountain, swimming pool, place for animals, patio/garden living room, or various outdoor work and hobby areas.

No matter which areas are most important to you, make sure you at least consider each of the six main types of spaces under roof. That way you can make intelligent decisions when custom designing your house, realizing the trade-off effects that having too much or too little of any particular space are likely to have once you move in.

A house ultimately expresses the unique personalities, goals, and lifestyles of its inhabitants. Thus a totally satisfying residence will provide you with deep-seated feelings of personal achievement and will remain a source of continuous pride.

Chapter 2

House Styles and Types

Your ideal house, a dwelling that's both handsome and practical, is much more than simply a collection of great rooms. Certainly, a truly well-designed house that offers plenty of living security, enjoyment, and pride of ownership, should feature memorable individual rooms. But that's not enough.

To work as a single unit, rooms should be arranged in such a manner that the house becomes the closest possible match to your living requirements and lifestyle. Your home should contain the correct amount of space to suit yourself and your family or future family. It shouldn't be too small, nor too large.

The best way to go about putting a house together in your mind is also the most logical. First, decide on the amount of space you want roofed over. Once that's established you can go on to the style and type of house that will best lend itself to a combination of your objectively arrived at space requirements and your subjectively arrived at preferences for appearance and setting. Naturally, both advantages and disadvantages exist for all the various types of houses, and these characteristics will be pointed out later in the chapter.

Somewhere in the back of your thoughts, while space requirements and house types are being talked about, keep in tune with your own financial abilities. If money or income is a major problem, pay particular attention to the chapter on planning for future expansion.

When deciding how much space will be roofed over:

1. Arrive at the number of rooms you want. The six types of space were discussed in the previous chapter. Figure out your ideal number of rooms: what they are, their sizes, and then, just to be safe, also determine the minimum number of rooms you can get along with (keeping in mind you can add more at a later date).

2. Determine a dollar-per-square-foot cost and decide how much you can afford or wish to spend. Consider how much of a down payment you can come up with, plus the cost of current mortgage rates in your area. To arrive at how much you can expect to pay per square foot of house space, attend new construction open houses that are built with similar materials and workmanship to what you'd consider acceptable. When at an open house, ask what the going price for the house is, *excluding* the lot, then divide those total dollars by the amount of square feet the house has. Square footage means the livable area of the house. It doesn't include the garage, basement, or attic. What you can

spend on a home mortgage is usually governed by ratios of indebtedness to income that are conservatively set by banks and savings and loans.

3. Decide on the style of architecture you'd prefer. Some styles are innately larger than others. Some styles are low key, while others literally exude a certain social status. Your choice is likely to be influenced by climate, geographic location, personal tastes, finances, and also by the dwellings already built in the area or neighborhood you decide upon.

4. The last consideration, and a consideration more important than style, is the *type* of house you want: a single-story, a 1 1/2- story, a 2-story, split foyer, or multi-level.

HOUSE STYLES

Style is a broad concern that will ultimately affect your choice of house. It most commonly indicates the decorative features of the exterior and to some degree the interior. An overview of the various house architectural styles reveals that individual styles are best suited to their own particular climates and locations. They're frequently constructed of local or native materials, with exteriors and even appropriate colors that complement their surroundings.

When looking at particular styles you will find that Early American, Cape Cod, Colonial, Georgian Colonial, and Southern Colonial are all styles that have withstood the whims of change. The first two are compact, informal houses well suited to the northern sectors of the United States. The Georgian and Southern Colonial are larger and more formal. The Georgian is adaptable to both the northern climates and the milder climates, and the Southern Colonial—actually a form of the Georgian—is especially suited to the warm and humid climate of southeastern regions. Meanwhile, the Mission or Adobe style is particularly good in the hot, dry climate of the Southwest. However, this is a prime example of using local materials; in the warm, wet areas of the Southeast, adobe would not stand a single rainy season.

A popular and far-reaching style that has been evolving over recent years is that of the Contemporary. This style is suited to the theory "anything goes" and has few rules to follow or break. Contemporary houses can be simple, basic, and inexpensive. Or they can be extremely liberal in their composition, consisting of any of a variety of singularly dominant characteristics, from long sweeping roof lines, to a half-dozen levels juxtaposed over one another. Contemporaries can be full of big open spaces, constructed with huge panes of glass, hand-hewn stones, posts and beams, and modern brightly colored man-made materials. Innovative features such as passive solar heating, central courtyards or greenhouses, interior balconies, and spacious wood decks are frequently part of the Contemporary plan.

But Contemporaries are not for everyone. They can mean skilled engineering, expensive plans and drawings, tricky construction, costly materials, and high utility bills from heating cathedral ceilings and similar extravagant spaces.

Although individual construction styles may vary greatly from one another, they all must answer to certain design guidelines. No matter which style you lean toward, consider the following points in regards to your own house plans:

1. Think about the design of your house in relation to the complexity of construction. If the house makes a turn anywhere, for example—L-shaped or U-shaped rather than running in a straight line—you introduce construction complications. A gable roof, which is simply two sloped surfaces meeting at a high point called a ridge, is economical and comparatively easy to build, *until* you choose to turn a corner with it; then the junction forms a V-shaped indentation and the affected rafters require compound angle fitting, a construction technique that takes much more time and skill. Instead of laying down simple angles where all the rafters are positioned in the same fashion, boards must be individually measured and cut to multiple lengths and angles.

Another suggestion that will keep construction costs lower is to build the outside walls free from a lot of ins, outs, jigs and jogs (Fig. 2-1). Most houses look better anyway if they don't have a sizeable amount of roof lines joined together. Otherwise a house will appear cluttered and haphazardly designed. In general, a simple square or rectangular plan gives you more house for the money.

2. A few additional comments about the roof and roof line can be made here: A continuous roof line gives an impression of greater size than a roof line broken up into several different planes. A roof should extend or overhang past the outer walls 2 to 4 feet. This not only lends a handsome and distinctive appearance to the house, but it also helps protect windows and outer walls from snow, rain, and sun. Another plus is that it keeps water away from the foundation and basement. A substantial overhang is a feature largely ignored by

Fig. 2-1. A house cluttered with many ins, outs, jigs and jogs.

many builders. It shouldn't be excluded. For the relatively small cost of an overhanging roof, there are too many benefits to go without one.

A roof over the main entranceway will shield you from the elements while you're fumbling for your house key, or will keep visitors out of the rain while they're waiting for you to answer the door. It also provides a nice spot for you to sit on a patio or lawn chair in comfortable shade during the summer.

3. Remember that good proportions are one of the first rules of a good-looking house. The exterior of your house should not have more than two (at the *most,* four) sizes of matching windows, apart from glass sliding doors. At the same time, the doors should not be of abnormal size or located in odd places. To arrive at a satisfactory scale and proportion of doors and windows, compare what's already been installed on existing houses in neighborhoods you'd feel comfortable in. Get a feeling for the styles of windows and doors you prefer.

The tops and bottoms of all equal-size windows should line up, conforming to one long horizontal line across the house. Smaller windows should line up with the tops or bottoms of larger windows. When the tops of exterior doors also line up with window tops, things

will look even better.

The exterior should not consist of more than two or at the very most three kinds of siding materials, so that the whole appearance works toward a single theme. If a wide variety of materials are present, the house will not be pleasing to the eye, plus the exterior will likely end up costing more than necessary and will be difficult to maintain properly. Never mix more than three materials such as aluminum siding, wood siding, bricks and stone, or glass. Each will compete with the others, resulting in an extremely "busy" and disconcerting effect.

Just as two many different building materials will clutter up the exterior of a house, so will too many colors. You've probably heard of Hollywood entertainers who paint their mansions a bright orange or obscene purple. If you merely want a practical, handsome exterior that won't draw curiosity seekers and will help ensure a good resale value, then the exterior of your house should be consistent with the exteriors on the rest of the neighborhood dwellings.

4. It makes sense to select a house style that will closely match the living pattern of your family. Consider your entertainment activities, hobbies, children's pastimes, gardening interests, maintenance desires,

and even the amount of time you like to spend away from home on vacation.

5. If you might want to add another room or section later on, when designing for expandability remember that building up or down is always cheaper than building out. Also remember that building out, if the only alternative, at a later date will be more cost efficient if you plan for it at the outset.

HOUSE TYPES

House "type" denotes the number and arrangement of a dwelling's living levels. The basic types of houses are the single-story ranch, the Cape Cod or 1 1/2-story, the 2-story, the split foyer, and the multi-level.

Again, an important consideration that will affect your choice of house type is your lifestyle. You want the house that best fits your needs and meets your ideas of personal acceptance and preference, as well as something that fits the setting you desire.

Some family activities that will affect the type of house you want are:

■ Entertaining—card playing, informal and formal dinners, outdoor barbecues, large cocktail parties, teenage parties and other pursuits all present different requirements.

■ Privacy—families and individual members differ in their desire and needs for privacy.

■ Hobbies—these can present special problems related to space, storage, and noise levels. For example, a drummer needs a different kind of space than a stamp collector. The woodworker needs room for bulky tools and materials, and will be a major generator of noise.

THE SINGLE-STORY RANCH

This type of house (Fig. 2-2) can be constructed in a wide variety of sizes, shapes, and designs. It can be built over a full or partial basement, a crawl space, or a concrete slab foundation (Fig. 2-3).

Before the advantages and disadvantages of the ranch are discussed, here are some general guidelines applicable to all single-story plans:

■ The single-story plan should provide access from both the front main entrance and a rear or side entrance into the house without routing people directly through the center of the living room or kitchen work area.

■ The living room should not be used as a corridor at the expense of carpeting and furnishings, but should instead provide the privacy for which it was intended. An entry foyer should distribute traffic so visitors don't have to step directly into the living room. This prevents congested cross-traffic and interruptions.

■ The kitchen, laundry, family room and any busy

Fig. 2-2. A ranch home.

Fig. 2-3. A ranch home cutaway.

work center of the house should be accessible from an outdoors patio or deck, if one is included with your plan.

■ The master bedroom suite should be assured of privacy by its remoteness from the family room, kitchen area, and outdoor living space.

■ If you choose to go with a ranch house, don't try to match it to a lot with a substantial forward slope. Ranches look best if they appear to be hugging the ground; a forward sloping lot requires an exposed foundation in the front, which detracts from the low look of a ranch (Fig. 2-4). Single-story houses are ideally suited to flat lots or sites that gently slope to the sides or rear, particularly if the plans call for a walkout from a basement or lower living area. This is also a way of economically increasing the amount of living space since the lower level is an extension of the foundation.

Advantages

1. It offers the greatest liveability for all members of the household regardless of age. Major rooms are located on a single level (unless a basement has been expanded into living space). The overwhelming advantage of this house is its suitability to families having senior citizens or young children. There are no second story stairs to climb up or tumble down.

2. Single-stories can be made very relaxing and informal through the use of outdoor space. They offer the most convenience for indoor/outdoor living, with plenty of possibilities for porches, patios, terraces, planters, and gardens that can be built adjacent to and integral with any room.

3. Single-story houses can have spacious basements. You might be one of those persons who thinks if you're going to have a cellar, it might as well be a large one. After all, basements are good for storage, workshops, heating appliances, billiards, ping-pong and other sports and hobbies, for future expansion of living space, and for laundry facilities if you can't have your laundry on the first floor.

4. As a rule, ranch houses are easy to build. They're close to the ground. You needn't resort to double-length ladders and scaffolding to reach much of the structure, as you have to with 2-stories and Cape Cods, for instance.

5. Because there are fewer living levels, heating and cooling systems don't have to negotiate additional floors and ceilings once the main living level is taken care of. Same with the plumbing system. It's much less maneuvering of ducts, pipes, and electrical wiring in a single-story.

6. A single-story plan is the easiest house to keep clean after it's built. Not having to climb stairs is a big time saver. In a ranch you don't have to worry about keeping cleaning equipment on more than one floor.

7. A single-story is easy to expand even if no preparations were made at the time of original construction. You can simply convert part of the basement into living area, or you can add on an exterior wing. Of course, if you know in advance that you want additional

space later (as discussed in Chapter 4), you can then make provisions in the walls where the expansion will be.

8. With a single-story house, you can consider a contemporary-looking sloped ceiling that follows the pitch of the roof so ceiling joists are not required and a feeling of spaciousness is created. Insulation and ceiling materials are applied directly to the rafters. This is a common building practice in the South, but it can also be used—with today's energy efficient insulation—in the North as well.

9. When designed with trussed rafters, a ranch can benefit from the popular open planning in which living, dining, kitchen, and family alcove sections are designed as part of one interconnected space. Because structural partitions are not needed except for privacy, areas can be arranged by furniture placements, room dividers, folding partitions, and even projecting or freestanding fireplaces.

10. A one-story house is much easier to inspect and maintain due to its proximity to the ground. The roof pitch is usually very low, so it's not difficult to climb onto, walk on, or repair. The outside walls, if constructed of painted materials, are simple to touch up or repaint, and it's easy to perform other routine housekeeping tasks such as washing windows and cleaning out gutters.

Disadvantages

1. Single-stories have been described as being typically informal due to the reduced amount of privacy found between their walls. The single floor layout increases the importance of effective interior zoning—for the careful placement of physical buffers between the living, working, and sleeping areas. The need for such buffers (along with an aversion to stair climbing) may be greater in a family that includes very old or very young members who need more rest than other family members.

2. Some people just don't feel comfortable sleeping on a first-floor level, for reasons of privacy and security.

3. Ranch houses cost more to build per square foot than other house types, because of their high ratio of foundation and roof to living space.

4. It's difficult and expensive to build upwards on a ranch.

5. Single-level houses usually require relatively wide lots and might be difficult to locate on smaller size

Fig. 2-4. A ranch home on a forward sloping lot.

parcels found in many neighborhoods.

6. Heating and cooling costs tend to be higher per square foot in ranches because all the ceilings and floors are essentially exterior surfaces. Exterior surfaces allow heat to leak out during the winter, and coolness to escape during summer. On multi-level houses, at least some ceilings and floors are interior surfaces.

7. Although it's easy enough when you have room to spread out on a building site, expanding a single-story house can be expensive and difficult on smaller parcels where the house has already been situated on the lot according to zoning restrictions. Local restrictions might prohibit expanding any closer to lot sides and setbacks.

THE CAPE COD OR 1 1/2-STORY

This traditional design (Fig. 2-5) derived its nickname from the place where it was first built. Originally a testament to pure function, it resembled a simple Monopoly-style square house capped with a broad low-slung roof all constructed around and over a massive stone chimney which stood erect through the dead center of the house.

Such a monumental chimney served several purposes. First, since the entire house had been built up around the chimney, each room had its own fireplace—either for cooking and/or warmth, and all fireplaces conveniently shared the same chimney. Secondly, the chimney also helped give the house stability against

fierce Atlantic gales and shifting seacoast sands.

Most of the original Cape Cods were 38 by 29 feet or smaller. They had low ceilings, rarely over 7 feet high. That was about the largest space that could be heated with wood. The entrance was centered at the front of the house, directly opposite a central stairway that led to the second floor. The second floor started at the roof line, and was often supplemented with "eye" dormers for additional room, light, and ventilation. Its resulting dormitory-style rooms were popular for storage and for children's bedrooms.

On the first floor, the front of the house consisted of two large rooms, one on each side of the central stairway. A large "Colonial" kitchen took up the entire rear of the dwelling, and was flush against the massive all-purpose fireplace hearth. The bathroom was out back. Way out back.

The Cape Cod's windows were small and shuttered to keep out windblown sand, hail, and driving January blows. The small panes or "lites" making up each window were used due to the limitations of the glass blowing industry in those days. Large panes were difficult to make true and clear.

The entire house usually faced squarely south to take advantage of every available ray of winter sunshine, plus, cleverly enough, to enable occupants to tell time: when the sun's rays came straight through the front window, hitting a marker on the floor in such a way, the people inside knew it was high noon.

As decades passed, and long after the original rea-

Fig. 2-5. A 1 1/2-story home.

Fig. 2-6. A 1 1/2-story home cutaway.

sons for the Cape Cod design had ceased to exist, the 1 1/2-story house again found supporters during the Depression days of the 1930s. Homebuilders liked them because Cape Cods were compact, thrifty houses to construct, especially when the huge fireplaces were left out. Cape Cods had another resurgence after the World Wars when builders mass-produced row after row of them. And the houses are still popular today, in larger more luxurious versions.

Here are a few general guidelines for the 1 1/2-story house:

■ It can be built over a full or partial basement, a crawl space, or a concrete slab foundation.

■ Although the Cape Cod typically still has a front center entrance, it's also a good idea to have a rear access as well.

■ As with the ranch, an entrance foyer should distribute traffic so visitors don't step directly into the living room from the outside.

■ In a Cape Cod, the master bedroom is often located on the first floor, with the other bedrooms upstairs. Whether on the first or second floor, the master bedroom should be assured of privacy.

■ Even if the second story will not be finished off initially, make provisions for future expansion (Fig. 2-6). Have the ceiling/roof area insulated instead of the second floor (more on this in Chapter 4).

Advantages

1. It's an economical house to build, with a low cost per square foot of living area. It requires proportionally less materials and labor to construct than other types, and because of its relatively small basement foundation and roof, a larger amount of living space can be had for a smaller financial outlay.

2. Its two living levels allow distinct zoning for privacy.

3. The Cape Cod is known for its low heating and energy costs, due to its efficient shape.

4. It can be constructed on a small lot.

5. The completed Cape Cod needs fewer furnishings, less interior decoration, and takes less time to clean and maintain than other types.

6. Its low roof line makes it fairly simple to build and maintain, but not so simple as a ranch.

7. The second story can be left unfinished and initially used for attic storage. Later, if more living space is needed, the area can then be employed for practical expansion. This is an economical approach to the problem of insufficient funds. You can initially run heat and air-conditioning pipes and ducts, plumbing pipes, and electrical lines to the second floor and then cap them off there. For air circulation you can install louvered

ventilators at each end of the attic from the start. In short, a design like the Cape Cod offers a great "finish it later" potential. This is good for young starters or newlyweds who at first need only a minimum of space but want more later on, as the family grows. Then, when the children eventually leave home, the second floor can be converted into an income unit or merely sealed off and used for storage.

8. Here, one person's advantage can be another's disadvantage. Many people contend that climbing stairs in a Cape Cod or other house having stairs is good exercise to help keep a person fit.

Disadvantages

1. Upstairs rooms beneath the roof tend to be hot during summer and cold during winter unless special care is taken when insulating and installing heating and air conditioning.

2. When poor planning is followed, the second floor can be cut up into odd size rooms, with sloping ceilings and dormer windows—causing awkward room layouts.

3. There's the need to provide space for a stairway leading to the second floor. This is space taken away from the first level. And stairs have to be ascended every time you want to visit the second floor.

4. If there's a basement beneath the house, the opening used by the downstairs stairwell can wreak havoc with the compact first floor, where every square foot counts.

THE 2-STORY HOUSE

All things considered, this is one of the best, most efficiently designed houses available (Fig. 2-7). It has become increasingly popular to families looking for a spacious yet economical and private dwelling. A 2-story can be built over a full or partial basement, a crawl space, or a concrete slab foundation (Fig. 2-8). Some general guidelines to consider are:

■ Once again, through the use of an entry foyer, access should be provided to all parts of the house from the main entrance. The entry foyer is usually the best place to start the upstairs stairway, so someone entering the house can go straight to the second floor if they choose.

■ As in the other house types, the living room should not be a traffic runway.

■ A rear entrance should be provided (or a side entrance), with direct access to the kitchen, laundry, family room, and all other working and living areas of the house.

■ The upstairs bedrooms should be planned so

Fig. 2-7. A 2-story home.

Fig. 2-8. A 2-story cutaway.

you don't have to walk through one bedroom to reach another.

Advantages

1. There's natural zoning between the upstairs and downstairs. The upstairs serves as an effective buffer between the sleeping areas and the downstairs living and work spaces.

2. Certain individuals prefer the feelings of privacy, security, and comfort brought about when the bedrooms are located on a second story, well above ground level.

3. Building up as opposed to out is cheaper per square foot. A 2-level square house with a given amount of floor space requires only half as much foundation and roof as it would if constructed as a single-level dwelling. Building down means a basement. One justification for a basement is its value as living area. But here the term "living area" is broadly used to cover basement recreation rooms, workshops, hobby areas, and similar spaces planned for special activities. If the basement will likely become a depository for junk, then it shouldn't be considered as possible living space.

4. Size being equal, it's cheaper to heat and cool a 2-story than a ranch. Cool air falls, and heat rises. At least one level in a 2-story reaps certain benefits, no matter what the outside temperature is.

5. The 2-story is adaptable to small lots. In fact, no other common type can match it for getting the most house on the least lot. It's a good choice on either high-priced land or on a tight little parcel. And a simple way to gain more space on the second level without affecting the foundation space requirements is to have the second-story walls overhang the first-story walls, Garrison style. This also aesthetically breaks up the high-wall appearance you might otherwise find unappealing.

6. A 2-story can be successfully expanded without much advance planning. A family room or wing on the side can be added, but at substantial cost. Due to its compactness on a lot, there's usually plenty of room to build an attached garage. Naturally, if you decide at the time of original construction that someday you will need more space, by building expandable features into the house you'll save money in the long run.

7. Physical fitness buffs will swear that a 2-story improves the cardiovascular system because occupants are forced to exercise by climbing up and down stairs.

Disadvantages

1. Stair climbing. Having to frequently go up and

down stairs makes housekeeping tougher and puts a strain on parents of young children and on the elderly.

2. It puts restrictions on a family that likes to spend a lot of time outdoors. Although the downstairs can be designed for easy access to the back yard, you might find the times needed to trudge upstairs to retrieve something from a bedroom will amount to a noticeable inconvenience.

3. Unless the attic of a 2-story is properly ventilated, the upstairs bedrooms will get uncomfortably hot during summer.

4. Without an elaborate network of wood decks and patios, the upstairs bedrooms are usually shut off from direct access to the outdoors and are not easy to escape from in case of fire or other emergencies.

5. Although the per-square-foot cost of a 2-story is much lower than that of a single-story, extra footage must be provided in the 2-story to compensate for the space lost to the second floor stairway. Plus the upstairs stairway limits the flexibility of the overall design.

6. Long ladders are needed to reach the roof, gutters, and second-story windows.

THE SPLIT FOYER

A split foyer (Fig. 2-9) is essentially a raised single-story house with the basement or lower level lifted halfway out of the ground and joined to an entry foyer. The lower level is usually finished off into part of the house's living quarters. An identifying characteristic of the split foyer is that the entry foyer is always located about halfway between the two living levels. In other words, once you step into a split foyer you have to go either up or down (usually a half flight of stairs) to reach a living level (Fig. 2-10). Split foyers have also been called, rightly and wrongly, mid-levels, raised-levels, and raised ranches. As with the other house types, there are general points to keep in mind when designing a split foyer:

■ Split foyers are meant to be constructed on lots having front-to-back or back-to-front slopes. It's silly to place them on dead level ground. Take a drive through practically any middle-class suburb. You'll notice that the worst-looking split foyers—the "no design" kind, have all been erected on flat lots. Built so, they appear ungainly and awkward. Instead, they should be closely fitted onto sloping parcels, so a natural marriage between the house and land results.

■ The two levels of a split foyer provide distinct zoning to help separate working and living activities. Further, room functions can be planned in a variety of ways. On a front-to-back downward sloping lot, the rear entrance is likely to be located in the upper living area, even if it means the construction of an outside wood deck or concrete patio with steps to the ground.

■ A split foyer's entry, like the other entry setups, must direct traffic through a hall or foyer arrangement to upstairs and downstairs rooms without marching visitors through the center of other rooms along the way. The rear entrance should also permit easy travel to the kitchen, laundry, and remaining areas.

■ The arrangement of living areas can vary greatly in split foyers. Some plans have all the bedrooms

Fig. 2-9. A split-foyer home.

Fig. 2-10. A split-foyer home cutaway.

on one floor, with the working and living space on the other level. Some plans combine living and sleeping areas on both floors. The garage can be included in the lower level, or attached to the side of the house with its own entry. In any layout, the bedrooms—especially the master bedroom—should be assured privacy from the rest of the house.

Advantages

1. This design offers easy entry from the outside to either interior level.

2. The entrance foyer midway between the upper and lower levels has direct access to at least one bath, thus reducing traffic near the main living areas.

3. Properly designed, a split foyer can look handsome and large, with only a short stairway from either level to the outdoors.

4. The split foyer can provide automatic interior zoning with the sleeping area on one level and the working and living areas on the other.

5. It allows greater window depth in the lower level, which yields improved lighting and ventilation. In turn, this helps the area normally referred to as the basement become more desirable for recreation rooms, baths, and bedrooms.

6. The split foyer uses its floor area to the maximum. It has a lower per-square-foot cost compared to a single-story ranch.

7. The simple floorplan and relatively compact design of this house results in construction convenience and savings.

8. With proper attention to correct insulation and careful positioning of heating and cooling appliances, the split foyer is easier to be made energy efficient than a dwelling that has many projections and corners, or one that is spread out.

9. Most split foyers are suitable for small lots.

10. The minimal foundation and roof areas help reduce construction costs.

11. When the upper living area faces downhill, you enjoy the full advantage of a view with height.

12. Again, with this design a physical fitness buff will proclaim that exercise on the stairs helps strengthen the heart.

Disadvantages

1. There's the inconvenience of frequent stair climbing.

2. Even though this dwelling can be energy efficient, because the stairway is more open than that of a 2-story, for instance, heat quickly rises to the upper level. If proper insulation, heating, and cooling steps are not taken, the lower level tends to be cold during winter. Special wall insulation below ground level is necessary, and a well-designed heating/cooling system is a must. Rooms over the garage can also be chilly if not well insulated. And during summer, without

enough insulation, rooms on the upper level will tend to be on the warm side.

3. The open stairway also reduces the effect of the zoning. Not visually, but because it allows odors and noise to freely move from floor to floor.

4. The split foyer, due to its structure and design, is not easily expandable at a later date.

5. Because the lower level of a split foyer is usually designed for living areas, there's not much room to have a basement workshop or storage facilities.

THE MULTI- OR SPLIT-LEVEL

These houses (Figs. 2-11 and 2-12) are essentially split in half vertically with two or more levels so the upper level is only half as high as the ground level floor. A typical layout for this design has the kitchen, dining room, and living room on one level (the ground level), with the bedrooms on another level a half story higher. The ground level may be over a full basement, or con-

structed on a crawl space in warm-climate regions, with the upper half story over a garage that's on a grade. Here are some general guidelines to keep in mind when planning a multi-level house, or split level:

■ Split levels are ideally suited for side-sloping lots on hilly terrain where the bottom level faces and opens toward the downhill side and the upstairs level opens toward the uphill side (Fig. 2-13). A split-level house placed on a flat lot will look awkward and will not be a very functional dwelling.

■ Arrangements of living areas vary in split levels. It's best to have the living facilities on one of the upper levels with the sleeping rooms on another for noise breakup and privacy zoning. As with other types of homes, the bedrooms, especially the master bedroom, should be carefully located for privacy.

■ The main entrance may be either on the upper or lower grade, as determined by the slope of the building site relative to the street.

Fig. 2-11. A multi-level home.

Fig. 2-12. A split-level home.

■ Matching a split level to a side-sloping lot allows close to a full height exposure on both sides of the dwelling, making each side fully accessible to the outdoors.

■ As with other house types, it's a good idea to have a front and rear entrance in a split level, even though these entrances may be located on different levels. In any event, the living and working centers of

Fig. 2-13. A split-level home cutaway.

the home, including the kitchen and laundry, should be easily accessible from either entrance.

■ An entry foyer should distribute traffic via a hallway to the rooms on its own level, and to the stairs leading to other levels.

■ Consider that the basement is usually the best place to put the heating unit, storage space, and perhaps a workshop if desired.

Advantages

1. It adapts well to lots with side slopes.

2. A split level has many of the same advantages a split foyer has. It produces automatic interior zoning. The sleeping area can be on one level, and the working and living areas on other levels.

3. A split level can offer easy entry from the outdoors to any interior level. On a side sloping lot you will have at least two main floors with access straight to the outside.

4. When properly designed, a good split level will look handsome and large, with only a short stairway from one level to another.

5. There can be greater window depths in the lower level, which will gain improved lighting and ventilation for possible expansion of the lower level into a recreation room, bedrooms, or any other living and working space that you may desire.

6. If the areas beneath the upper level are being used for living space, or even for a garage, that means at least three-quarters of the floor area is actively put to good use, thus making the price per square foot reasonably low.

7. Frequent short bursts of stairclimbing are considered a plus by the health conscious.

8. When the upper level faces downhill, there's the scenic advantage of a view with height.

9. With the ability to have access from the outside on at least two levels, there should be no problem locating a bathroom near each entrance.

Disadvantages

1. Split level room arrangements can be jumbled and disjointed if care is not taken during the initial planning stage. You could end up with a house that requires you to climb steps no matter where you want to go, or what you want to do.

2. The heating and cooling requirements of split levels can be very demanding. The lowest level tends to be cold during winter and needs special wall insulation for any area below ground. A well-designed heating/cooling system is a must. Rooms constructed over a garage tend to be chilly, and the rooms in the upper level are frequently too warm in summer unless proper precautions are taken. This is largely due to the number of open stairways present, similar to those in a split foyer.

3. The open stairways reduce the zoning effect by letting odors and sounds travel freely from one level to another.

4. Because the split-level house has so many different levels, it can be tough to build. This runs up the cost per square foot, especially if substantial bulldozing of the lot is needed for the foundation and landscaping, or if retaining walls must be constructed.

5. Another drawback to split levels is that they're often difficult to expand due to limitations of the lot and also because tampering with the original plan can easily harm the dwelling's appearance. Too many jogs and angles in the roof lines or exterior walls make them look cluttered and unplanned, as if put together piecemeal.

6. The split level is a poor choice for individuals who prefer large basements for woodworking shops, or want surplus storage space and footage to eventually turn into a basement recreation room.

7. With the two sections of the side-to-side split level framed under two distinct roof lines, the builder must skillfully integrate the two sections in an artistic manner to avoid the appearance of two separate houses joined together to make a single dwelling.

Chapter 3

Traffic Planning and Zoning

After you've reviewed the first two chapters, you should be ready to begin sketching the general layout and design of your house. As you do this, give careful consideration to the overall traffic plan and interior zoning best suited to your own particular situation.

First let's look at the space in a typical house from another angle. What about the place you're living in now? Chances are, you can (or you wish you could) identify three kinds of space, by function. Excluding storage areas, there's *private space*, or areas needed for sleeping, dressing, lovemaking, bathing, and studying. There's *social space*, or areas for being with others, entertaining, relaxing, and recreating. And there's *transitional space*, or places, depending on the circumstances, where either private or social activities can occur.

To discuss interior zoning in relation to those areas, we must study traffic that enters and exits the house, traffic that moves within the house itself, plus room-to-room relationships.

TRAFFIC ENTERING AND EXITING THE HOUSE
Service Access

It might sound snobbish, but there should be a definite entrance to be used by servicemen, repairmen, and individuals delivering items they have to carry into the house. Such an entry should be wide enough, direct, and as short to service areas as possible. The kitchen, laundry, basement, and utility rooms should be the prime considerations here, since those areas are the places most frequented by servicemen and vendors.

This entrance should be a logical alternative to protect the living room from unwanted intrusions by casual visitors such as messengers, salesmen, or unexpected visitors for whom a proper reception has not been prepared, such as your clergyman or parents in-law. Ideally, this access, or a sidewalk that leads to it, should be visible and obvious from the front of your house so people who have never been to your place can determine which entrance to use by themselves. Otherwise, a small tasteful sign can be strategically placed at a sidewalk that leads to a side or rear entrance.

The most important idea of a service access is to reduce cross traffic through the living areas of the house whenever possible. Why invite a meter reader or other serviceman to pass through your dining room while enroute to the basement? It's wise to eliminate as much of this type of cross traffic as you can.

Guest Access

This entrance is traditionally for friends and guests of the family. It is the main front door, the entrance that usually faces the front street. It should be easily accessible from the driveway and the front street, and from all rooms inside the house so occupants are able to quickly answer the door when someone arrives.

This entrance should provide guests exterior shelter from the elements while they're waiting at the door, and should have a place to remove coats, a closet to hang them up in, and an area in which visitors can adjust to the surroundings (entrance hall, foyer, vestibule, gallery).

It should provide efficient access to those parts of the house guests are most likely to frequent. This includes the living room, dining room, recreation or party room, an office, a patio, the bathrooms . . . each family's pattern of living will determine the needs for guest accessibility. This access should give a pleasant impression to visitors and a sampling of the quality and character of the house.

Day-to-Day Indoor/Outdoor Movement

Children need a good access to repeatedly go in and out while playing. Because very young children need almost constant supervision, a back door to their play yard may be a necessity. Excessive running through the house can be minimized further by locating toilet facilities near the door most often used by children.

Guests invited for outdoors activities usually enter the house by the front door then proceed to the location of the activity. This path should be fairly direct and avoid, if possible, passing through a room.

Because the outdoors living areas are likely to be the setting for picnics, barbeques, and similar events, the exits should be located near the kitchen. Here again, the availability of bathroom facilities for adult outdoor activities (in addition to the convenience for children) comes into play.

Removal of waste materials should involve a minimum of travel through the inside of the house. Containers for staging garbage and trash are usually located near the service areas, screened from public view.

If you have a basement planned, consider an outside door for the convenience of children who are playing downstairs. This lets them run straight outside without tracking through the rest of the house. And it's a great energy and time saver if your laundry equipment is situated in the basement, where you can quickly

walk outdoors to hang up wet clothes in nice weather. Also, when you're working outside in the yard, you can enter the basement, change clothes and get cleaned up. It's nice to be able to bring large items in through the basement door instead of carrying them through the house and down the basement stairs. This means a lot when transporting clothes washers and dryers, freezers, pool tables, and other large objects. A direct basement access will also be a timesaver when you store outdoor equipment such as screens, storm windows, garden tools, and lawn mowers in the cellar.

In addition to the main front entrance, a separate side or rear access usually serves as the family entrance for grocery shopping, for children going to and coming from school, for family members taking out the laundry from a first-floor utility area, plus other informal activities.

General Guidelines for Entries

For safety's sake, install a peephole or window in your front and rear doors so you can see who is ringing or knocking.

Again, it's not a good idea to have a front door that opens directly into the living room. A main entrance center hall or foyer should both shield you and your visitors from an inrush of wind, snow, or rain, and keep your living room privacy intact. Although the main entrance should lead to the living room area, it shouldn't encourage people to pass through the living room on their way to the rest of the house. If so, it causes interruptions, wear and tear on the carpet, and other inconveniences. A good floorplan will provide access to all main living areas through hallways or foyers rather than directing traffic from one main room to another. This also means that you should provide the ability, even in open-style plans, for direct access to the kitchen and bedrooms without intruding on living room activities. Thus children can come and go without interrupting a conversation you are having with guests in the living room.

The access from the front door to the kitchen should be easy and direct, as this path is frequently used.

There should be a convenient sheltered entrance to the kitchen from the garage, carport, or driveway, so groceries can be brought into the house and put away without a lot of effort and fuss.

You'll want good guest circulation with the ability to move the guests from the front door to the coat closet, bathroom, and living room. A clothes closet is essential near the front and side/rear entrance, not only for guests but for family members as well.

The remaining components of traffic planning and zoning in a house deal with internal movement and room-to-room relationships.

INTERNAL MOVEMENT

Here are five points to consider when developing your floorplan:

1. The living, sleeping, and work areas should be separate from one another. Yet they should be positioned properly in relation to each other and to additional factors such as orientation to the street, the sun, and even to scenic views. It's important to weigh your feelings toward "bedrooms" versus "work and play noise," or "entertaining guests" versus "bothering sleeping children." In other words, how much of a buffer zone between the bedrooms and the rest of the house do you think you'll need?

A 2-story house contains natural zoning, with the kitchen and living rooms on the first floor—a full story below the bedrooms. In a single-story house the living and sleeping areas should generally be located at opposite ends of the house, neatly connected by the kitchen and utility room.

2. Try to separate quiet rooms from noisy ones by distance. Keep bedrooms as far away as possible from the living, food preparation, and utility areas. Isolate study rooms from play areas, hobby rooms, party rooms, and workshops. Adequate sound-absorbing features become particularly important in moderate-size houses with open kitchens, combined living and dining room areas, and all-purpose family rooms, and also in children's play areas and adult workshops.

3. In the interest of silence, though, don't get carried away. Make sure no key area is completely isolated (laundry rooms and bathrooms especially), and also see that there's a safe place for children to play while you're entertaining in the living room.

4. In houses where stairs are necessary, the head of the stairs should be centrally located. This not only minimizes the need for halls, it also frees exterior walls for windows and adds natural lighting and inexpensive ventilation. In 2-story houses the stairway to the basement is usually positioned beneath the stairs to the second floor. The problem with this setup is one of arranging your plan so the head of the basement stairs is located near the service or rear/side door so items

can come and go from the basement in an efficient manner.

5. For ease of internal movement, many people favor an open plan. Open planning attempts to achieve a feeling of spaciousness; the interior of a house is made to appear larger than it really is through the elimination of solid walls between activity centers and by substituting partial walls, screens, or open room dividers.

A small dining room will appear larger if no solid wall stands between it and the living room. The uninterrupted expanse of ceiling visually increases the appearance of the dining room, and it has the same effect in the living room. The two rooms "borrow" space from each other.

Similarly, a kitchen will appear larger if it opens to a family room or dining space. This can be accomplished by using a breakfast bar to separate the two spaces. The open feeling is not lost even if you opt for a cabinet over the bar. Space between the bar and cabinet, and the cabinet and the ceiling, as well as the absence of a door between the two areas will be sufficient to retain the openness.

Of course, there are disadvantages to the open plan: a lack of privacy, conflicting activities can be distracting, and incorrectly grouped furniture might "float" without unifying walls.

ROOM-TO-ROOM RELATIONSHIPS

Although you might have already read about some of these points in the first two chapters, many are so important that they bear repeating. A main consideration to keep in mind is that your traffic patterns should not take people through the middle of several rooms (or even one room) while enroute to another.

The Kitchen

This is considered to be the most important room of a house. Certainly it's the heart of a house, a place used by the entire family for a variety of activities. It's most often placed adjacent to both dining and living rooms and close to a patio or deck where a barbecue can be located. If you have a first-floor family room, this too should be within easy reach of the kitchen, so that the dishes used for snacking are not far from the sink.

The kitchen should be centrally located to have "control" over the entire house. Here are other considerations:

■ It should be easily accessible to the front door so guests can be received, and to the family entrance at the rear or side so a car can be unloaded and deliveries accepted.

■ From the kitchen you should be able to keep an eye on children playing either inside or out.

■ The basement stairs should be close by, especially if you have any food storage down there, or if your laundry is in the cellar.

■ Having a bathroom within a few steps of the kitchen saves a lot of time.

■ Consider that an open area from the kitchen to the dining or family rooms allows for free conversation and visiting while you're cleaning up or preparing something to eat.

■ Ideally a kitchen has no more than two doors, one to the dining area or front of the house, and one to the service/family entrance or garage. Within the house there should be alternate ways to reach those areas without going through the kitchen. If there are three or more doors in the kitchen they should be located in one passageway that doesn't break up the kitchen work traffic.

■ After the kitchen's location is decided upon, appliances should be arranged so that distances between the central cooking area (range), the preparation and cleaning area (the sink), and the food storage area (the refrigerator and pantry) are no more than 7 feet each (Fig. 3-1).

■ If possible, the kitchen should have at least one, and preferably two outside walls in which to place windows. Window screens will enable grease fumes and cooking odors to be removed. Windows also serve as pleasant distractions for the busy cook by providing scenic and interesting views when available.

■ Although the kitchen should have a convenient central location, watch out for any traffic patterns that route individuals from the living room to the bedrooms by passing directly through the kitchen or dining areas.

The Laundry Area

This feature is usually placed in one of four areas within the house:

1. In or next to the kitchen, which allows much of the required housework to be done in the same general vicinity, without wasted motions.

2. In the basement. But this forces you to constantly go up and down stairs when doing the wash, and if you don't have a direct outdoors access from the

Fig. 3-1. A kitchen work triangle.

basement, it makes hanging up clothes in the yard cumbersome.

3. Near the bedrooms. This makes it handy for changing out of your soiled clothes and putting away clean garments.

4. In warmer-climates where basements aren't included with many houses, a popular place to put the laundry is in a mud room attached to part of the garage or in a breezeway.

No matter where the laundry room is, you'll need easy access to the outdoors, you'll need space for a clothes washer and dryer, soaps and cleansers, and you'll need enough room nearby to set up an ironing board.

The Utility Area

This is where the furnace, air conditioning unit, water heater, and possibly a humidifier are set up. From the standpoints of maintenance, efficiency of operation, and cost of installation, this area should be somewhat centrally located in a basement, but not out in the middle of the cellar floor. It should be close to a wall to avoid breaking up a large portion of otherwise usable space. Although the furnace, water heater, and related appliances are typically located in the basement, when no basement is included they're placed in a mud room, garage, or integrated into their own area right inside the house.

Bathroom Facilities

Here are some considerations:

■ Avoid having bathrooms visible from other rooms, especially from the living room or head of the stairs.

■ Be certain that guests can easily get to a bathroom without being under direct observation from the living room.

■ Position at least one bathroom near the bedrooms.

■ A bathroom should be simple to get to from the rear or side entrance for children playing or adults entertaining outdoors.

Bedrooms

Bedrooms should be placed together in one part of the house, protected from outside noises as much as possible and also convenient to bathrooms. This makes

parental supervision easier and simplifies the problem of maintaining a quiet sleeping area.

■ Don't situate bedrooms one after another in a series requiring passage through each other.

■ When dressing, people should be able to move between the bedrooms and the main bathroom without being seen from the living areas.

■ There should be a buffer zone between the living and sleeping areas so parents can entertain while children sleep. The placement of bathrooms, hallways, closets, bookshelves, utility rooms, fireplaces or other interior masonry walls can serve as sound barriers between the various quiet rooms and the noise-producing ones. Avoid "closet" walls that have only a thin plywood or fiberboard backing as the barrier between two rooms.

The Dining Area

The dining area should be adjacent to the food preparation space. For quick meals many people use a bar that is part of the kitchen work counter. Others rely on a table in the kitchen, or position a table in the family room or breakfast room. Still others choose to have a separate formal dining room next to the kitchen.

The Living Room

The living room should be near the main front entrance. It's used for such activities as reading, visiting, and entertaining guests. If your house won't have a separate family room, you can use the living room for watching television and listening to music. In any case, this room should be at a dead-end location to discourage unnecessary traffic from interrupting conversations and other activities.

Storage Areas

■ There should be ample storage in all rooms of a house.

■ Each bedroom should have its own closet(s).

■ It is convenient to have storage space in bathrooms either under the sink or in a separate closet.

■ The hallway leading to the bedrooms should have a linen closet for clean sheets, pillowcases, blankets, and other bedding and bathroom linens.

■ Plan closets near all of the entrance doors: front door, side door, and rear door. These closets are needed to keep guest and family member coats, hats, boots, umbrellas, and similar personal clothing and care items.

■ If there's no formal laundry area, consider a laundry closet near the kitchen or bedrooms.

■ In the kitchen a pantry closet for storing canned foods and appliances is a true plus.

■ A broom closet in the kitchen is needed to store dust mops, brooms, sweepers, ironing boards, and other unwieldy items.

GENERAL GUIDELINES

Make sure the size and arrangement of your rooms allows for flexibility of living arrangements. Check that your house plan has enough windows to provide plenty of natural lighting and cross ventilation, yet that the windows will not severely limit the wall space for your furniture and decorations.

COMMON MISTAKES CHECKLIST

Before you go to the next chapter, on size planning and future expansion, here's a checklist of things people often overlook or fall for:

—No separate entranceway or foyer to receive visitors.

—No window or peephole in the front and rear doors, so the occupants can't see who's knocking or ringing the doorbell.

—No roof overhang or similar protection over the front door.

—An isolated carport or garage with no sheltered direct access from the car to the house.

—No direct access route from the driveway, carport, or garage to the kitchen.

—No direct route from the back or side yard to a bathroom so children can come in and out with a minimum of bother.

—Are gas, electric, and water meters planned for the inside of the house? In the garage or basement? If so, move them outside to eliminate the need for meter readers to clomp through the house every other month.

—A fishbowl picture window in the front of the house exposes you to every passerby.

—Accident-inviting basement doors that open inwards toward the cellar steps.

—Walls so cut up by windows and doors that furniture placement is extremely limited. Plan ahead to accommodate your furnishings. Is there sufficient wall space? Develop a sense of scale and dimensions as you evaluate room sizes and window and door locations. The height of the window sills are important factors. Desks, bureaus, chests, dressers, and buffets all require wall space. If the window sills are high enough, some of the furniture can be placed beneath windows. Many major furniture pieces are between 30 and 32 inches high.

—Windows in children's rooms that are too low for safety, too high to see out from, and too small or difficult to escape from in case of a fire.

—A hard-to-open double-hung window over the kitchen sink is a big pain in the neck. An easy-to-crank casement window is best here, and a sliding window second best.

—A window over the bathtub causes problems. It can result in cold drafts as well as rotted window sills from condensation.

—Bathrooms located directly in the line of sight from living areas, or directly in view from the top of the stairs so everyone knows when others are using the bathrooms, makes for embarassing situations.

—Having only one bathroom is especially tough in 2-story houses and split levels.

—No light switches at every room entrance/exit.

—No lights or electrical outlets on a porch, patio, or terrace.

—No lighting outside to illuminate the approach to the front entrance.

—Noisy light switches that go off and on like pistol shots. Silent switches cost only a little more.

—Child-trap closets that can't be opened from the inside.

—Small closets that are hardly large enough for half your wardrobe. Also watch out for narrow lost doors that keep much of the closet out of easy reach unless you happen to use a fishing rod. Be careful of basketball-player shelves too high for a person of normal height, and clothes poles fastened so low that dresses and trousers can't hang without hitting the floor.

—Avoid a situation where there is no room for expansion. Sometimes, due to how the dwelling is placed on the lot, or because of building or zoning restrictions or construction methods, a house simply cannot be expanded.

—Watch out for rooms that are too small to be practical. There's competition among builders and developers to get the largest number of rooms in a given square footage, at the lowest price. A dining room is too small if you cannot walk around the table and chairs.

—A floorplan that provides poor circulation in and

out of the house and from one room to another.

—No interior zoning: the living, working, and sleeping areas are all jumbled together, each infringing on another's integrity.

—No consideration to the number of floor levels—one, 1 1/2, two, or multi-levels—that offer the most advantages and greatest living conveniences to *your* family.

—A house interior that's dark and drab from a lack of ample window and glass placements. Strategic windows and glass sliding doors can go a long way to make your house bright, cheerful, and attractive. But don't overdo them; hang on to your privacy as well.

—A kitchen that's situated at one end of the house, not centrally located.

—A poorly designed kitchen. An inefficient work triangle, skimpy counter space and storage, no place to eat in comfort, and a lack of outdoor access. If any room of the house deserves the most attention to detail, it's the kitchen.

—Inadequate storage space throughout the house might not become apparent for a few years, but when it does, that lack of space will be most frustrating. Make sure the closets are large enough for storing household items, linen, and laundry, as well as for personal possessions—seasonal and routine items.

Chapter 4

Size Planning
and Future Expansion

Before you begin drawing your final plans and prints, within the best of your abilities try to determine how large your house should be and whether or not you can realistically afford it just yet. If your dream house is financially out of reach, face the facts. You'll have to decide exactly what's necessary and what can be scaled down, completely eliminated, or added on at a later date. If you elect future expansion as the best option, then plan for that expansion long before your new house is begun.

To start you out on the right track, this chapter covers two important subjects:

1. How to arrive at the correct size house to suit your present and future needs.
2. How to prepare your house for future expansion.

DECIDING ON THE SIZE OF YOUR HOUSE

The size of a house is generally expressed, as mentioned earlier, in total square footage of finished floor area. This is the key figure used today for determining building value.

Often a higher-priced house offers more square footage for the money, or more value in space per dol-

lar than that offered by an inexpensive dwelling. Thu cheaper houses are frequently more costly per squar foot. Of course, it also depends on special items an options built into each home but, all things equal smaller houses tend to be more expensive per squar foot of living space.

That holds true because smaller houses incur over head costs comparable to those of larger houses, anc in smaller dwellings those same overhead costs mus be spread out over a lesser amount of square footage. In a sense, it's like buying groceries in bulk. The smaller packages usually cost more per ounce or per pound than their larger-size counterparts.

To get a handle on how the costs in a typical house and lot are broken out, those expenses can be separated into three relatively equal categories (Fig. 4-1): one third of the costs result from the land and improvements to that land. Improvements can include a water well, septic system, utility connections, landscaping, a driveway and sidewalk, and even road construction if needed. Another one-third of the costs go to the house structure from foundation to roof, including the house shell and entire framework. The remaining one-third of the costs come from the vital organs of the house, which include the plumbing, heating, and electrical systems,

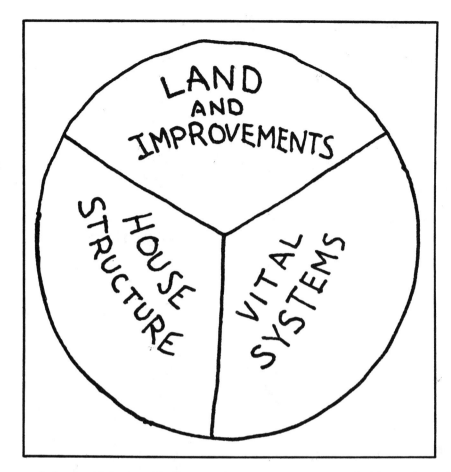

Fig. 4-1. A breakdown of house/lot costs.

and also the kitchen and bathrooms.

If you study that division of costs, it becomes evident that when the first and last categories remain fairly constant in price, you can considerably add to the size of the house by increasing the dimensions of the basic structure at a substantially reduced cost per square foot of living space.

Here are some pointers to weigh when size planning:

1. When considering the overall size of your house, the number of bedrooms can be used as an acceptable guide. At least three bedrooms are recommended even if you have no children or are planning no children in the future. A 2-bedroom house is for the most part more difficult to sell because fewer people are looking for them. Anyway, you can always turn a spare bedroom into a study, sewing room, or game area.

2. To economize on materials and labor, view the space inside your house in terms of actual requirements. Determine what each individual room will be used for, and then decide on their size and shape. You might realize that certain rooms such as walk-in closets are a waste for you.

3. When designing and planning rooms, consider your present dwelling. Are the rooms there too small? Or too large? Think about rooms in other houses you've been in. Focus on ones that are close to what you want. If you can, measure the ideal rooms to leave no doubt in your mind. And don't leave out the work area in the kitchen that will have to accommodate appliances, cabinets, and closet space.

4. The housing requirements of a typical family change about every five years, as time marches on. You might want to keep that in mind while laying out the size and shape of your new house. How do you fit into the following typical household scenario?

—Using a young couple just starting out together

49

as the initial family unit, all their living requirements could be contained in a small dwelling, from a 1-bedroom kitchenette/bath apartment to a 2-bedroom house.

—Between 5 to 10 years, with the addition of one or two children, the needs go up to at least two and possibly three bedrooms, larger living spaces, and more storage.

—Between 10 to 15 years, the typical family, if three children are present by then, needs more sleeping space, a second bath, and more living and storage area.

—From 15 to 25 years, this 10-year span while the children are maturing is likely to be stable unless the wage-earner's business requires a change of locale or an upgrading of living standards.

—At 25 through 40 years, during each 5-year period it's likely that one child will fly the nest—for marriage or at least for college or career.

—Over 40 years, the family is once again down to two people, and large amounts of space can become a hinderance rather than a help.

Planning For Future Expansion

For any of a variety of reasons you might not be able or be willing to start out with the size house you'd like to eventually have. If you want to avoid jumping from house to house every few years as your housing needs increase, consider the following:

1. Initially plan a house that can be easily expanded at a later date.

2. Review the house design you want in relation to the ease or complexity of making additions. Allow space and structure adaptability to add a garage, foyer, family room, bedrooms, second bath or whatever it may

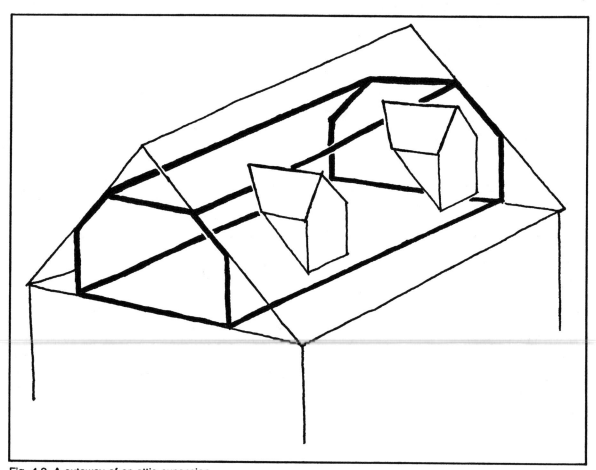

Fig. 4-2. A cutaway of an attic expansion.

be that you didn't initially build into your house.

3. When more space is required of a single-story ranch or 2-story house, the construction of an addition to the rear or side of the house may be the best (but not the cheapest) answer. In both types of houses, it's a good idea to begin with dwellings that are able to accept additions without having their overall architectural character ruined by those additions. Consider how roof slopes, window types, exterior finishes, dormers, and similar features will be affected by the modifications you're planning. Also pay close attention to property line construction setbacks, so they won't rule out your ability to expand in other ways at some future date.

4. If your expansion plans call for knocking out a wall or walls later to enlarge part of your house, plan for that *before* the house is completed. Leave plumbing out of the walls that will be removed. Support the above ceiling or floor independently of the affected walls. Build the framework for the new entrance right into the wall so all you need to do is knock out the rest of the wall when the time is right. Electricity is not a problem; light switches and outlets are easily moved.

5. When working with large unfinished places that will eventually be rendered into living areas, place doors and windows in such a way that easy-to-install partitions can be erected to turn big, open spaces into several individual rooms, as desired.

6. Although it might not be the way for you, there's no doubt that building up or down is cheaper than building out.

When Expanding Upwards

If you start out with a Cape Cod, a 2-story, or even a ranch house having unfinished upper levels or attics, there are a number of technical points to be addressed when planning the initial structure.

To permit maximum expansion without resorting to exterior modifications, the roof slope must be steep enough (or a gently sloped roof raised up enough) to provide adequate head room for new living areas. For example, on a house having a width of 24 feet or more, make sure your roof slope—if the roof is not stepped up or raised up substantially—has a minimum of 9 inches of height for every 12 inches of travel or run toward the peak. Also make certain the floor joists of the unfinished floor or attic are large enough to carry typical floor loads.

Whatever your plan, you'll have to comply with local building codes that usually require one-half of a room's ceiling area to have a minimum floor-to-ceiling height of 7 1/2 feet. Thus, if your attic space has windows at the gable ends, sufficient head room, and properly sized floor joists, the construction of one or more rooms is a relatively simple and inexpensive project, with no need for exterior remodeling. Extensions of existing heat, air conditioning and return air ducts that had already been installed to and capped off at the area to be finished, and activation of the already rough plumbing and wiring should result in a satisfactory expansion of your existing house at a nominal cost.

To attain increased floor area and more wall space for windows, dormers are recommended. Two types of dormers may be used: the shed dormer, and the window or gable or "eye" dormer (Fig. 4-3).

The shed dormer is the most practical because it adds a great deal of both floor and wall space and is relatively simple to construct. However, because it's not as pleasant to look at as the window dormer, it's generally built at the rear of the house. A window or "eye" dormer offers less space, but it's still a major improvement over nothing. The illumination, ventilation, and increased floor space it brings can transfer an otherwise little-used space into a cozy bedroom or study. This addition of dormers can often be accomplished without removing the existing structure's entire roof.

When Expanding Downwards

If you start out with a house having a basement, and you plan on eventually using that basement for additional living space, then keep the following points in mind:

1. Make sure you have enough clear area to allow for basement expansion. You wouldn't want your water heater, furnace, stairs, and other items to be laid out so they break up the entire cellar.

2. Use larger-than-normal basement windows for extra light and ventilation, and consider an outside access so you don't have to walk through the main part of the house every time you want to go in or out of the cellar.

3. If you plan to add a bathroom to your basement in the future, be sure to locate the sewer drain pipes below the cellar floor. If the sewer pipes enter the basement midway up the wall, you're in for extra expense and inconvenience when you have to add a pump system to push the refuse upwards.

In conclusion, you'll find that interior expansion is always more economical than exterior expansion. But

Fig. 4-3. Shed and gable or eye dormers.

exterior expansion provides extra space without sacrificing attic storage or basement areas.

No matter which direction you plan to expand in—whether it be up, down, or out—make sure your furnace/air-conditioning units will have the capacity to handle the extra load.

All in all, a small increase of effort in the initial planning, plus a little additional cost, can make enormous differences in the ease and expense of expansion at a later date.

Chapter 5

Prints and Drawings

Now that the first four chapters have touched upon most of the major options involved when a house is planned, such as dwelling styles, types, floorplans, individual rooms, and sundry accessories, you can see how inefficient it is to just go to a builder and describe what you want in vague generalities.

WHY YOU SHOULD CREATE THE PLANS

To walk into a skilled contractor's office and hint merely at what you think you need, and then let him charge ahead and build it, using his own discretion, is in a sense like going into a clothing store and—without even browsing through the racks—telling a salesperson who knows neither your size nor your taste in clothes to please select an entire suit of clothes, right down to the shoes, that you'll be sure to like. The salesperson would be astonished. No one ever leaves that many decisions up to a total stranger, however knowledgeable about clothes the salesperson might be. Naturally, the salesperson would start asking questions, inquiring what *type* of clothes? Sports, leisure, business, or social? What *size* slacks do you take? Made of which materials, in what colors? For winter or fair-weather use? And what price range are you looking in?

To take the analogy a step further, what about the purchase of an automobile? A new car is a major expenditure to most of us. You decide what style, type, and size vehicle is best suited to your needs, desires, and pocketbook. Few people will stroll into a dealer's showroom and request a "four-door sedan," and fewer yet will ask a salesperson to simply suggest something out of the blue. Try it sometime and see what happens. It's human nature for a salesperson to try to convince you to purchase whatever is being pushed at the time by the dealer. And salespersons are good at that. That's why they're in sales.

Builders are a different breed. Builders are accustomed to dealing with people who walk in cold, having only a foggy idea of what they want. By far, most builders will play fair with their customers, will try to help a potential homeowner arrive at a suitable house design. Some, however, especially during times when plenty of work is available, become irritated by customers who come expecting counseling services to help plan an entire dwelling from scratch. These builders will invariably suggest a stock plan—a house they're thoroughly familiar with, perhaps one they've built within the past few months.

Other builders, when faced with an undecided client, will ask dozens of questions such as "How large do you want the bedrooms?" and "What size garage?" Then they try to piece together an appropriate plan from your answers.

Somewhere along the line each builder is going to have to know how much material he's going to have to order. How else can he bid the job? To know that, he's going to need a set of plans, a set of blueprints and drawings to use as a guide. And just as a cook needs to follow recipes in the kitchen, so does a builder need prints and drawings to follow in the field.

If you're still not convinced that you should participate fully in the creation of the prints and drawings by making many of the optional construction decisions yourself, then consider that the builder wants you to be pleased with the finished product, but rarely does that concern go too far past the technical level. Do you really think it will register on a contractor if you say "Gee, you should have made the family room 2 feet longer so we could fit our overstuffed sofa in front of the fireplace." What else can he say but "I'm sorry. It was on the plans you approved." To mention that the basement has heaved up and it's full of water . . . now *that* will register. Builders can handle complaints centering on the actual construction—the carrying out of the original agreed-upon plans. But to chastise a contractor after he erects the house that the garage should have been located on the other side of the dwelling to act as a shield against prevailing winter winds—your efforts will likely fall on deaf ears.

FINAL SET OF PLANS

No matter where you get them, your final set of prints and drawings should in some way address the following:

Floorplans. One blueprint for each floor. These are working drawings showing overall dimensions, room and hallway sizes, location and sizes of doors and windows, location of interior partitions and wall thicknesses, location of electrical switches, plugs, and appliances, plumbing fixtures, water supply, drains, and other information needed to complete the house (Fig. 5-1).

Lot Plan. The lot plan shows the original contour of the land, the proposed finished contours, original trees and the ones which will be left standing, the location of a water well and septic tank if applicable, the driveway, electrical service, and the placement or orientation of the house on the building site (Fig. 5-2).

Foundation Plan. The foundation plan is a draw-

ing that shows the height of the sill (where the house frame rests on the foundation or top of the basement) above grade, the chimney location, the extent of excavation and grading required, plus the location of water and sewer hookups and easements (Fig. 5-3).

Elevation Drawing. The elevation drawing depicts the lines of the house from its four or more sides. You might need a front, rear, and two side views, depending on the complexity of the house. It will show where windows and doors go, and their respective sizes. Exterior material types can also be indicated here (Fig. 5-4).

Interior Sketch. If the interior has any unusual or distinctive features, you might want artist's renditions to help you visualize what the interior will look like. An example would be an elaborate bathroom or special kitchen cabinet setups (Fig. 5-5).

Perspective Sketch. The sketch shows what the completed house will look like from the outside, including landscaping if desired.

Finally, you should request at least four and possibly five complete sets of plans—one for the builder, one for the lending institution, one for the local building department, one for yourself, and, just in case, an extra set for your records.

PREPARING YOUR PLANS

There are a number of ways you can complete the plans you'll need before going to contractors and soliciting bids.

1. Have an architect do them for you, with your help.
2. Purchase a set of stock plans through a mail-order company.
3. Prepare them yourself with help from books, magazines, and friends.
4. Use packaged planning kits.
5. Employ a personal computer and special home designing software.

Architects

The truly great houses in the world have practically all been designed by architects. Indeed, architects are wonderful at custom engineering houses that require complex construction techniques or unique materials. If you've got a building site in the middle of a swamp, or on a ledge of solid rock, an architect will be able to figure out how to build on it. And due to the nature of

Fig. 5-1. A floor plan.

Fig. 5-2. A lot plan.

Fig. 5-3. A foundation plan.

Fig. 5-4. Front and side elevations.

Fig. 5-5. An interior sketch.

their business, they stay abreast of the latest energy-saving and other technical innovations. Whether or not an architect is right for you depends on the complexity of your design and building site, and your bank account. Like any professionals, architects are well compensated for their efforts.

Stock Plans

If you'd prefer not to draw your own initial plans, and hiring an architect is out of the question, consider stock prints and drawings that have already been prepared by professionals and are available through mail-order companies. Even if you can't find exactly what you want, plans can always be modified to some extent by your builder/architect. Remember that care must be taken, because major changes can cause structural problems—especially when altering a foundation to match a lot or when changing load-bearing walls.

One advantage to using stock plans is simple economy. They're inexpensive themselves, and their dimensioning of rooms and spaces for suitable structural members are made with a minimum amount of waste by using standard lumber. This is because stock plans are prepared by experienced home design experts who have accumulated a broad range of planning ideas over a considerable length of time. Some are licensed practicing architects.

Do-It-Yourself

To help you commit your ideas to paper, there's a wonderful invention available at nominal cost: graph paper. Graph paper is an ideal medium for expressing room sizes and relationships of a floorplan. A good scale to work with is 1/4-inch equals 1 linear foot. You'll also need a roll-up steel or plastic tape measure at least 25 feet long, plus some home design books and magazines that will give you average dimensions and proportions of various features in houses such as hallway widths, door heights, and wall thicknesses. Many of these dimensions can be found within this book.

Once you finish your plans the best way you know how, the builder can use his knowledge of construction materials to point out where adding a few feet here, or taking away a few inches there, can result in substantial savings. If plywood sheets come in 4- x -8-foot sheets (and they do), it would be silly to spec out a room that would require coverage by ten sheets of 4 1/2- x -8-foot sheets. That would result in sizable waste.

Builder/architects can take your estimates and customize them—still keeping your plan's original integrity—to dimensions that lend themselves to standard building material sizes in order to reduce waste and give you more for your money. Builder/architects have a realistic feel for how to stretch materials to the maximum. They'll be able to take your rough drawings and show you where you can pick up efficiencies, and where you could use substantially less materials—perhaps by merely altering a particular dimension in some minor way.

Packaged Planning Kits

If you'd like to explore a simpler method of creating your own house plans, packaged planning kits are available from a variety of companies and publishers. They consist of scaled grid sheets or boards and appropriately-sized furniture, construction parts such as windows, doors, and walls, and even landscaping trees and shrubs. They're simple to use and they can help plan how much furniture your house will need, and where it should go.

The kits come with instructions and most are designed with a 1/4-inch scale. They can help you reduce the cost of professional drawings; the designer can do the drawings a lot faster if he or she just copies from your layout.

Most kits are manufactured in two-dimensional style, but some offer three-dimensional planning—a great help if you have a difficult time conceptualizing and visualizing whether a room design you are planning would fit the space allotted. The 3-D kits tend to be twice the scale of the others to give you a better grasp of the cardboard, plastic, or foam furniture replicas. These life-like objects include plumbing fixtures, televisions, stereos, sofas, toilets, even pianos. Some kits use decals, and plastic or cardboard printed in 3-D style. Other kits supply foam furniture that you can paint or even upholster to try out color schemes.

Computer-Assisted Plans

For the ultimate in home design at your fingertips, turn to the personal computer. Architects have been using CAD (computer-assisted design) graphics and engineering packages for years, but now that the personal computer has become so affordable and easy to use, it's certainly an interesting option.

Software is available to help you draw your own designs on a monitor or computer screen. And computer screens are great to work with because they're so easy to change, with no harm done if you make a mistake. With some programs you can look at the drawings in 3-D; you can rotate them right on the screen to see all angles of a room or entire floorplan. You can save the drawings on floppy discs and can print them out on a printer whenever you need a copy.

The details on these programs are incredible—down to the patterns on wallpaper. You can ask for overhead and side views, and you can even fast forward a newly planted landscaping scheme to see how it will look years in the future, when fully grown.

These accurate designs can help reduce the effort and cost of having professionals complete your set of plans.

BEFORE APPROACHING THE BUILDER

It's important that your plans are as detailed as possible before you meet with contractors; otherwise you'll be at a disadvantage in regards to the bids. In order to protect themselves against work that is not clearly spelled out, they might pad their bids. At the same time, some could attempt to make you do without items you are taking for granted which are not initially specified.

Part 2

How to Build It

If you never even lift a hammer throughout the entire construction process, it still pays to know how a house can and should be put together. How else could you make educated decisions when arriving at the building specifications that are so crucial to establishing a dwelling's quality of construction and overall durability, safety, and comfort? Often the difference between mediocre and excellent construction involves a ridiculously small materials cost. Knowing construction methods and materials will also assist in your dealings with whichever contractor you choose.

While it's not necessary to be able to recite good specifications from memory, it's important that you have a sense of how the contractor should go about fulfilling his obligations. Chapters 6 through 24 cover what should happen from foundation to rooftop, and point out essential decisions that need to be made and should be made with your input.

Chapter 6

Footers and Foundations

No house will be solid enough to reach its full life expectancy unless it's carefully constructed on adequate footers and foundations.

FOOTERS

A footer, as the name implies, is that lowest part of a house upon which the rest of the dwelling is placed. And like anyone's feet, if they're not firmly planted on the ground, proper posture or position of the rest of the body (or in this case, the house) becomes difficult if not impossible to maintain. A house footer serves to firmly situate the building onto and into the ground. And because it plays such a basic and critical role, it's important to make sure that the footer is done correctly. It's largely inaccessible once covered and landscaped over, and if ineptly constructed, major problems will result in huge expenses and inconveniences to homeowners.

Few people realize exactly what a house sits on, and how the house is joined to the earth. There are right and wrong ways to construct a footer; the main idea is to evenly spread or distribute the weight of the house over a large enough area of soil so that settling or moving will never occur. The chief enemies are gravity and

time. Downward pressure that's not evenly supported from below will ultimately result in cracked floors, foundations, and tell-tale symptoms such as doors that will not close, cracks in plaster, and worst of all, an obvious tilting of the house.

Construction Guidelines

Consider the following when planning your footer:

1. It must be built on solid earth if conventional building practices are to be followed. Sure, some houses are constructed on swampland—but they require extra engineering, expense, and effort. The standard-type house should rest upon solid ground. This means shy away from a building lot that has been filled in and graded to bring its surface up to a respectable level.

2. If you realize the land consists of recently filled loose soil, and you still desire to build on it, several options exist. You can have the soil compacted through mechanical means. Contractors can use a heavy-duty tamper that hammers the soil down, plus huge rollers that pass back and forth over the surface, compacting as it moves. Together these machines press the soil until

the proper "load level" or ability to support weight is reached that's similar to that of virgin earth (undisturbed ground).

If the depth of uncompacted soil is too great to be efficiently compacted, and a deep cellar might lend itself to the house's structure, you could excavate and remove the disturbed ground until you hit virgin soil, then build from there.

A last and most radical alternative could apply if virgin ground is entirely too far below grade to fix by either compacting or excavating. Here the solution is to architecturally design a one-of-a-kind footer of the floating nature to adequately support the house you plan. In most cases, though, the expense is too great to bother with. It is better to find a more suitable site.

Either of the first two options will also cost more than an ideal lot, but they're necessary if you are to prevent the house from shifting and being subject to the difficult-to-correct ailments described earlier. Unless concrete is poured onto undisturbed or properly compacted soil, its weight, when coupled with that of the foundation and the rest of the house, will slowly press down the loose soil below. This will result in cracks, heaving, and tilting which naturally will disturb the framework of the rest of the house. Suddenly the windows won't go up as easily, the doors will no longer fit their frames, plaster walls will crack—all indicating that stressful pressures are at work that will probably brand the house as shoddily constructed.

3. The frost line must be taken into account. The earth is an insulator, and in northern sections some of the top ground freezes and offers protection to the unfrozen soil below. The frost line is an imaginary undulating plane located at some depth below the surface, or the average depth of ground that can be expected to freeze during winter, year after year. In the northern parts of the United States for instance, it ranges from about 2 to 5 feet below typical ground level. If a footer is not placed below that frost line, the alternate expansion and contraction of the earth above the frost line might cause the footer to move—to heave upwards or list downwards, causing cracks to occur in rigid concrete footers and foundations, with their accompanying ill effects. In general, national building codes recommend that footers should be located at least 12 inches below the frost line (Fig. 6-1).

4. The type and condition of the soil must also be taken into consideration. For example, it's an unwise practice to build on organic type soils such as peat:

they haven't the proper load-bearing strength to support the weight of a house. Groundwater content likewise influences the ability of the soil to support weight, and greatly affects the installation of proper drainage to prevent water from seeping into lower levels. Also be careful not to build over a spot where a large tree root system still exists: the roots will slowly disintegrate, leaving voids that will undermine the footer and lower level floor. Major tree roots should be removed and the holes left by them filled in and compacted.

5. Naturally, the contour of the building site and the distribution of the house's weight can have a major effect on the footer's construction demands. A 2-story dwelling with one floor directly above another, even if it weighs the same as a multi-level that's spaced out over more area of ground, will distribute its weight in a different manner—thus the need for a different footer than that of the multi-level's. In many cases, footers must be custom-designed for lots having substantial slopes. "Steps" are commonly included to compensate for grade differences (Fig. 6-2). When preparing for a block foundation, as a general rule the depth of each step should be in a multiple of 8 inches, which happens to be the height of the standard concrete building block. That helps build uniformity into the foundation so you won't end up having to add a half-course of block somewhere along the top of the foundation, with the accompanying waste and bother.

6. When a full basement is specified for a house there should be an excavation of proper width, length, and depth to accommodate the foundation wall, columns, and chimney stack footers, the basement floor, and an adequate drainage system (Fig. 6-3).

7. Pour separate footers wherever steel-support columns will be located (concrete columns that support the house's main steel beam or beams). This helps relieve downward pressure and will help prevent the basement floor from cracking.

8. A bed of gravel must be laid under and around the planned footer and foundation wall, no matter which type of footer is used (Fig. 6-4).

9. Footers for single-story and 1 1/2-story houses should be at least 8 inches thick and 16 inches wide; for a 2-story dwelling, 12 inches thick and 24 inches wide. Larger footers are needed for homes constructed on unstable earth or on filled land. In general, a rule of thumb is that footers be at least as deep as the foundation wall is thick, and twice as wide (Fig. 6-5). With concrete slab construction, contractors often simply in-

Fig. 6-1. A footer and foundation below the frost line.

crease the thickness of the slab under the load-bearing walls instead of pouring separate footers. In all cases the footer must meet the local building code minimum specifications.

10. As mentioned before, if the earth on a potential building site is unstable, you would do best to avoid such a lot in favor of another with virgin soil. If you decide to build on a filled lot anyway, in addition to having the ground tamped, increase the footer's thickness to 20 inches, and the width to 40 inches, double the amount of steel reinforcement rod used, and obtain a new or used solid 1-inch-thick steel cable and lay it through the center of the footer, embedded in the concrete the entire length around the house for extra strength.

11. The concrete used in footers should have a strength of at least 3000 pounds per square inch.

12. Reinforce a footer for extra strength. This applies to all footers, including those for fireplaces and support columns or piers. In normal situations, embed at least two steel reinforcement bars, preferably three, lengthwise throughout the footer (Fig. 6-6). The steel

bars should be 1/2- to 5/8-inch thick, and elevated from the ground during and after the concrete pouring through the use of "foundation chairs" concreted right in, about every 6 feet. The bars should be situated so they will be covered by a minimum of 3 inches of concrete at all points. One-foot overlaps should occur wherever the bars meet.

13. Allow concrete used for footers two days to set to gain most of its strength before anything is done on top of it.

14. Depending on the type of foundation your house requires, there are a number of items that might have to be prepared for while the footer is being installed. These include drains and sewers, plus water, gas, electric, and phone lines. If the necessary holes or trenches are dug for these items while the back-

hoe/shovel is present for excavating the footer and basement, they can be completed at less cost. The backhoe/shovel won't have to come back a second time, nor will the contractor need costly labor to dig them by hand.

15. Special attention to the sewer or septic lines prior to pouring the footer will prevent basic sewage problems. If your house will be connected to a street sewer, this connection should be made at the excavation/footer/foundation stages of construction. The sewer usually runs under a wall footer and basement floor to the main stack location. If a septic system will be used, the same sanitary sewer pipe installation must be made from the septic tank location to the stack.

16. Lastly, make sure the contractor grades and stones the driveway while the footer is going in—before

Fig. 6-2. A step footer.

Fig. 6-3. A straight footer, wall, post, pilaster, and chimney footers.

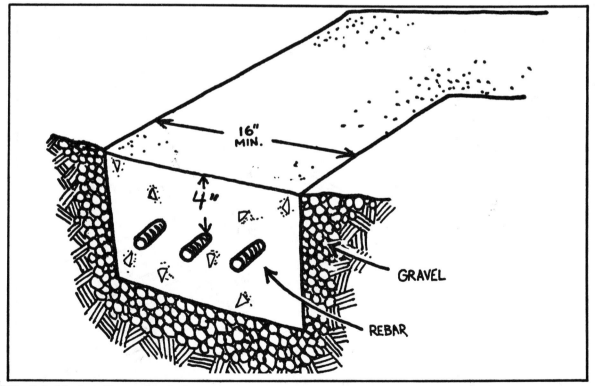

Fig. 6-4. A footer with gravel and rebar.

W = WALL THICKNESS

W

1/2 W   1/2 W

W

2W

Fig. 6-5. Footer design.

REBAR

FOUNDATION CHAIR

Fig. 6-6. Footer reinforcement—rebar on chairs.

the construction of the foundation and rest of the house begins. An early graded and stoned drive where the finished driveway will go is convenient for receiving material deliveries and for simply getting onto and off the site in bad weather. It will also encourage the heavy cement trucks needed for the footer, foundation, and basement floor to pack down the gravel and earth driveway long before the finished driveway will be poured or asphalted. All of the heavy equipment and delivery traffic will result in a stronger driveway base.

FOUNDATIONS

To put it in simple terms, the foundation of a house is what sets directly above the concrete footer and below the wood-frame living levels. Another way of describing most foundations is to call them basement walls. In houses without basements, a slab foundation also incorporates the dwelling's footer in one continuous piece of concrete.

The foundation must be strong enough, whatever its construction, to support the house sills (heavy horizontal timbers or planks attached to the upper part of the foundation to serve as a starting point for the house walls) and other related members of the house structure, as required.

From a structural standpoint, the foundation performs several key functions:

■ It supports the weight of the house and any other vertical loads such as snow.
■ It stabilizes the house against horizontal forces such as wind.
■ It acts as a retaining wall against the earth fill around the house.
■ In some cases, a basement might be needed to act as a barrier to moisture or heat loss.

No matter which foundation type you must have, or elect to have, some general points apply. All foundations, whether slab type, crawl spaces, basements, or any others, should extend above the final grade enough so that wood members of the house are some distance above the soil. That distance might be regulated by local building codes. However, keep in mind as the specs of your foundation are laid out that if local building codes are being used as a guide, they'll help, but they could still fall short of what many people would consider optimum construction. Local specs lay down minimum rules for safety and health, but that's about all. In other words, you might want to go a few steps

beyond what they recommend. In this case, it's better to position the top of your foundation slightly higher, away from the ground level, for added protection against moisture and insects.

When excavating is necessary—and that's likely for any of the foundation types you will consider—have the valuable topsoil scraped off and saved. It should be pushed into a pile and kept out of the way until the house is completed and the landscaping is roughly finished, so you can spread the topsoil around to provide a fertile base in which to plant grass seed or to lay sod. Don't let the topsoil become "lost" amidst the rest of the soil that's excavated and used to fill in around the foundation.

Slab Foundations

The slab foundation, as mentioned earlier, generally involves a combination of footer and foundation into a single slab of concrete (Fig. 6-7). They're popular wherever basements are impractical or impossible to have, in certain parts of the Southwest for instance, or in areas with high groundwater levels.

The "floating" slab is unique in that the finished concrete floor, foundation walls, and footers are reinforced together with steel mesh and metal rods and poured as one integral mass over a bed of gravel (Fig. 6-8). The entire slab thus floats on top of the ground while functioning as the floor of the house. However, the depth of the concrete should not be equal throughout its overall area: the slab must be thicker beneath support walls if the footer is considered part of the slab.

In cases where the terrain is not relatively flat, a footer must be poured separately from the slab, and terraced or stepped down. Then a foundation, usually of concrete block, is constructed to a certain level height so a concrete slab floor can be poured. A slab foundation can be very troublefree and economical when built in adherence to the following construction points:

1. First the site must be properly graded and compacted, and the footing trenches dug.
2. If the slab will be placed directly on soil, make sure the area is free of biodegradable tree roots and debris.
3. Granular fill makes a better base than soil. Bank or river gravel, crushed stone, or slag can all be used, in sizes from 3/8 to 1 inch thick. A 4-inch-thick granular bed of fill will suffice.
4. At this stage all of the following underground utilities should be installed: plumbing, drains, sewers,

Fig. 6-7. Slab foundations.

heating service lines and ducts, radiant pipes, electrical work, and any other public or private utilities. Remember that if your house will have a slab foundation, you're going to end up with one continuous slab of concrete. If anything is done wrong and not corrected after the concrete is poured and before it dries, just think of the trouble and expense you'd have to go through to simply get at the problem, let alone to fix it. Take precautions to see that the utilities are all accounted for, and installed in a safe and correct fashion.

When run beneath a concrete floor, water lines should be laid in trenches deep enough to prevent freezing (below the frost line).

Pressure check the plumbing for leaks before the concrete is poured. The water should be turned on with all faucets and shutoffs closed to make sure there are no leaking seams, cracks, or holes through which concrete could seep into, to solidify and plug water or drainage lines.

Make sure copper pipe is wrapped in a rubber or plastic tape wherever it will come in contact with concrete. An undesirable chemical reaction occurs when bare copper meets concrete. It's another simple precaution that can save a lot of time, expense, and inconvenience at some later date.

5. The subgrade for the slab should be dressed

Fig. 6-8. A floating slab foundation.

up or smoothed out in preparation for the concrete pour. Whether the subgrade is gravel or slag, it must be thoroughly compacted. It should end up higher than the surrounding grade so water will drain away from the house and so the top of the slab is comfortably higher than ground level.

6. At this stage, a vapor barrier is placed over the subbase to stop the movement of liquid water and water vapor into the slab. Among materials used as successful vapor barriers are heavy-duty sheets of roofing material, polyethylene plastic, and construction paper. They act as both an insulation and moisture control, holding dampness in the ground rather than permitting it to penetrate cracks that could form in the slab foundation.

7. The slab should now be reinforced with steel rods. Although the steel reinforcement will not assure the prevention of cracks, it might reduce the magnitude of cracks that would otherwise occur. For best results the steel is placed horizontally through the middle of the slab, and held in position until the concrete dries. A common practice of placing the reinforcement rods on the subbase, then pulling it up to the center of the slab with a rake or hook should not be permitted. It's virtually impossible to accurately control the location of the rods or wire fabric with this method. Instead, steel bridging or "chairs" can be anchored into the subbase. The chairs will stick up to about the midpoint of the slab and will correctly hold the reinforcement when the concrete is poured. Naturally, the chairs will be concreted right into the slab, along with the reinforcing steel.

8. In addition to steel rods, welded wire fabric or mesh is required for slab foundations. Again, this material makes the concrete less likely to break loose from itself. Wire joints should be overlapped 6 to 8 inches or at least the width of one of the openings, unless the entire job can be done with a single piece.

9. In an application where a foundation is required to hold up a concrete slab (with the interior filled with bank gravel to support the slab) then the top block—if it's a concrete block foundation—or the top ledge—if it's a poured foundation—should be a header or shoe block form in which a portion of the block/concrete has been cut/left out to provide a base for the concrete slab to be fastened to or supported on (Fig. 6-9). Vertical reinforcement rods should also come up from

Fig. 6-9. A foundation that supports a concrete slab.

the foundation walls and be bent into the slab. This will help hold the walls to the floor slab. More about this type of construction can be found in the crawl space/basement section of this chapter.

10. Before the concrete is poured, a means by which the house sill—the wooden horizontal planks that support the main upper structure of the house—can be secured to the foundation must be arranged. Anchor bolts can be positioned about every 4 feet around the perimeter of the slab so they'll be embedded into the slab when the slab dries and cures.

11. When the slab is framed and ready to be poured, it should hold the top of the floor about 8 inches above the ground, and the surrounding grade should be sloped way from the foundation to keep water running away from the house. If an elevated floor slab is used, it should be a minimum of 4 inches thick at any part.

12. If a smooth finish is desired, specify that the concrete should be troweled by hand or by power-driven machines. A textured finish can be obtained by dragging a broom across the surface before the concrete is fully set.

Advantages:

1. They're very economical to build, especially when compared to crawl space or full basement foundations. Most slab foundations take much less labor and time to construct.

2. They're worm- and rot-proof.

3. They can't catch fire.

4. They're basically wearproof and are certainly more secure than any other kind except solid rock.

5. They can store heat from the earth and are naturals for use in a passive solar heating system.

6. They require little insulation from the elements.

7. As foundations, slabs are outstanding because they simultaneously act as one gigantic footer. Consequently, they impose the lowest soil loading per square foot of all foundations.

8. They experience less problems from ground moisture . . . there's no leaky basement from a slab.

9. The slab is more adaptable to filled or unstable soils where conventional foundations would settle unevenly and crack.

10. The slab-on-grade foundation eliminates the need to frame a floor on the first level.

11. They are not affected by underneath drafts.

12. Vinyl flooring and wall-to-wall carpeting can be installed directly onto the top surface of a slab.

Disadvantages:

1. If a problem occurs with a utility that's concreted into or positioned beneath a slab, it's extremely expensive and troublesome to access the malfunction, to make repairs, and to restore the foundation the way it was.

2. Slab foundations can be efficiently used only on relatively flat lots. They require substantial sitework when employed on uneven ground, whereas a crawl space or basement foundation readily adapt to hilly terrain.

3. Floors constructed over a crawl space or basement foundation are easier on the feet and legs.

4. Plastic and other moisture barriers must be punctured for pipes and electrical wires to pass through, thus allowing some underground dampness to rest against the bottom of the slab.

5. Because the slab is mostly below ground, no ventilation reaches its lower surfaces. The slab tends to adjust to room temperature very slowly and instead follows fluctuations in ground temperatures whenever they occur.

Concrete and Block Wall Foundations

Many houses built today sit on foundation walls that form either a crawl space under the house, a partial basement, or a full basement. These types of foundations are more common in northern locations where deep frost lines are encountered, but are also found, conditions permitting, in the South. They're constructed of either solid poured concrete or concrete blocks, and if either has the edge over the other, it's concrete blocks. Both foundations will not only support a dwelling, they'll also protect it from water, frost, and insects while providing (when desired) a basement to be used for storage or expanded living space.

If proper foundation construction is not followed the result will be a below standard foundation and possibly a house that tilts, floors that sag, walls that crack and leak, doors that won't fit their jambs, and windows that won't open or close—all of the same defects that can also be attributed to a poorly executed footer.

Although crawl spaces are frequently left with soil or gravel floors, both the solid concrete and the block wall foundation floors should be poured with concrete to provide cleanliness, to prevent moisture and insect encroachment, and to supply a useful floor for storage or additional living areas.

The National Building Code requires a foundation to start at least 1 foot below the frost line. Local building codes can tell you how deep the frost line is in your area. Should bedrock (solid rock) be encountered before the prescribed depth is reached, digging can stop because bedrock will not move no matter what happens or how cold it gets. Even if you build where there's never freezing temperatures, still see that the foundation is situated at least 1 1/2 feet below grade (ground level) to assure a firm and level base for the framing structure.

Crawl Space Foundations

These foundations are cheaper to construct than basements, and acceptable when the storage, utility, and living spaces otherwise found in basements are neither needed nor desired (Fig. 6-10).

Advantages:

1. A crawl space foundation is cheaper to construct than a partial or full basement.
2. Because a crawl space is relatively low to the ground, there's not much risk of cracked walls.
3. The crawl space foundation takes considerably less time to build, thus speeding up the overall construction time of the house.
4. A crawl space provides ventilation below the first floor, separating the living areas from contact with the ground and letting the floor follow suit to the temperatures maintained by the living spaces (unlike a slab, which is more affected by ground temperatures).

Disadvantages:

1. A crawl space foundation is practically useless for storage or living space (for humans).
2. A crawl space can't accommodate large or tall appliances such as water heaters or furnaces.
3. Crawl spaces will attract a variety of small furry creatures (notably rabbits, squirrels, chipmunks, skunks, possums, and mice).
4. Water or sewer lines that run through crawl spaces must be insulated extra well to prevent pipes from freezing.

Partial and Full Basement Foundations

There's a convincing argument that since the house will be placed over a footer anyway, you might as well put the potential space below the regular living level to good use, too (Fig. 6-11).

Advantages:

1. The functional living areas of the house can always be expanded into a basement's lower level. A recreation room can be installed there at a minimal cost.
2. The partial or full basement can easily accommodate a water heater, furnace, and other major appliances such as freezers, washing machines, and clothes dryers.
3. A basement can include a separate entranceway into the house on the lowest level.
4. These foundations provide considerable storage and workbench areas.
5. Like the crawl space, a basement also insulates the main living areas from the ground.

Disadvantages:

1. Basement foundations are more expensive to construct.
2. Because of their height, the walls of a basement foundation are more likely to crack and develop problems.
3. A basement foundation can at times be virtually out of the question in areas where the water table is high.

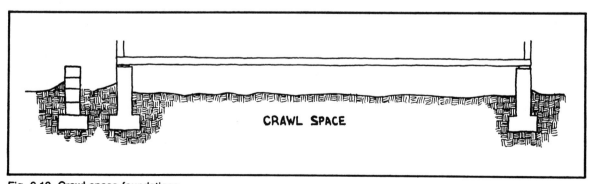

CRAWL SPACE

Fig. 6-10. Crawl space foundations.

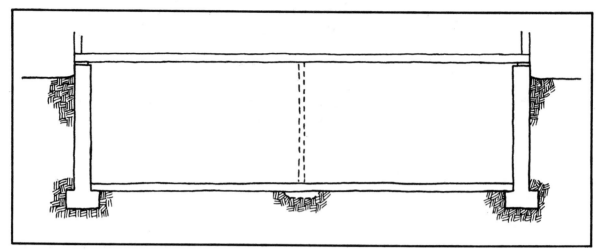

Fig. 6-11. A basement foundation.

4. Basement foundations take a relatively long time to construct, and string out the entire housebuilding process.

FOUNDATION CONSTRUCTION

Once you decide on the type of foundation that you think best fits your requirements or is intrinsically suited to the house you are planning, you then must decide how that foundation should be constructed. Two options are concrete block or poured concrete. Either can be used to provide crawl space foundations (usually built up from a footer having walls 18 to 24 inches above ground level), or basement foundations that are typically 7 feet high on the inside.

Concrete Block Construction

You probably already know what concrete blocks are. They're rather rough-feeling, heavy, gray, and have several rectangular holes running through their insides, vertically (Fig. 6-12). When installed, they're laid one course or row on top of another, staggered so their vertical joints don't coincide with each other. The staggering of joints makes for a stronger interlocking bond. At the same time, the center rectangular holes are large enough to overlap so the courses can be tied together with reinforcement rods that are inserted vertically and then filled with wet concrete.

On a foundation, the concrete blocks begin at the footer and are laid to form the desired height of the house foundation or basement. On houses using brick exterior finishes, the brick may also start with the footer and follow the house blocks right up. This will help support the brick veneer and strengthen the entire foundation. An alternative for brick exteriors is to go with a wider foundation block that provides a ledge at the outer top of the foundation walls on which the brick can be conveniently laid or begun from.

Advantages:

1. Concrete block walls permit a stop-and-go schedule during construction.

2. Concrete block walls are easier to repair than poured concrete walls are.

3. Concrete block construction is often preferred by builders costing out the foundation job because block construction eliminates the need for concrete forms.

4. Concrete block walls absorb sound better than solid poured walls do.

Disadvantages:

1. Concrete block walls are strong, but not as strong nor as impenetrable as solid poured concrete walls.

2. A concrete block wall is more likely to develop small cracks that can allow air infiltration, moisture, and even insects inside the foundation.

Solid Poured Concrete Construction

Solid poured concrete foundations, even though not as common as block foundations, are desirable in a number of situations.

8" CEMENT BLOCK

16"   8"

8"

10" CEMENT BLOCK

16"   10"

8"

Fig. 6-12. Concrete blocks.

77

Advantages:

1. They have a slight advantage in strength and load carrying capacity over block wall construction.

2. They offer the best protection against air infiltration, moisture, and insects. They're also less likely to result in wet basements.

3. They can often be cast integrally with the footer, at a substantial time and cost savings.

Disadvantages:

1. They are usually more expensive than block construction.

2. It's somewhat difficult to be sure that you'll get as good a mix of cement as specified. There's an element of risk involved. It means, again, dealing only with reputable contractors for the concrete. Established builders will do so, but vanishing or marginal builders might not. A defective mix might not be detected until the house is up and the contractors are long gone. By then it will be a nightmare to correct major flaws or problems.

Foundation Construction Points

Here's a collection of various construction specs and procedures used in proper concrete block or poured concrete foundations:

1. The minimum thickness for any home in the most ideal situations requiring minimum loadings is 8 inches. Ten inches, however, is preferable and safer to go with. For extra-deep basements, use 12-inch concrete block. The minimum thickness should be increased to 10 inches if the walls will be subjected to any lateral pressures such as large snowdrifts, if the walls are more than 7 feet below grade, if the walls are longer than 20 feet, or if the house is going to carry a heavier than normal load. This can be the case if you elect a 2-story home or if you plan on having considerably heavy furniture and items such as a grand piano, pool table, water beds, or exercise equipment.

2. Before the foundation floor is concreted, all necessary plumbing and sewer pipes should be installed. Make sure all pipe cleanouts are present in the foundation walls and floor. Once the rough (underground) plumbing is situated in the foundation, the plumbing inspector may wish to: check that the drainage system, when full of water, will hold up without leaking; check for proper pipe slope or fall; check the cleanouts; inspect the piping for proper sizing.

3. Once the foundation walls are up, the floor of the crawl space or basement should be filled with 6 inches of 3/4-inch stone, with 3-inch drain pipes running through the stone. Over the drain pipe and stone a plastic vapor barrier should be placed to seal out dampness. This is especially important if the foundation is a crawl space. A concrete floor at least 4 inches thick will hold the water and dampness down into the ground, and the drain pipe will direct the water to a sump hole. Should the sump hole ever fill up, a sump pump can be installed to pump out the water and pipe it away from the house.

4. If concrete block is used, the first layer of blocks against the footer should be special drain blocks with grooved or weep holes along their bottoms. This will prevent a buildup of water pressure against the walls. Instead the water will flow into the 6-inch gravel bed beneath the concrete floor. The drain pipe running through the gravel will convey the water to the sump hole.

5. Along the outside of the foundation, 4-inch drainage tiles having openings or perforations along the top should be placed end to end on a gravel base along the footer, pitched toward the spot where the water will be piped away from the house. With this setup, water flowing against the wall of a home has a place to go. These drainage tiles need to be continuously sloped toward the discharge end; otherwise sediment might build up at a low point and completely block the line. The water table should be maintained no higher than the elevation of the tile under the entire basement or crawl space so water pressure is held at a minimum. This perimeter drainage tile system should be covered with approximately 12 inches of 3/4-inch stone.

6. A sump pump should be installed in the foundation floor when necessary (Fig. 6-13). Some foundations, either crawl space or basement types, might never need one, while others may have to employ one year round. Many building codes have been making it mandatory to include a sump hole so a pump can be added later, if needed. For the most part, sump pumps are set up so if water backs up to a certain level and threatens to flood the cellar, the pump will automatically kick in and pump out the water into an outside drainage line that will carry it away from the house. The sump hole itself should be 24 inches in diameter or 20 inches square, and should extend at least 30 inches below the bottom of the basement or crawl space floor.

7. One aspect of the foundation that deserves special attention is waterproofing. Damages due to wa-

Fig. 6-13. A foundation interior with sump pump and drainage.

ter and moisture are among the most serious causes of home deterioration. They cause wood rot, unsightly paint peeling, mildew, rusted appliances, and other maladies. They can even effect the health of the occupants. The prevention of damp and wet basements calls for good waterproofing techniques, proper drainage, vapor barriers, correctly graded lots, appropriate landscaping and positioning of shrubs and trees, generous roof overhangs, and plenty of gutters and downspouts.

8. If you're going to have a basement, have the septic disposal system line or the sewer line located below the basement floor if possible. Otherwise, wastewater and solids generated in the basement have to be pumped up to the level of the main lines for disposal. If the disposal lines cannot be lowered, the simplest solution is often to completely avoid any sanitary drains in the basement (that means no toilets, wash basins, showers, or laundry equipment). Then the house sewer or septic lines can be suspended beneath the first-floor joists and run through openings cut in the foundation walls. The same thing can be done with a crawl space foundation if necessary, as long as precautions are taken to prevent the lines from freezing.

FOUNDATION REINFORCEMENT

Reinforcement of a foundation's floor and walls is a critical part of the housebuilding process that can be easily slighted by marginal builders. Most homebuyers don't realize what's involved and depend solely on the recommendations of contractors who can, at times, underemphasize the specs that are needed for a sturdy foundation. This cost-cutting philosophy can lead to big problems later, at the homeowner's expense.

Here are some guidelines for the reinforcement of foundation floors and walls:

1. A concrete floor should be strengthened with reinforcement bars, or rebar. The bars should be elevated to the middle of the floor's thickness on steel bridges or chairs set so they're positioned evenly across the floor before the concrete is poured. If welded wire fabric is used instead of or along with the rebar or reinforcement rods, its joints should overlap 6 to 8 inches, or at least the width of one of the openings if the job can't be accomplished with a single piece.

2. If a foundation has long walls or walls subjected to above average stresses, they can be strengthened with pilasters (Fig. 6-14). A pilaster is a vertical block or concrete column poured or constructed adjacent or adjoining a foundation wall, located at about the middle of the wall's length. More pilasters are often needed along unusually long walls. By having thicker walls at selected points, this extra support lessens the overall stress on the walls and helps prevent cracking. Pilasters are normally not required for walls under 18 feet in length.

3. Another more modern and commonly used method of reinforcing the walls of a foundation is to place long pieces of 1/2-inch rebar through the rectangular openings of every other concrete block, vertically, and then to fill in those reinforced holes or "cores" solidly with mortar or concrete (Fig. 6-15). The same

Fig. 6-14. Foundation block pilaster.

Fig. 6-15. Vertical concrete block wall reinforcement.

bars can also be embedded in the footing pour as a tie from wall to base, especially when high walls must withstand considerable pressures from slopes, water, or backfilling. All house foundations should have either the pilasters or the reinforcement rods for vertical support.

4. For horizontal support, reinforcement wire should be placed in the mortar bed joint of every other course of blocks (Fig. 6-16).

5. Block walls that are quickly erected and back-filled might require bracing within the crawl space or basement for temporary support until the concrete and mortar dries, to make sure the structure is tightly knit before stresses are applied.

6. Load-bearing foundation walls should not be joined or tied together with a masonry bond unless the walls join at a corner. Instead, steel tie bars vertically spaced not further than 4 feet apart will form a strong bond (Fig. 6-17). If the walls are concrete block, strips of lath or steel mesh can be laid across the common joints in alternate layers or courses (Fig. 6-18). If a non-

bearing wall will be constructed at a later date, ties should be incorporated into the first wall, to be left half-exposed so they'll be available when needed for the second adjoining wall.

Masonry Joints

If using brick or block for foundation walls, you have an option to select any of the following mortar joint types that would best go with your style of construction: flush, struck, V, concave, raked, beaded, extruded, and weathered (Fig. 6-19). The V and concave versions are the most popular since they look neat and do not form a mini-ledge that could accumulate water. Beneath the grade, or the ground level, where appearance does not count, the joints are typically left flush.

Concrete Forms

If you're using poured concrete walls, make sure you realize that the quality of the concrete forms directly affects the finished appearance of the walls. Concrete

Fig. 6-16. Horizontal concrete block wall reinforcement.

forms must be tight, smooth, defect-free, properly aligned, and well-braced to resist lateral pressures created by the poured concrete.

FOUNDATION FLOOR SUBBASES AND FLOORS

The floor subbase should consist of a compacted layer of 3/4-inch stone 4 inches thick. Plastic sheathing that comes in 4 mil or 6 mil thicknesses or other suitable vapor barrier materials is placed over the subbase to form an effective insulating moisture barrier before the floor concrete is poured.

The concrete floor should be a minimum of 4 inches thick with welded wire mesh and rebar running through it for strength, and the concrete should be sloped to the floor drains. If you desire a smooth finish on a basement floor it will be necessary to specify that you want the concrete steel-troweled. Should you desire a textured finish, it can be obtained by having brooms dragged across the surface before the concrete is set.

FOUNDATION WALL TOPS

On top of all foundation walls anchor bolts (Figs 6-20 and 6-21) should be installed or partially embedded at

TIE BAR

METAL SCREEN

CORES FILLED WITH MORTAR

BEARING WALL

Fig. 6-17. A steel tie-bar.

approximately 4-foot intervals, with protruding bolt lengths long enough to securely fasten the sill plates (wooden planks that join the upper framing structure to the foundation).

Have foundation walls constructed of concrete block capped with a course of solid masonry blocks that will act as an insect barrier and help distribute the weight of the house's upper structure. When solid blocks are not used, the cores or rectangular holes in the top course can be solidly filled with mortar or concrete. To do this, a strip of thin metal lath must be placed in the mortar joint under the top course. The strip, which

is just wide enough to cover the block cavities, forms a base for the concrete that fills the top course cavities.

WALL COVERINGS AND INSULATION FOR THE FOUNDATION

Exterior

The foundation will be less susceptible to frost damage, moisture transfer, leaking, and insects if the walls are insulated on the outside (Fig. 6-22). Concrete block walls should be pargeted (plastered) with a half inch of cement mortar.

83

Fig. 6-18. Steel mesh reinforcement.

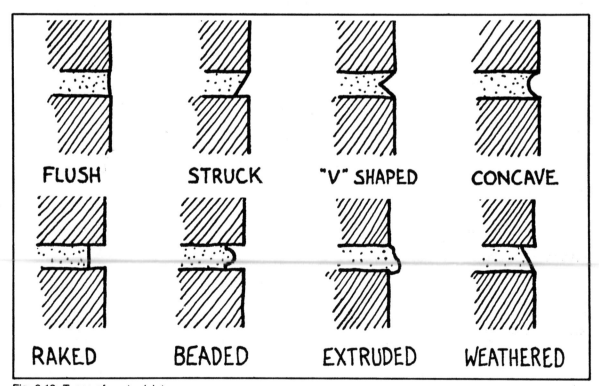

Fig. 6-19. Types of mortar joints.

Fig. 6-20. Anchor bolt in a poured concrete wall.

For any foundation wall, request two coats of tar or bituminous waterproofing material to be troweled on. Troweled tar is the best, but brushed on or sprayed on waterproofing is better than nothing. If local weather conditions are severe, protection can be further improved with a layer of sheet polyethylene or asphalt-impregnated membrane. Bentonite clay paneling can be placed against the foundation before backfilling. Alternatives to bentonite are rigid plastic or glass fiber sheets.

Above-grade foundation wall exteriors can be protected with stucco or treated plywood.

Interior

To reduce heat loss, to prevent moisture and water leak-

age, and to further prevent the possibility of insect penetration, foundation interior walls can be insulated and covered.

A thick waterproofing white paint that's brushed onto the interior surface will reduce moisture penetration and discourage insects from infiltrating interior walls (Fig. 6-23). Foundation wall interiors can be insulated by first putting up furring strips and then applying blanket insulation in the usual way (Fig. 6-24).

FLOOR SUPPORT BEAMS, OR GIRDERS

If the side walls of a house are more than 16 feet apart, one or more girders (load-bearing beams that help support the first-floor joists) should be installed. Steel I-beams make the most reliable girders, but wood girders constructed of 2- x -10 inch planks joined together by bolts or nails are also used (Fig. 6-25). Steel I-beams are rather consistent in quality. The wood girders will vary in quality because they're dependent on the quality of their individual component planks. Girders either rest in pockets formed in the tops of the foundation walls, or on top of corresponding masonry pilasters. At any event, the steel girder tops are generally made flush with the top of the wooden sill with wood planking that's laid along the top of the entire steel beam (Fig. 6-26).

FOUNDATION VENTILATION

Ventilation in foundation walls can be accomplished either with windows or with vents. Basement windows are discussed in the chapter on windows. If vents are decided upon, have them installed near or at the top course of concrete blocks, or as high as possible in the foundation walls at the rate of one every 50 linear feet,

each being about the size of a concrete block: 18 x 16 inches (Fig. 6-27). They would be the type that can be closed during cold weather. At least one vent should be positioned at each corner of the house, with cross ventilation arranged for a minimum of two opposite sides. If the house is located in an area experiencing high humidity during much of the year, increase the number of vents to one for every 15 linear feet.

On a crawl space foundation, in addition to vents there should be at least one access door of not less than 18 x 24 feet installed. If it's a large crawl space of 2,000 square feet or more, or an unusual shape, more than one access door should be included.

BACKFILLING

A formal inspection should be made by you once the foundation work has been completed and before backfilling occurs. Backfilling is simply the pushing back of excavated soil around the house to fill in the construc-

Fig. 6-21. Anchor bolt in a concrete block wall.

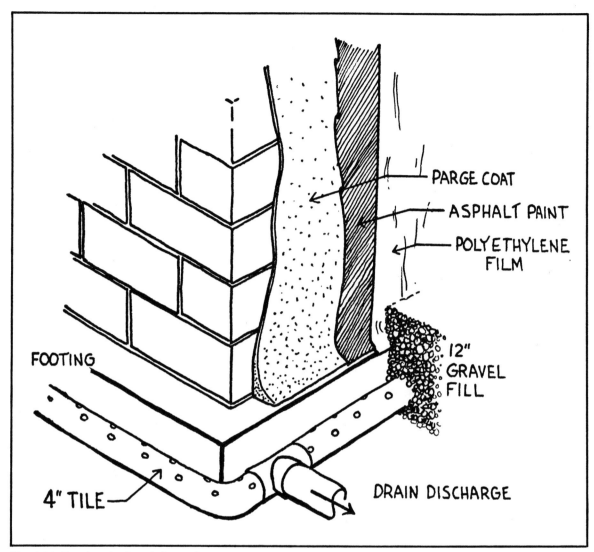

PARGE COAT

ASPHALT PAINT

POLYETHYLENE FILM

12" GRAVEL FILL

FOOTING

DRAIN DISCHARGE

4" TILE

Fig. 6-22. Exterior of a foundation wall, cutaway.

tion ditches.

To avoid subjecting a "green" foundation that isn't fully cured to pressures that could damage it, plank or timber bracing should be installed inside the crawl space or basement, supporting the walls at about 15-foot intervals. The house's central longitudinal support beam or girder, plus the first-floor joists and floor decking should also be erected to help strengthen the foundation walls before backfilling takes place.

As a general rule, backfill height from footer to grade should not exceed about 7 feet.

The waterproofing must be protected during the backfilling, since rocks and other hard materials in the backfill could scratch and penetrate the waterproofing and allow moisture seepage. If the soil that will be pushed back contains large rocks, the contractor should apply 4- x -foot or 4- x -10-foot sheets of impregnated sheathing or equal material for protection.

ALTERNATIVE FOUNDATIONS

Pole and Pier Foundations

Another less often used foundation is the pole and pier

arrangement (Fig. 6-28). It's a good setup for small homes built in steep terrain, or for vacation homes and cottages. Poles or piers are firmly implanted into solid ground so they're stationary (frequently cemented right into the ground) and then the frame of the house is constructed on top of them. For this type of foundation concrete tubes (concrete-filled cardboard cylinders) and telephone poles are very popular.

Advantages:

1. It lends itself to steep terrain where there is con-

siderable variation in the height of the piers and where a regular masonry foundation is impractical.

2. Grading isn't required. There's a minimum of site preparation involved.

3. It's an inexpensive foundation and easy to build.

4. There's plenty of natural ventilation between the ground and the living levels.

Disadvantages.

1. With the underside floor surface fully exposed

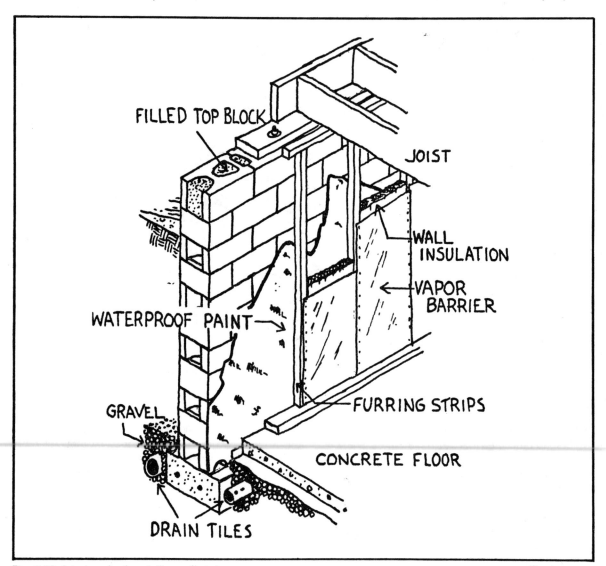

Fig. 6-23. Interior of a foundation wall, cutaway.

INSULATION

MASONRY WALL

FURRING STRIPS

POLYETHYLENE
VAPOR BARRIER

Fig. 6-24. Foundation wall furring strips and insulation.

to the elements, in cold climates the floor must be exceptionally well insulated.

2. It allows wind and small creatures to get beneath the building.

3. It is practically useless for any kind of storage or future expansion.

Wood Foundations

This type of foundation, admittedly, isn't very popular. But it is an alternative. It employs gravel and treated lumber to construct footers and foundations. For a contractor who is familiar with them, wood foundations

usually require less time to build and should cost less than their masonry counterparts (Fig. 6-29).

Here are some specs regarding wood foundations:

1. The exact to-the-letter proper construction of this type of footer and foundation is particularly important if the plan calls for a below-grade living area or basement.

2. Wood used in construction must be salt and pressure treated, plus all below-grade fasteners should be made of non-rusting metal such as stainless steel. The pressure treatment for all lumber that's to be used

Fig. 6-25. A wood girder.

Fig. 6-26. A steel I-beam girder.

Fig. 6-27. Foundation wall vents.

below grade should be .60 CCA, which is 50 percent higher than wood recommended for normal ground contact.

    3. A key element in this type of foundation is the control of moisture. Roof runoff must be directed away from the foundation by properly installed gutters, downspouts, splash blocks, and underground pipe drainage systems.

    4. Plywood joints must be caulked full length and then covered with a 6-mil polyethylene or other plastic film to direct the water down into the footer drainage system. As an added precaution, the foundation can be coated with plastic film after two layers of hot or cold tar with alternate layers of building paper are applied.

    5. Footers may be gravel, crushed stone, or sand at a minimum thickness of 4 inches for walls of a single-

story dwelling. For 2-story houses the thickness should be increased to 6 inches. There should also be gravel placed around the base of the foundation, and a drain pipe system leading to a sump hole or drainage outlet.

6. Follow good backfilling techniques, as mentioned previously. Proper sealing techniques must also be used. Consider horizontal insulation to help keep the footer area warm.

7. In general, many of the guidelines provided for concrete foundations should still be adhered to when constructing an all-wood foundation.

MISCELLANEOUS ITEMS

1. In locations with large populations of termites or carpenter ants, an effective way of providing additional protection is to have professional exterminators poison the soil around the foundation with chemicals that will have residual effects for many years. Additional

treatments can then be reapplied when needed.

2. Specify if you want windows in your foundation: their size and make, how many, and their location if in a full or partial basement.

3. Specify if you want a rear or side outdoor entrance to the basement. Use concrete or steel lintels for door openings. Precast lintels are simple to install because they're made to match the height and width of concrete blocks. They should be long enough to allow at least 8 inches of overlap on the bearing block. Another method involves forming and pouring a lintel on the site, using wood forms and reinforcement rods.

The outside stairwells should begin at a point lower than the basement floor, to prevent water from entering the house. A 6-inch sill protecting the basement door is standard. Or, a built-up curb could be constructed as a partial remedy. A decision should also be made if you want the stairwell opening covered by steel doors.

Fig. 6-28. A post and pier foundation.

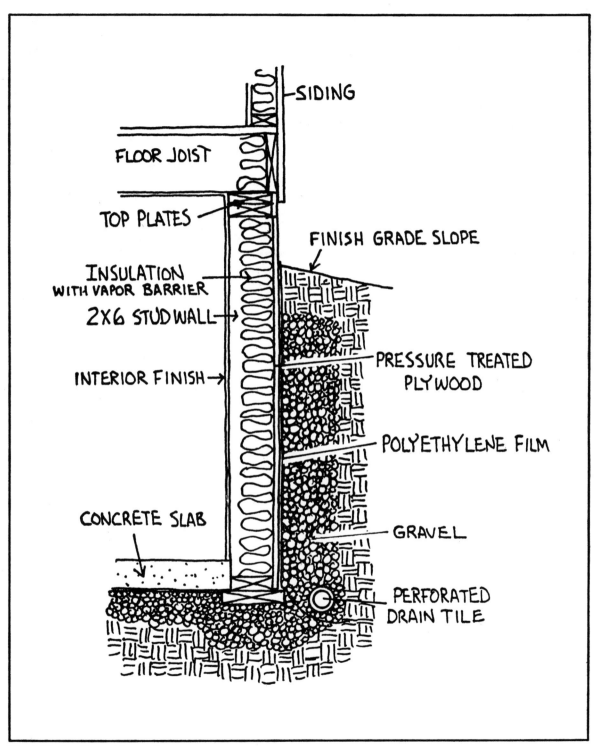

SIDING

FLOOR JOIST

TOP PLATES

FINISH GRADE SLOPE

INSULATION
WITH VAPOR BARRIER

2X6 STUD WALL

PRESSURE TREATED
PLYWOOD

INTERIOR FINISH

POLYETHYLENE FILM

CONCRETE SLAB

GRAVEL

PERFORATED
DRAIN TILE

Fig. 6-29. An all-wood foundation.

4. Don't let the contractors confuse the topsoil that was pushed to the side when the construction first began, with the soil that was excavated for the foundation. The topsoil should only be used for finished site grading, not as backfill around the foundation.

5. It's not a common request, but poured concrete walls and floors can be colored by the addition (when wet concrete is being mixed) of mineral pigments sold by ready-mix concrete and block producers, lumberyards, and building material dealers. Reds, greens, yellows, browns, grays, and other colors are available.

6. If you're going to have a full basement and you plan to eventually put in a recreation room there, try to locate it as closely as possible to the basement stairs. Opt for larger windows than normal so the room will have adequate light and ventilation during the day. Check building code requirements related to any partitions you plan to install around the furnace and water heater; you might have to meet clearance and surfacing requirements.

Avoid placing partitions, present or future, too close to plumbing-heating-electrical elements such as sump pumps, water meters, shutoff valves, and waste cleanouts. Give them an extra inch or two clearance to permit quick access in case of trouble.

When possible, plan a workshop area close to an outside access an also nearby and unexcavated sheltered area where a crawl space can be used for lumber and other long materials storage.

The typical cellar wall should be 11 courses high (or 88 inches). But if any rooms on the floor above it will be sunken, consider an extra course (12 courses total) in case you someday decide to finish off the basement. If you go the extra block, you'll still be able to have a normal ceiling height when finished.

All basements should have at least one water drain in the floor, and those cellars larger than 900 square feet should have two drains. The floor should taper toward the drains, and the drains should transport water to the house's sump hole or exit drain.

The vertical steel columns supporting the house floor should be at least 4 inches wide, and the steel beams or girders, 8 inches.

Chapter 7

Floor Framing

The wood frame of a house has been compared to the skeleton in a human body, in that it forms the shape and size and provides the strength to a dwelling. Even with a brick house (in today's modern construction this means brick veneer), the framing actually supports the brick, not the reverse as many people think. It's critical that the framing be erected correctly; it's not an area in which to compromise to reduce costs. Any errors discovered after the framing is complete are likely to be expensive to correct.

As a rule, the sills, girders, floor joists, and subflooring are the first members of the wood-framed structure placed on the foundation walls. These are followed by the outside wall studding and corner posts.

LUMBER

All the lumber used should be air- or kiln-dried and of No. 2 grade or better. Regular lumber, while seasoning after it's nailed in place, can shrink away from the shank of a nail, reducing friction between the nail and the surrounding wood and causing nail popping—the heads of nails protruding from lumber they had previously been nailed flat against.

Besides an increase in strength and nail holding

power, air- or kiln-dried lumber holds paint and preservatives better and is less likely to be attacked by fungi or insects. Only pressure-treated lumber should be used for framing members that are exposed to moisture.

Here are some guidelines:

■ .25 CCA (pounds salt solution per cubic foot) for above-grade applications, such as decks and railings.
■ .40 CCA for lumber that comes in contact with the ground, such as columns supporting decks.
■ .60 CCA for lumber located below grade, such as in an all-wood foundation.

SILL PLATES

The sill or sill plate is the timber (usually made of 2- x -10-foot wood planks) that's secured to the top of the foundation walls to form a link between the foundation and the home's upper structure.

Some sill plates are laid in a bed of wet mortar. Most, however, are installed in the following manner: Anchor bolts are partially embedded in the concrete foundation wall tops, then a resilient, waterproof layer of sealer (similar to felt or styrofoam) is pushed over the

Fig. 7-1. Securing and sealing the sill plate.

protruding bolts down onto the top surface of the foundation walls. The sill plate, having holes drilled in it to correspond with each anchor bolt, is then placed over the bolts onto the fiberglass layer, pressing the fiberglass flat against the top of the foundation walls. The sill plate is secured along the foundation top with sturdy lock washers and nuts tightened onto each anchor bolt (Fig. 7-1).

SUPPORT BEAMS OR GIRDERS, AND POSTS

Floor joists that traverse long spans between foundation walls must be supported by longitudinal beams or girders as mentioned in the foundation chapter. These structural members can be 8-inch-thick steel I-beams (the best alternative) or three or four widths of 2-x-10-inch or 2-x-12-inch wood planks nailed or

bolted together. The ends of these longitudinal beams or girders rest either on pockets in the foundation walls, or on concrete block or poured pilasters. The girders should bear (or rest on) a minimum of 4 inches of each foundation wall and should clear the walls of a pocket by 1/2 inch on both sides and the end. The top surface of a longitudinal beam or girder must be made level with the top surface of the foundation wall via a wood plank set on top.

Vertical posts of steel at least 4 inches thick help support the beams or girders through the interior of the span (Fig. 7-2). To prevent sinking, the vertical posts should be located over piers (concrete-filled holes about 18 inches across and 8 inches deep) set in a gravel floor or beneath a concrete floor so the weight the posts support will be distributed over a broad area.

FLOOR JOISTS

Floor joists are horizontal structural planks placed on edge against a house's sill plate in an orderly fashion to distribute the weight of the wood framing to girders and sills and to provide a base to which the floor deck-ing can be fastened. Joist ends usually rest both on the sill plate and on interior longitudinal girders of steel or wood (if intermediate supports are needed) (Fig. 7-3). The joists should be of sufficient strength, stiffness, and number to support the floor loads over the area

Fig. 7-2. A vertical support post.

STUDS

SUBFLOOR

FINISH FLOOR

CROSS BRIDGING

BLOCKING

DOUBLE JOISTS

STEEL POST OR COLUMN

STEEL I-BEAM

FLOOR JOISTS 16" CENTERS

MASONRY WALL

HEADER

SILL

SILL SEALER

Fig. 7-3. Floor joists support.

spanned, with no perceivable deflection or "give" that could result in cracked plaster or pulled-apart drywall seams.

In good construction, joists are placed at least 16 inches on center, and are made of planks a minimum of 2 x 10 inches. For a stronger-than-usual floor, the same 2-x-10-inch joists can be installed every 12 inches, or 2-x-12-inch planks can be used instead. Planks running around outside wall perimeters that the floor joists are fastened to are called headers (Fig. 7-4).

Double-thickness floor joists—two planks fastened together—should be used in certain situations, for example, wherever the first- or second-floor walls run parallel to the floor joists. This will occur at openings around a stairway (Fig. 7-5), near a fireplace chimney, or at any major change in joist direction such as where partitions are built to provide clearance for hot air ducts and returns, under cast-iron bathtubs, or to provide additional support where joists cross a girder (Fig. 7-6).

HEADERS

Again, a floor framing header is a wood plank that the floor joists are nailed perpendicular against on two op- posite sides of the house, and parallel against on the other sides. It's erected vertically on its long edge, resting on the sill plate along the exterior of the foundation walls.

FLOOR DECKING

A number of structural materials, mostly wood or composition wood products, are used to cover floor joists and to construct the first-floor platform on which all subsequent house framing rests. These are mainly walls that are bearing units for the upper floors and the roof structure, and walls that form interior partitions. The most popular materials are plywood, hardwood boards, and particleboard, in thicknesses that range from an average of 1/2 inch to 1 inch.

Plywood is one of the strongest and most convenient floor materials available. It not only makes an effective floor, but also serves as a strong structural tie between the floor joists and wood beams or girders.

Hardwood boards are available in many sizes and thicknesses. They are often used as floor coverings, and alone over the joists as the main floor structure as well.

Particleboard is popular because of its low cost and

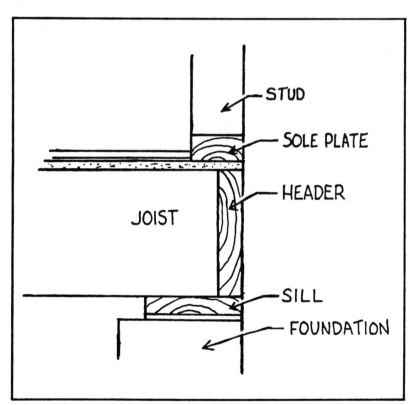

Fig. 7-4. Header-joist construction view.

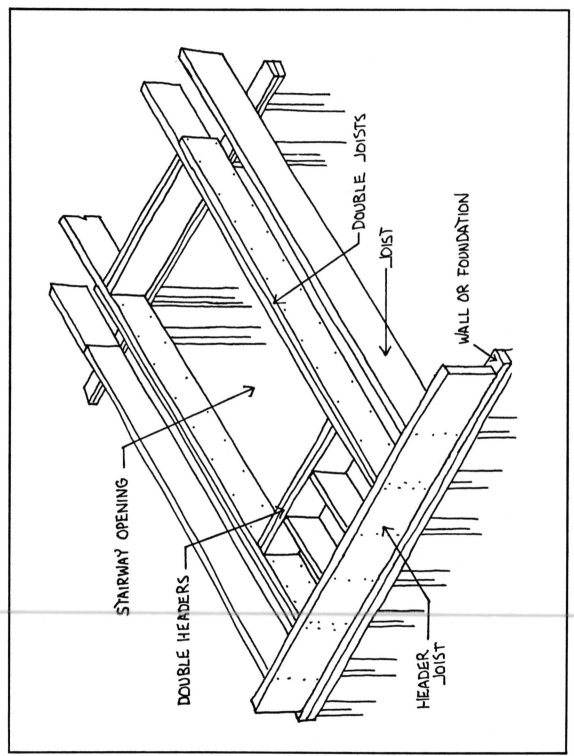

DOUBLE JOISTS

JOIST

WALL OR FOUNDATION

STAIRWAY OPENING

DOUBLE HEADERS

HEADER JOIST

Fig. 7-5. Framing a stairway opening.

100

Fig. 7-6. Double thicknesses of floor joists.

lack of sheet curvature (it's stiff and straight). When used with an underlayer of plywood, what results is a thick, sturdy floor.

An excellent all-around floor construction consists first of 1/2-inch-thick (or thicker) plywood sheets glued and nailed or stapled to the floor joists. The glue helps eliminate squeaks and nail popping, and increases the stiffness of the plywood-to-beam bond from 10 to 90 percent. In fact, the adhesion of mastic-type glue is so strong that plywood and joists tend to behave like integral T-beam units. Nailing the flooring to the joists is superior to stapling. Threaded sinker nails have proven the most efficient; they'll help prevent springiness, uplift, horizontal shifting, warpage, and nail-head popping. When used along with glue, they'll provide a trouble-free floor.

The plywood and other sheet materials should always be attached so the joints don't line up in a regular fashion (Fig. 7-7). On the first layer of flooring it's best to arrange the plywood sheets so their lengths run at right angles to the joists for maximum strength. Once the 4- x -8-foot sheets of plywood are laid down, they

should be covered with 15-pound asphalt saturated felt paper. Then a second structural layer of particleboard (5/8-inch) sheets can be fastened to the plywood—again, staggered so none of the seams coincide (Fig. 7-8). If additional strength is desired, the second layer can also be plywood sheets (5/8-inch). In either case, the result will be a sturdy, quiet floor.

If you select plywood, make sure that the contractor uses a good grade, one that contains exterior glue lines on both sides. If the bottom plywood layer is nailed with an automatic nail gun, it should be renailed along the vital joints—where one piece of plywood faces another—with screw nails to prevent squeaking. Gaps of 1/16 inch should be left between the sheet edges and end joints in all plywood layers. Plywood expands when it absorbs moisture and will buckle if it doesn't have enough room.

Even correctly installed single-layer floors will squeak and give if the sheets are too thin. Use 1 1/8-inch-thick tongue and groove plywood over joists erected at 24 inches on center, and 3/4-inch-thick tongue and groove plywood over joists erected at 16

Fig. 7-7. Joints should be irregular.

inches on center (Fig. 7-9).

CAULKING

Make sure that parallel beads of caulk are run between the bottom of the floor decking and the sole plate as well as between the top of the floor decking and the outside wall framing. This will reduce the loss of heat due to air infiltration.

BRIDGING AND BLOCKING

To stabilize the floor joists, bridging or blocking must be used. They help keep the joists properly aligned so the floor decking has a continuously level base to rest upon (Figs. 7-10, 7-11, and 7-12).

Bridging consists of pieces of wood (or sometimes metal), usually 1- x -3 inches, nailed crossways between the top and bottom of adjacent joists at about the center of each joist span. There should be no more than 8 feet between individual rows of bridging, and nailing is easier if the bridging is slightly staggered along individual rows (Fig. 7-13).

Blocking refers to 2- x -10-inch or 2- x -12-inch wood blocks fit and nailed firmly between joists at the center of each joist span. Again, there should be no more than 8 feet between individual rows of blocks. Blocking not only holds the joists parallel and plumb, but will also act as a firestop that will retard the horizontal spread of flames. In homes where bridging is used to support the floor joists, a row of blocking can also be installed for fire protection.

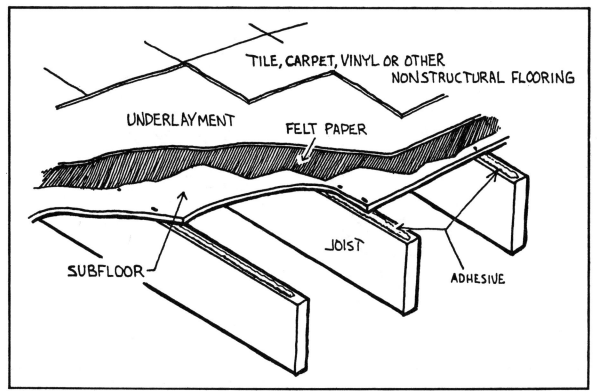

Fig. 7-8. Covering the floor joists.

Fig. 7-9. Plywood tongue-and-groove sheets.

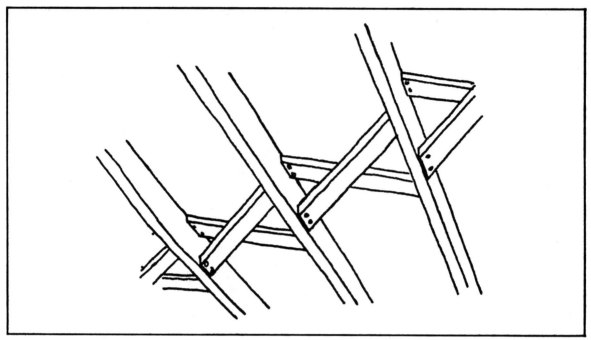

Fig. 7-10. Cross bridging of floor joists.

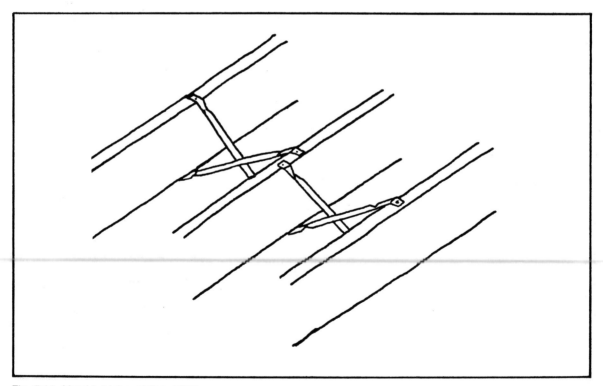

Fig. 7-11. Metal bridging of floor joists.

104

Fig. 7-12. Solid bridging or blocking of floor joists.

Fig. 7-13. Staggered bridging.

STONE AND TILE FLOORS

If your plans call for stone or tile floors in areas other than bathrooms, allow additional space where needed for the setting of stone or tile in a thin bed of cement by dropping the floor decking.

FLOOR EXTENSIONS

Give careful consideration to house designs that call for floor extensions that protrude outside the basic perimeter of the walls. They might be needed for such features as porches, second-story overhangs or decks, or bay windows. A sizable protrusion could require supports outside of the foundation wall or even an actual adjustment to the foundation.

TERMITE CONTROL

Termites frequently enter a housing structure near the first-floor level. In addition to poisoning the ground along the foundation, toxic treatment of framing lumber may be warranted. Another method is to install a continuous strip of thin metal between the foundation and sill plate so the metal strip extends out from the foundation, bent downward about 2 inches at a 45-degree angle.

FLOOR INSULATION

The usual procedure for installing an effective vapor barrier is to place 15-pound building paper between the first and second layers of floor decking (if there are two layers), or else between the only layer of structural floor decking and the floor covering (Fig. 7-14). In addition, if required by your climate and house type, blanket or batt insulation can be installed between the floor joists beneath the floor decking. An integral vapor barrier should also be situated against the underside of the floor decking. This will prevent the movement of moisture from the living area into the insulation.

Insulation from noise can be an important feature in a home, especially in single-story dwellings in which basement recreation rooms are planned. An example of a soundproof type of floor, starting at the top, consists of a layer of carpet, the carpet pad, a 5/8-inch particleboard top floor decking, a layer of 15-pound felt paper, a 1/2-inch plywood bottom floor deck, 3 1/2-inch-thick fiberglass batt insulation without a vapor barrier between the joists, and finally, a layer of 1/2-inch fire-rated drywall nailed to special resilient channels attached to the joists.

Fig. 7-14. Floor insulation cutaway.

Chapter 8

Wall Framing

The wall frame is the next logical part of the house to erect after the first-floor decking is attached to the foundation. Wall framing performs three basic tasks. First, it supports the home's upper floors, ceilings, and roof. Secondly, it acts as a base on which outside and inside coverings can be fastened. Thirdly, it provides space for and conceals essential wiring, pipes, heating ducts, and insulation. In its most general sense, wall framing also includes room partitions that are constructed within the outer perimeter of a dwelling.

TYPES OF WALL FRAMING

There are four types of wall framing worth noting: platform, balloon, plank and beam, and pole.

Platform Framing

In this type of framing each floor is built separately, one on top of the other, with the first floor providing a work platform for the second level, and so on (Fig. 8-1).

Balloon Framing

In balloon framing, the studs or vertical members of the exterior walls are continuous from the sill plate of the first floor to the top plate of the second floor. These long

studs are more expensive than the studs used to frame single-floor levels, and the labor to erect the longer studs, due to scaffolding required, is higher (Fig. 8-2).

Plank and Beam Framing

This type of wall framing uses long, thick structural members—often rough-hewn posts and beams. There are fewer framing pieces required, and those pieces must span lengthy open spaces. Wooden planks are used for the floors and roof. In fact, they supply the sole support over long spans in both, being nailed at their ends to wood beams. Consequently, plank and beam framing doesn't employ joists for the floors or ceiling. This type of framing is very popular where exposed beams and beamed wide-open ceilings fit the particular style or decor of a home (Fig. 8-3).

Pole Framing

In certain situations this uncommon type of framing has an advantage over studded wall framing in that long wooden poles, if properly pressure treated, can be embedded deep into the ground to provide a total bracing effect for the walls against the force of strong winds (Fig. 8-4).

Fig. 8-1. Platform framing.

Fig. 8-2. Balloon framing.

Fig. 8-3. Plank and beam framing.

STUDDED WALLS

By far, studded walls support most of the floors, ceilings, and roofs of modern dwellings. This conventional wall framing consists of a combination of header, studding, and top plate, which should be doubled (Fig. 8-5).

Exterior Studded Walls

The header for the exterior walls runs on and along the top outer edge of the sill plate and against the perimeter of the first-floor joists and decking. It should possess greater width than as individual wall studs—usually 2 × 10 inches. Specify the quality of header and studding lumber to be a minimum of No. 2 BTR SPF.

Many contractors will suggest the use of 2- × -4-inch

studs in the wall framing. Don't be swayed by a builder who advocates such construction. By going with 2- × -6-inch studded walls, you'll end up with stronger walls that will accommodate thicker blankets of insulation—that will, in turn, yield greater energy efficiency for a proportionally small increase in construction costs. Some contractors claim that increased insulation value provided by styrofoam panels will make up for the loss of stepping down to 2 × 4s. Don't believe them. It's not entirely true and it will result in substantially weaker walls.

Have the wall studs spaced 16 inches on center to provide a sturdy base for exterior and interior wall coverings to be attached to. Use a double-width top plate of 2- × -6-inch lumber to hold the top of the walls together.

PURLIN

OUTSIDE POLE

FLOOR JOIST

BLOCKING

LEDGER

Fig. 8-4. Pole framing.

111

Fig. 8-5. A studded wall frame, with double top plate.

Interior Studded Walls, or Partitions

Conventional interior wall framing consists, as does the exterior wall, of a combination of sole plate, studding, and top plate (often doubled) to receive the weight of the ceiling joists. The sole plate generally runs across the floor decking, parallel or perpendicular to the floor joists, depending on the direction of the partition. The ceiling joists are typically positioned directly over the supporting studs applicable.

The partition tops should also be capped with two pieces of 2-inch-thick boards which are lapped or tied into exterior walls wherever they intersect.

If your house is a 1 1/2-story or 2-story design, it will require the installation of at least one load-bearing partition before the structural framing work is completed. Later interior partitioning consists of fitting the other various room and intersecting partitions to load-bearing partitions and exterior walls.

In a conventionally framed 2-story home, the load-bearing first-floor partition should be placed directly over the main longitudinal support beam or girder that rests on the foundation walls. Whenever possible the load-bearing partition on a second floor should likewise be positioned over its corresponding member on the first floor, so as much weight as possible will bear down upon the main girder beneath the first floor. If a particular floorplan will not permit the second floor to be partitioned in such a manner, the main load-bearing second-floor partition may then rest across the second-floor decking.

Remember, due to their size and shape, combination bathtub/shower units must be installed during the wall partition and framing activity (Fig. 8-6). Select the models and colors well in advance so they can be ordered and received in time for the plumbers to erect them when the framing crew is ready.

Corners

The corners of wall frames need extra support because they must provide stability to both intersecting walls. They involve the assembly of "posts" with insulation, and corner bracing for strength, also with insulation (Fig. 8-7).

Posts

The posts are block assemblies of 2- x -6-inch planks used at corners of wall frames and where interior partitions abut an outer perimeter wall. They should be constructed to provide a good nailing surface for exterior and interior wall coverings, and because they're hollow, they must be filled with fiberglass or other insulation so heat is not wasted through the thermal break they'd otherwise create (Fig. 8-8). The insides of posts must be insulated as they're being put together. It can't be done later, when the posts are covered over.

Corner Bracing

If plywood at least 1/2 inch thick is used as the outer sheathing on the wall frame corners, other bracing is not necessary (Fig. 8-9). But if the outer sheathing is a material that's not very strong, such as rigid board polyurethane or polystyrene, additional bracing will be needed. An effective corner brace to specify is the diagonal support that's "let into" or inset into the outer corner studs (Fig. 8-10). This type of bracing is achieved with either 1- x -4-inch or 1- x -6-inch boards that fit snugly in notches cut at appropriate angles at the proper height of each wall stud crossed. The inset boards on both sides of a corner rise to meet each other in the shape of a triangle (Figs. 8-11 and 8-12).

OPENINGS IN WALL FRAMING

Remember that it's the outer walls of a home that protect the occupants from winds and inclement weather, noise, and unsightly views. Openings should be strategically placed as well as being soundly framed. Here are some pointers to consider when planning the openings in your wall frames:

1. It's important to know far in advance of the wall framing, the size, type and brand of each window and door you want so the rough opening dimensions can be secured from the manufacturers and passed along to the carpenters who will be erecting the wall framing.

2. In good construction, board cuts and joints should be accurate and tight at the junction of the roof rafters and the ridge board, at the headers of windows and doors, and at all intersections. Cuts and joints that result in large, obvious gaps create weaknesses in the framing that you should not accept.

3. Wherever an interior door or window will be hung there must be double studding around the opening to make up for the support studs that would otherwise go in place of the door or window opening. Double studding also presents a needed place to nail door and window trim against. A double horizontal lintel or header is used to support short studs that reinforce the top plate

Fig. 8-6. A bathroom tub/shower unit and framing.

of the wall above the opening (Figs. 8-13 and 8-14). The planks for the headers should be 2 × 10s.

    4. If any of your exterior walls will have brick veneer coverings, you'll need steel lintels to support the brick veneer over the tops of window and door openings (Fig. 8-15).

    5. If you desire, blocking can be installed to the right and left of each window between the studs to pro-

vide a solid backing to which curtain rods can be mounted. Anywhere you plan to hang heavy or often-used objects such as large pictures, trophies, or hanging bookcases, arrange for blocking between the studs to take the weight. Keep this in mind in the bathroom, for fastening towel racks, soap dishes, and toothbrush holders to the wall.

    6. If your bathrooms call for any flush-mounted

medicine cabinets in which the storage portion of the cabinet is recessed into a wall, additional framing must be installed to accept them. Select the cabinets well in advance so you'll have the correct dimensions of the rough openings to give to the framing crew.

7. Make sure the framing crew remembers to install nailers to which drywall or plaster lathing can be attached, especially in odd corners and spaces where regular framing isn't used.

VENTILATION

Plan the openings—especially window openings—for cross ventilation as far as possible. Good air flow occurs when the air inlets and outlets are approximately the same size. A better air flow results from a larger outlet than inlet.

SHEATHING

One of the last operations of the wall framing is the in-

Fig. 8-7. Exterior corner post construction.

INSULATION

DOUBLE TOP PLATE

3 · 2 X 6 STUDS

EXTERIOR WALL

INTERIOR WALL

SOLE PLATE

INTERSECTION
OF INTERIOR WALL
WITH EXTERIOR WALL

Fig. 8-8. Post construction.

stallation of the exterior sheathing that's attached to the wall studs. Four materials are typically used, mostly in 4- x -8-foot sheets: CDX plywood, particleboard, fiberboard insulating sheathing, and styrofoam insulating sheathing (Figs. 8-16, 8-17, 8-18, and 8-19).

Exterior sheathing performs three functions:

1. It braces the structure. Plywood is the strongest, most rigid of the four, followed by particleboard. Plywood is especially important in plank and beam con-

struction where a lot of bracing is needed. If plywood is used, at least on the corners, it will eliminate the need for diagonal bracing with its accompanying high labor costs. Diagonal board with structural insulation board sheathings such as particleboard are also acceptable.

For best results the sheathings should be applied vertically in 4- x -8-foot or longer sheets with edge and center nailing.

2. It provides insulation. If insulation, not strength, is the main concern, styrofoam board is the winner here.

It insulates superbly, but is quite expensive.

3. Sheathing provides a weathertight base for the exterior siding. Plywood and particleboard sheathings are a strong base for exterior siding. Be aware that there are two types of plywood: exterior and interior. Definitely, only exterior grade plywood should be used for the outside of a house. Exterior plywood is made with high-quality veneers and is bonded with waterproof glue. It offers the best durability and its glue won't weaken with age or with long exposure to foul weather.

TOP PLATE

PLYWOOD SHEATHING USED AS CORNER BRACING (INSTALLED VERTICALLY)

Fig. 8-9. Plywood corner bracing.

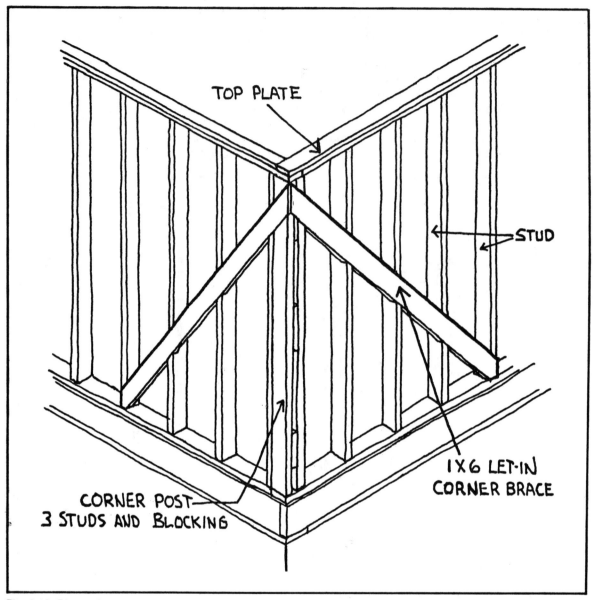

TOP PLATE

STUD

1×6 LET·IN CORNER BRACE

CORNER POST— 3 STUDS AND BLOCKING

Fig. 8-10. Diagonal corner bracing.

NAILS

As minor as they might seem, nails are what hold much of a housetogether. Only galvanized nails should be used on exposed materials, both inside and out. Unless the house is roofed over immediately, a sudden downpour could cause regular steel nails to rust and streak the surfaces in just one night.

If your walls will have studs with 24-inch O.C. (on center) placements instead of 16-inch O.C. as recommended, nailers will have to be installed so drywall or plaster lathing can be securely attached. Nailers are small 2- x -4-inch or 2- x -6-inch blocks attached perpendicular to the studs, in between pairs of studs. The nailers will also give extra support and stiffness to the walls, and will act as a fireblock to discourage flames from spreading throughout a wall.

It's a good idea to specify that exterior wall sheathing be applied by hand nailing only. Some of the more fragile sheathings such as polyurethane board can be easily torn and damaged by power nailing equipment.

For optimum efficiency, nails should be spaced 6 inches O.C. along the sheathing edges and no further than 12 inches apart on intermediate studs. On 3/8-inch plywood, size 6d nails should be used, and 8d nails for 1/2-inch-thick materials.

ENERGY

Energy is of great interest no matter which part of a house is being discussed. But here, with the exterior walls, insulation takes on a special importance. Naturally, you want to reduce the consumption of energy for heating and cooling, and to increase the level of comfort in the home by muffling the effects of the elements. That's another reason why openings should be

Fig. 8-11. Corner bracing.

DOUBLE PLATE

STUD·CONTINUOUS BOTH FLOORS

JOIST

FIRESTOP

BLOCKING

1 X 4 RIBBON (LET·IN)

CORNER POST 3· STUDS

1 X 6 LET-IN DIAGONAL BRACE

1ST FLOOR JOIST

SILL

Fig. 8-12. Corner bracing.

DOUBLE HEADER

DOUBLE STUD

DOOR OPENING

Fig. 8-13. Door opening reinforcement with double header.

carefully planned to prevent unwanted heat loss or heat gain, and to allow for natural ventilation and heat gain when desired.

The major obstacles to well-insulated, sealed walls are doors, windows, and electrical outlets. Again, eliminating as many potential problems as possible in the design stages is the first and most logical step. Place wall switches and outlets on interior walls when possible, and use as few windows, glass sliders, and doors

as practical. Have reasons for everything that goes into the exterior walls. When you can, incorporate features that must be inset into a wall, within interior walls, where the interruption of insulation is not a factor.

Because wall insulation is so important, the use of 2- x -6-inch studs is stressed over and over. Old-fashioned 2 x 4 studs were fine in their day, when energy was inexpensive, but only 3 1/2 inches of fiberglass batt or blanket insulation will fit into such walls.

121

Fig. 8-14. Door and window reinforcement.

122

Fig. 8-15. Masonry steel lintel reinforcement over window.

With 2 × 6 studs, 5 1/2 inches of the same kind of insulation can be laid (Fig. 8-20). That makes a big difference in energy use.

SOUND INSULATION

There are two ways to arrange effective sound insulation for interior walls. The first employs staggered 2-×-4-inch studs erected on a 2-×-6-inch plate. The studs should be positioned 16 inches O.C. Because the studs are staggered, this technique eliminates the touch-

ing of drywall or plaster on both sides of the wall by any single stud, thus reducing the wall sound transmission capabilities. In the second procedure, the air voids between the studs are "woven" with 3 1/2-inch fiberglass blanket or batt insulation, which will further deaden noise transmission.

Interior walls are constructed with 2-×-6-inch studs and 5 1/2-inch or 6-inch batt or blanket insulation. The additional wall thickness makes up for sound transmission deficiencies.

TOP PLATES

SPACE NAILS
12" O.C.

SPACE NAILS
6" O.C.

Fig. 8-16. Plywood sheathing construction.

VAPOR BARRIERS

In addition to being insulated, living areas should also be sealed with a moisture-proof layer or vapor barrier that's applied to the inside of the wall studs to prevent the movement of moisture from the living areas into the insulation. Insulation will lose some of its thermal qualities if it becomes damp or wet, and if the moisture within the living spaces is retained within the house, the occupants will still feel comfortable with less heat, due to the inside humidity.

There are several methods for applying a vapor barrier to exterior walls: install insulating batts or blankets faced with vapor barrier backings such as treated Kraft paper or aluminum foil. The vapor barrier should always be placed closest to the living area that's heated. If unfaced batts or blankets are used, a vapor barrier or polyethylene film not less than 3 mils thick (or an equivalent) should be applied.

If you elect loose blown insulation, establish a vapor barrier by stapling or nailing polyethylene sheet ma-

124

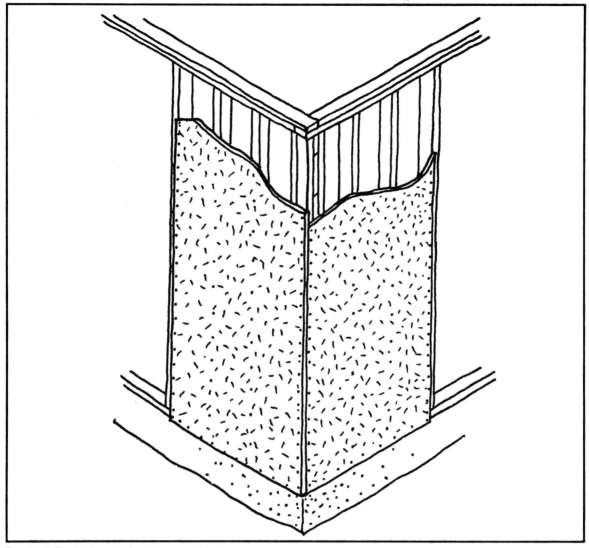

Fig. 8-17. Particleboard sheathing construction.

terial to the interior of the wall studs. You can also install aluminum foil-backed drywall before the insulation is blown between the studding.

BEFORE THE WALLS ARE CLOSED

Before the walls are closed up, the following items should be completed, if applicable:

—Sink drains
—Vents
—Water supply for sinks
—Cold water for toilets
—Water for showers and tubs
—Hot water for a dishwasher
—Hot and cold water for a clothes washer
—Gas lines
—Appliance vents
—Built-in vacuum system
—Electric wires and doorbell
—Intercom system
—Phone lines
—Alarm systems
—Heating and cooling ducts

Fig. 8-18. Fiberboard.

WOOD CORNER BRACE

METAL CORNER BRACE

VERTICAL APPLICATION

HORIZONTAL APPLICATION

Fig. 8-19. Styrofoam board sheathing construction.

FOIL BACKED INSULATION
STAPLED FLUSH WITH STUDS

STUD

SOLE PLATE

UNBACKED INSULATION

STUD

FINISH WALL

POLYETHYLENE FILM
STAPLED TO STUDS

Fig. 8-20. Wall insulation views.

Chapter 9

Roof Framing

There's no getting out from under it; the roof of a house is often all there is between you and the sky. And as such, the roof will protect you from such inconveniences as snow, sleet, hail, rain, sun rays, wind, dust and dirt, acid rain, insects and animals, moonlight, and noise. It will, when constructed and insulated properly, keep cool air in the house during summer, and warm air outside. Then conversely, during winter, will keep cold air at bay and warm air inside.

The typical roof also serves, however infrequently, as a platform for contractors to walk on when they're performing maintenance and repairs—to renew the flashing on a chimney, for instance, or to dislodge a stubborn bird nest from a false flue.

A roof must be securely fastened to the rest of the house, not merely "tacked onto" the upper level. It has to be able to resist updrafts of wind that would otherwise yank a roof right off.

The roof and ceiling frame of a home's upper level needs to be sturdy enough to support whatever covering or options are planned, including heavy tiles or slates, solar panels, and skylights. At the same time, however, it should be noted that any irregularities or impediments to simple roof lines provide opportunities for water or moisture to seep through the roofing. This includes chimneys, vents, roof lights, and anywhere the roof has a valley or dormer where one roof line intersects another.

Other practical considerations to be made with the roof frame are the size and shape of the area between the upper level's ceiling and the rooftop, commonly known as the attic space. In certain types of houses this space can be constructed in a variety of ways. It tends to follow patterns dictated by the shape of the rest of the house, tempered by basic aesthetics. Although it should be in correct proportion to the surrounding architectural components, certain decisions must be made regarding roof pitch, materials, and construction techniques.

The roof overhang—the part of the roof that protrudes beyond the exterior walls—protects the exterior sheathing or siding, windows, and doors from the elements, especially rain and sun. A wide overhang will block the sun on summer days when the sun rises high overhead, then will let the sun rays enter during winter, when the sun travels a much lower route in the sky.

ROOF STYLES

Six styles of roofs account for the lion's share of roofs,

129

old and new: Gable, Gambrel, Hip, Mansard, Flat, and Single-Pitch or Shed. The first three are the most widely used (Fig. 9-1).

Keep in mind that the following styles are not restrictive of each other. A single dwelling can and will often have a combination of several roof styles.

The Gable Roof

The gable roof is the single most popular roof style built today. It consists of two usually equal-sloped roof planes that meet at a topmost ridge. In fact, it's the ridgeline running the entire length of the house, or at least on the parts of the house having that style of roof, that most characterizes the gable style. The gable also means the upper triangular area formed on each end of such a roof. Gable or shed dormers are frequently added to plain gable roofs to break monotonous lines or for the practical purpose of providing natural light, air, and additional space to make an attic area more useful.

The Gambrel Roof

In general, this is a gable roof having two separate roof slopes on each side of the topmost ridge, the flatter or least-sloped of the two being above the level of any dormer windows. An advantage this roof has over the gable roof is that it increases the usable attic space, and when dormer windows are installed it's almost equivalent to having a second story.

The Hip Roof

With the hip roof, the ridge does not run the full length of the house. Instead, hip rafters extend up diagonally from each corner to meet the ends of the ridge. Essentially, the sides or slopes of this roof angle up in four planes from the outside walls. It's an exceptionally strong roof design.

Although by far the majority of houses built today have one of the above three roof styles, there are an equal number of lesser-used styles that should also be mentioned.

The Mansard Roof

This is a variation or modification of the gambrel roof, also referred to as a hip version of the gambrel. Its advantages lie in the space added to the attic and in the additional strength of its construction.

The Flat Roof

Not often used in typical residential construction, the flat roof frequently employs the use of rubber roofing materials covered by fine gravel. Naturally, such a roof would have to be constructed extremely strong for any building located in a climate expecting substantial snowfalls.

The Single-Pitch or Shed Roof

This simple style features a single roof surface or plane that's usually gently sloped in a single direction. It's not a bad roof, but its appearance is rather dull and uninspiring.

ROOF PITCH

When describing a roof's configuration, "pitch" is the measure of its steepness or the degree of slope the roof or part of the roof has. It's expressed in two corresponding numbers: a value of rise per a value of run (Fig. 9-2). "Rise" means just what it says—a vertical distance. "Run" addresses the horizontal travel it takes to reach a given rise. A 4-12 pitch means there are 4 units or measures of rise for each 12 units or measures of run (Fig. 9-3). It can as easily be expressed with inches or feet. Whatever measures are used, the ratio remains constant. Generally, when it comes to roof pitch, builders tend to do their thinking in inches. A very low-pitched roof will have a substantially smaller rise than run: 1-12 or 2-12, for example. Medium-sloped roofs range up to about 6-12—about the steepest slope that a novice can comfortably walk around on. A roof having a 45-degree angle is considered steep, having a pitch of 12-12.

The pitch of a house roof is influenced by a variety of factors, including expected snowfall, the heaviness or lightness of construction materials, how much storage or living space you want beneath the roof, how much insulation is planned, and of course, how the house should look so it's architecturally balanced. Different climate conditions and different house types will call for different roof pitches. Here are the main characteristics of a low-pitched roof:

1. It will shed water well in warm climates, and will, if built soundly, retain snow for added insulation in colder climates.

2. The area directly under roof will consistently

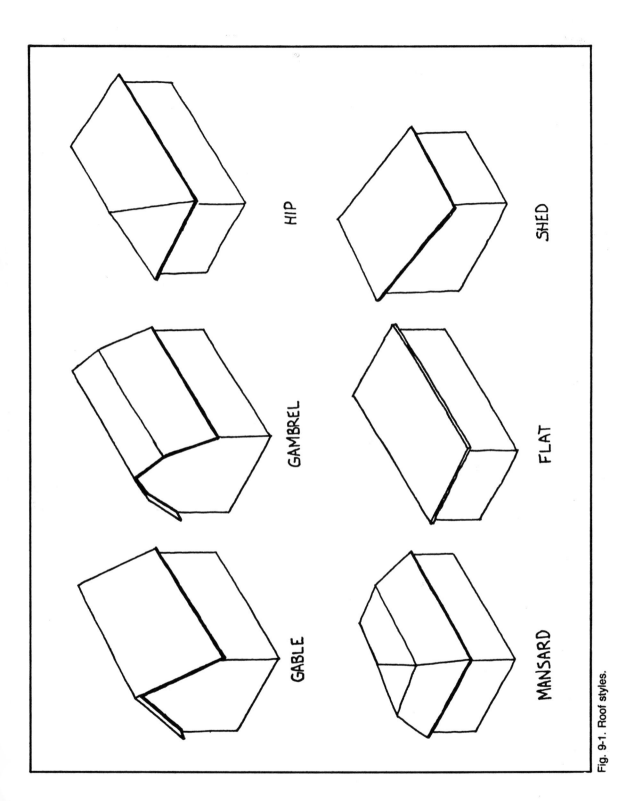

Fig. 9-1. Roof styles.

131

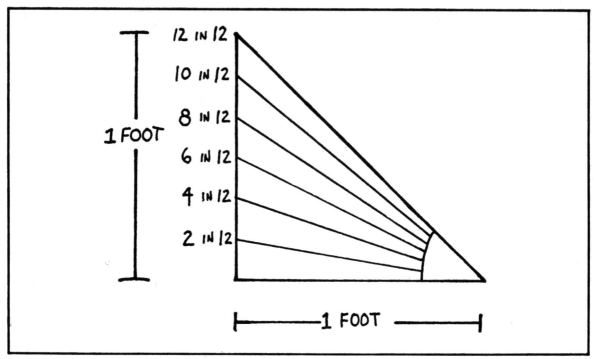

Fig. 9-2. Determining roof pitch or slope.

be a wider, more open space, providing extra room with pleasantly sloped ceilings.

3. The low pitch will make both interior and exterior maintenance safer and less expensive.

4. Since its initial construction requires less material due to its minimal surface area, it's also less costly to build than a steeply pitched roof.

To balance out our roof picture, some characteristics of a steeply pitched roof, from 6-12 to 12-12 and above:

1. A steeply pitched roof sheds just about everything in a hurry. Rain races off its surface, and snow is less likely to accumulate there. A steeply pitched roof rarely leaks. It doesn't give the moisture a chance to penetrate.

2. Although certain individuals can benefit by lofty

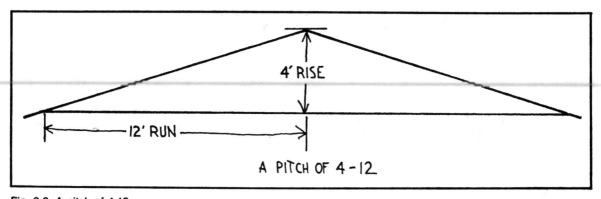

Fig. 9-3. A pitch of 4-12.

storage spaces provided within a steeply pitched roof, care must be taken to avoid ending up with inaccessible, unpleasant attic rooms. Let no one sway you into believing that every inch of an attic constructed beneath steeply sloped roof planes can be gainfully employed.

3. A steep pitch makes maintenance and repairs tasks for professionals, especially when the exterior must be accessed. That means inconvenience and high expenses for both.

4. Since more surface area is involved, a steeply pitched roof requires more materials at an initially larger cost outlay.

5. Steeply pitched roofs are good for placing solar collectors on in cold and warm climate locations.

Keep in mind that the pitch of your roof might limit or even dictate what roofing materials you can use. Wood and asphalt shingles, wood shakes, and tile or slate can require a pitch of 4-12 or steeper. Roofs sloped less than 4-12 are uncommon, and might need to be covered with an industrial-type flat roof of rubber or tar and fine gravel.

BUILDING METHODS

There are two common methods of constructing roofs:

■ The stick-built system uses individually erected rafters, ridge boards, ceiling joists, and collar beams assembled on the job.

■ The prefabricated truss system is a newer method in which trusses are made to your roof's specifications by a fabricating company that specializes in this work.

Stick-Built Roofs

Figure 9-4 shows an example of this roof.

Ceiling Joists. When all exterior and interior walls are framed, plumbed, and nailed, and after the top plate has been fastened in place, the ceiling joists go up to tie the walls together and to form a structural base for the erection of the roof. In most cases, these joists must span the width of the house, one overlapping with another, supported by a load-bearing collar beam toward the center of the house. The size of lumber to use for joists is determined by the distance to span, the type of wood used, and the load that will have to be supported above. The dimensions of the lumber used for ceiling joists should be specified in the plans. Try not to settle for anything less than 2- x -6-inch planks.

In a 2-story home, the ceiling joists at the second level become floor joists for the attic, and it's reasonable to assure they must be as sturdy as those used below, and constructed in a similar fashion.

The spacing of the ceiling joists may vary, but, as with studs, 16 inches O.C. is considered standard good construction. The doubled top plate of the wall frame supplies sufficient strength, and joists can be positioned anywhere along it. It makes most sense, though, to locate them over the wall stud positions (Figs. 9-5 and 9-6).

Ceiling joists serve several purposes: they resist the outward thrust imposed upon the walls by the roof rafters; they provide nailing surfaces for the ceiling and the upstairs or attic flooring; they support any weight placed on the upper floor. Because of their place in the framing scheme, joists must be securely nailed to the top plate of every wall that their ends rest on and to every load-bearing partition wall that they cross or join on.

Note that ceiling joists are not used with houses having truss framing. The bottoms of the trusses, normally constructed of 2 x 4s, become the ceiling joists.

Rafters. Rafters are wood planks or boards that span the distances from the top of the exterior walls to the roof ridgeline or peak to form the skeletal structure that the roof deck is fastened to. The size of the rafters vary with the distance they must span and the steepness or pitch of the roof. It's often computed by referring to tables that show the load-bearing capacities and deflection qualities of various woods and boards. For a roof having a 6-12 pitch with a span of about 20 feet (so each individual rafter would be 10 feet long), 2- x -8-inch lumber is sufficient, even in locations where substantial snowfall is expected.

The rafters should be spaced in the same manner as ceiling joists are erected, then each rafter can be tied or nailed to a companion joist as well as to the plate.

Advantages:

1. Dormer expansions are relatively simple to make with rafter framing. If you think you might want to expand into an attic, it's a must to use rafters instead of prefabricated trusses. And if expansion is likely, then beef up your floor joists to 2- x -10s.

2. The rafter or stick design is also the better choice when adding dormers because it's much easier to tailor the roof to accept the dormer framing.

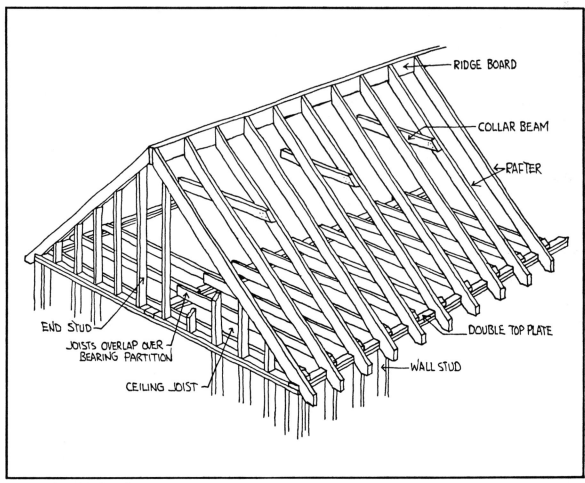

Fig. 9-4. A stick-built roof.

Fig. 9-5. A stick-built roof construction view.

Fig. 9-6. Positioning the ceiling joists and rafters.

3. Rafter-built roofs allow you to have cathedral ceilings.

Disadvantages:

1. The main disadvantage of rafter roof framing is that load-bearing interior walls must be relied upon for support.

Truss-Built Roofs

Roof trusses, unlike ceiling joists used with rafters, span the entire width of a structure (Fig. 9-7). They are triangular with wood interior bridging (W-shaped) for strength. Mathematically, the principle that gives what appear to be flimsy components the strength to span such long distances is the inherent rigidity of the triangle. Trusses simply rest upon and are fastened to opposite exterior walls. The individual truss units may be spaced 24 inches O.C., even when 2- x -4s are used for its makeup—although 16 inches O.C. is preferable

to avoid the risk of a sagging roof deck.

Advantages:

1. Ceiling joists aren't needed in the attic.
2. Trusses are built using smaller dimensional lumber than is used by the stick-built rafter system.
3. Despite the small lumber and the fact that trusses are usually installed 24 inches O.C., the truss design still provides adequate strength to a roof.
4. Savings in framing costs can usually be realized due to the reduction of materials and labor involved with truss installation.
5. The greatest advantage of a trussed roof is that it eliminates the need for load-bearing interior walls. Trusses are engineered to span entire distances between opposite exterior walls without relying on intermediate support. Thus, complete design freedom in the planning of interior space is possible.

Fig. 9-7. Truss-built construction.

Disadvantages:

1. The diagonal members used to reinforce the truss design greatly restrict the amount of usable attic space. To many homebuilders it comes down to the question of free use of living areas versus the importance of storage space.

THE ROOF DECK

The roof deck is what gets fastened to the exterior of the rafters, and what the finished roofing shingles, shakes, tiles, or other materials are fastened to (Fig. 9-8). It consists of a structural sheathing and a moisture-resistant underlayment.

Roof Sheathings

Most roof sheathing is done with plywood sheets in thicknesses suitable to properly strengthen the rafters and to correctly prepare for whatever roof covering will be used. Here are some construction points to remember about plywood sheathing:

1. The type of plywood to specify is CDX. This comes with a clear (C) or smooth side to be seen, and a rough side (D) to be hidden. The X means it's all held together with exterior-type glue.

2. The plywood thickness for roof sheathings should be at least 1/2 inch, and preferably 3/4 inch. If standard plywood roof sheathing is used on 24-inch O.C. roof framing, metal "H" clips made for this purpose should be fastened between the edges of the plywood sheets to reduce the potential of the sheathing to sag and to give extra support between the framing rafters or trusses. Thicknesses of plywood greater than 3/4 inch will not require the clips. The added thickness gives superior resistance to high winds and affords better penetration for shingle nails.

3. For the greatest overall strength, the 4- x -8-foot or larger sheets of plywood should be laid crossways to the rafters or trusses—similar to their use over floor joists—to tie the greatest number of framing members together as possible with a single sheet.

4. The joints should also be staggered by at least one rafter or truss so there is no continuous joint line from a cornice to the roof ridge board. No adjoining panels should butt together over the same rafter or truss.

Roof Underlayments

The second step toward completing a roof is to place a layer of underlayment or saturated roofing felt paper on top of the plywood sheathing. The roofing felt should be at least 15-pound (preferably 30-pound) material, which means that the weight of the amount of felt paper that would cover 100 square feet in a single ply is 15 or 30 pounds, respectively.

Roofing paper should be applied with 6-inch end overlaps and head overlaps along the edges of 2 or 3 inches. Many felt papers have white stripes on them for indicating the correct overlaps, and such stripes can also be helpful guides for coursing the shingles.

There are three basic reasons that felt paper is placed over the sheathing before the final roof topping material is laid:

■ It provides additional weather protection for the roof.

■ It's a resilient padding between the shingles and the wood sheathing.

■ It keeps the sheathing dry until the final roofing material can be applied.

It's a good idea that the roofers try to paper a roof

on a mild-weather day. If the temperatures are too cold, the paper becomes brittle and tears easily. If the weather is too hot the paper becomes soft, and likewise tears easily. Of course, building paper must be applied perfectly flat to avoid bulges on the finished roof.

VENTILATION

When closing a car door from the inside, have you ever experienced pressure on your ears because the car is practically airtight, and there's barely any means for the air to escape? Or how about storing fresh mushrooms in an airtight plastic bag? Any cook knows that to do so invites spoilage: if air cannot freely circulate around the mushrooms to remove "expired" moisture, then that moisture will quickly condense onto the mushrooms and will cause them to deteriorate, even if they're kept cold.

The same principle holds true with houses, espe-

cially when it comes to roofs. A house/roof combination that's too "airtight" is unhealthy. Moisture that's given off from a variety of our appliances and fixtures such as toilets, showers, clothes washers and dryers, dishwashers, cooking surfaces, and even from our own breathing—not to mention periods of high-humidity weather—becomes an agent that will, if not removed, rot wood, wreak havoc with insulation, and even go right through to the underlayment of a roof and affect asphalt or wood shingle roofing materials themselves—all from within. Materials stored in moist environments will also tend to be ruined by mildew over the long haul.

Thus an airtight, self-contained roof/house combination is not only undesirable, but is downright dangerous to have. Many old houses were constructed without vents. Instead, the owners relied upon large double-hung windows and screens positioned at the gable

FELT OVERLAP

PANEL CLIP IF REQUIRED

ROOFING FELT

PANEL SHEATHING
INSTALL WITH LONG DIMENSION ACROSS SUPPORTS, STAGGER THE VERTICAL JOINTS

Fig. 9-8. A cutaway of the roof deck.

Fig. 9-9. Types of roof ventilators.

ends, opening them for a cross breeze. That worked fine until the windows rotted or "froze" shut and couldn't be opened, or were closed during rainstorms and times of high humidity. That's why you'll find so many old houses with rotting wood roofs and musty-smelling attics that are extremely hot in the summer and freezing in the winter.

Nowdays, all good builders realize that attic or roof vents are necessities, and the builders supply one or a combination of several vent types to let the house breathe and rid itself of unwanted moisture. Consider, too, that roof ventilation also allows heat that rises to the upper reaches of the roof interior to escape, and does the same for dangerous gases or fumes that could collect there in the event of an accident or emergency.

How much ventilation is needed? Your builder should have a good feel for that. Sometimes it can be affected by the direction of prevailing winds, the amounts of shelter from the sun, or even from the positions of neighboring buildings or parts of the house itself. Some roof lines can channel wind toward a certain part of the roof, while others might hinder particular airflows from having desired effects. Roof ventilation is especially important to houses exposed to continuous sunshine and equipped with air conditioning.

There are four popular types of ventilators that can be used on new houses (Fig. 9-9).

Ridge Vents

Because the ridge line is the highest part of a roof, it

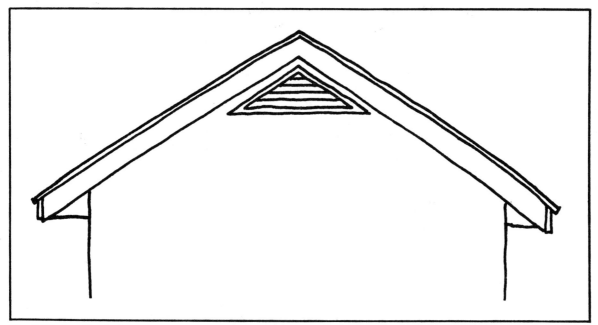

Fig. 9-10. A gable vent.

offers an efficient location for a ventilator. A ridge vent typically consists of a long metal channel or "crack" covered by an upside-down gutterlike piece of metal that permits air to escape from the house and prevents rain from entering.

Gable Vents

These are the most widely used roofing vents (Fig. 9-10).

They're the triangular slatted arrangements you'll notice on practically every house that has gable ends. They're tucked into the very top of the upper gable peaks, and are frequently made of galvanized metal. Because they're located close to the roof ridge, they do a good job of getting rid of heat. Depending on the location of the house, a cross-breeze will sometimes result.

Fig. 9-11. A roof-slope turbine vent.

Fig. 9-12. A roof-slope fan vent.

Roof-Slope Turbines and Fans

Roof turbines and fans are galvanized or aluminum vent units fastened to a side of a roof, on one of the slopes, positioned to act like a wind-powered turbine or fan when the wind blows, and like a free vent when the air is still (Figs. 9-11 and 9-12). They can be electrically powered to assist in removing hot air during summer, to augment an air-conditioning system so it doesn't become overloaded.

Soffit Vents

Soffit ventilators are similar to gable vents, except they're positioned at the eave or soffit areas of the roof (Fig. 9-13). Fresh air flows in through the soffit openings into the attic and then stale air is expelled through the gable vents, ridge vents, turbines, or fans.

Window Fans

When any or all of the above vents do not provide

Fig. 9-13. A soffit ventilator.

enough air circulation, an attic window fan might be necessary to draw out hot air and rid the attic of unwanted moisture. If you plan to install an attic ventilation window fan, make sure the builder includes large enough gable or other vents to permit the passage of enough air so the fan can operate efficiently.

Chimney Vents

Although they have little to do with attic airflows, chimney vents also pass through the attic and roof, or are positioned directly adjacent to them. Especially if it's a manufactured metal chimney that has a built-in air space, it should be located at least 2 inches away from any nearby wood frame. The top of a chimney should either clear the roof ridge by 2 feet, or should be lo-cated at least 10 feet away from it, laterally.

INSULATION

Insulation of the attic and roof is a task that must not be taken lightly. In cold-climate locations the heat generated by a furnace or fireplace rises, and it's the attic insulation that prevents it from escaping. In warm climates, heat from the sun beaming down on the roof tends to make the rooms that are directly beneath the attic too warm. Again, it's the attic and roof insulation that will prevent heat from passing through the attic floor into the living areas.

Attic and roof insulation are discussed in Chapter 20.

Chapter 10

Roof Exterior Finishing

The exterior roof covering is an important milestone in the house construction process because it brings the job's progress to the point of being closed in against the weather, or in the terminology of many builders: under roof.

The reasons for reaching this stage as quickly as possible are to protect the already completed construction from extensive damage due to rain, snow, and exposure, and to provide cover and enclosure so that further construction can proceed despite inclement weather.

ROOFING MATERIAL SELECTION

Good reasons exist as to why you should explore the various types of roofing materials available for your house. Indeed, your selection may be influenced by the following:

1. The desire for fire protection. At one time, the combustibility of a house's roofing material substantially influenced the fire insurance rate charged. Certainly, if you're miles away from the nearest hydrant you might want to think twice about wood shakes or shingles.

2. The effect that weather elements have in the area you live in. Certain roofing materials hold up better in certain climates than others will.

3. The life expectancy of the roofing. The price of the labor needed to replace a roof is high. Therefore, it's important to pick out a roofing material that will last. Don't make the mistake of selecting a material that will need replacing in less than 10 years. A quality product should last 20 to 30 years of normal use.

4. The type of house and how the house is positioned on the lot. If large expanses of sloping roof will be visible from the ground, try to choose a material that will contribute to the overall attractiveness of the home. Too often an owner will select expensive siding materials only to downgrade the building's appearance with a cheap roof. Instead, give careful attention to your roof; use materials that add color, patterns, or textures as desired.

TYPES OF ROOFS

There are basically six types of roofing that cover about 95 percent of the residential roofs: asphalt shingles, fiberglass shingles, wood shingles and shakes, slate shingles, tile shingles, and metal shingles.

Fig. 10-1. An asphalt three-tab shingle.

Asphalt Shingles

Most roofs today are being covered with asphalt shingles (Figs. 10-1 and 10-2). By far, asphalt shingles are the most common roofing material in warm and cold climates. Sometimes they're also referred to as composition shingles.

These durable shingles, depending on their weight, have a life expectancy of 15 to 30 years. They're made of a heavy paper known as felt, which is coated with hot liquid asphalt then covered with fine rock granules.

Asphalt shingles are manufactured in many different colors by many different companies. These shingles vary in weight from about 165 pounds per roofing square (a roofing square equals 100 square feet or roof area) to about 340 pounds per square. The heavier shingles are more expensive, have greater textures, and longer life. They also take more time and effort to put up, and most roofers will charge a higher rate for applying them.

There are many different kinds of asphalt shingles.

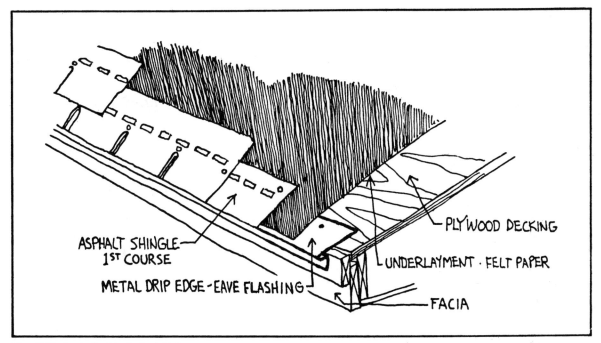

Fig. 10-2. Asphalt shingle installation.

The minimum grades weigh about 165 to 235 pounds per square. The medium grades run about 235 to 250 pounds per square, and the top grades weigh above 250 pounds per roofing unit. Don't go with anything less than a medium grade. The heavier the shingle, the longer the life expectancy. A 300-pound or heavier asphalt shingle should last between 20 to 30 years. Besides being more durable, the premium shingles are offered in better colors, colors that do not fade as quickly as the less expensive models do. The 250- to 340-pound shingles are also less susceptible to wind damage than the lighter shingles—due to their heavier, stiffer construction. But even if you live in a severe windstorm or hurricane area and decide to use the heavier asphalt shingles, it's still a good idea to have their tabs cemented down so curling doesn't result.

Winds can play havoc with asphalt and other shingles. To prevent such damage from occurring, cement down the tabs on each shingle as mentioned, plus opt for the interlocking types having tabs and slots used to hook each shingle together with the adjacent ones.

Lastly, although asphalt shingles are the most popular selection for roofing materials on new construction, they should be used warily on roofs having pitches of less than 3 to 12. With very low slopes, water seepage can occur under the shingles, especially during times of high winds. However, if professionally installed over a 2-ply underlayment of building paper, asphalt shingles may then be used on roofs having pitches of only 2 to 12.

Fiberglass Shingles

These shingles are similar in appearance to the asphalt variety, but are more resistant to fire. Ask your insurance agent about the possibility of reduced premiums for your homeowner's policy if you elect to go with fiberglass shingles. As with the asphalt type, the heavier fiberglass shingles are the more durable. Stick with self-sealing or interlocking shingles for protection against wind and curling. Along with the asphalt shingles, fiberglass ones should not be used with a pitch of less than 3 to 12 unless a properly installed 2-ply underlayment is laid first.

Wood Shakes and Shingles

Wood shakes and shingles are available in several species of wood, with red and white cedars the most popular, followed by cypress and redwood (Fig. 10-3). The term "shingle" means that the wood has been sawn, whereas "shake" indicates that the wood has been split. The shake is usually thicker and has a more rustic appearance.

Supply and labor costs to install wood shakes or

SHINGLE
(SAWN)

SHAKE
(HANDSPLIT)

Fig. 10-3. A wood shake and shingle.

shingles can be four to five times that of installing standard asphalt or fiberglass shingles. Homeowners, though, consider wood shakes and shingles a step up in quality and beauty.

"Hand-split" wood shakes and wood shingles have been popular for quite a while in the western United States, but did not reach the Midwest and East in appreciable numbers until the late 1960s. Their increasing popularity is attributed to their textures, deep shadow tones, longevity, weather resistency, and compatibility with Colonial, Modern, and Contemporary house styles.

The main drawback to wood shakes and shingles is their flammable nature.

Slate Shingles

Slate is one of the finest roofing materials available, and one of the most expensive (Fig. 10-4). Certainly it's one of the most durable shingles you can cover a roof with. But it also can weigh over 3,000 pounds per square (compared with 165 to 350 pounds for asphalt and fiberglass, and 200 to 450 pounds for wood), so a slate roof frame must be designed strong enough to support such an ambitious load.

Roofs made of slate shingles can add considerably to the value of a house. Pieces of slate are available in smooth commercial grades or rough quarry runs, in different colors and variegated shades depending on where they come from. They make a beautiful roof and if cut from a good mineral bed, will last 100 years and more.

It's unfortunate that slate has been given a bad name from older homes where tree-damaged roofs, or roofs undermined by rotting wood supports caused by a lack of proper ventilation result in loose and fallen slate roofing. People have heard horror stories about the high costs of repairing old slate roofs, and have unjustly grown overly wary of the slate shingles themselves.

Tile Shingles

Clay or cement tile shingles are especially popular in the sunbelt areas (Fig. 10-5). They come in all kinds of decorator shapes and colors and textures. They're simple to install but physically taxing because of their incredible weight—from 800 to 2600 pounds per square. As with slate shingles, a tile roof needs to be well braced to support its own weight. The tiles are apt to be expensive, especially in areas where they aren't frequently used (outside of warm-climate locations), but they are durable and have a long life expectancy.

They should be used sparingly for flat-sloped roofs, and should generally be applied where the pitch is steep enough for water to run down quickly to avoid water backup and leaking roofs.

Metal Shingles

Metal roofs, especially ones made of high-quality copper, terne (tin/lead alloy) or aluminum are very durable (Fig. 10-6). At the same time, they're relatively expensive and can be noisy to the point of aggravation in a rainstorm.

PREDRILLED NAIL HOLES

Fig. 10-4. Slate shingles.

Fig. 10-5. Tile shingles (clay or cement).

Aluminum shingles are lightweight when compared with other roofing materials (about 40 to 60 pounds per square) and come in many modern colors, shades, and styles, mostly in a shake-type texture. They'll last indefinitely if fastened securely with aluminum nails to avoid any electrolitic reaction that would occur if steel nails are used.

ROOFING COLOR

If your roof is a complex one, with many dormers, val-

FORMED ALUMINUM ROOFING

Fig. 10-6. Metal shingles.

leys, and varying planes, dark roofing shingles will tend to pull it all together in a nice way.

Check other houses that are already completed in the area or on the street you're planning to build. If several or many of them are using similar colors, it might be wise for you to select something different so your home doesn't "blend in" with the rest of them, to break the monotony.

Be aware that a light-colored roof reflects heat and is more desirable in areas where air-conditioning is the greatest energy user. Although it's not *critical*—you can choose a medium color if its better for the overall house appearance—you should stay away from black in warmer regions. If your heart is still set on a black roof, and you live in central Texas, just make sure you engineer enough attic insulation and ventilation, and that you have a four-star cooling system installed.

In colder climates, the reverse is true. Black absorbs heat from the sun, so a darker color is clearly the more practical choice.

VENTILATION AND SOFFITS

The roof and attic should be assured of adequate ventilation to allow for the escape of heat and humidity (Figs. 10-7 and 10-8). All ventilation units installed on the roof or in gable ends should be designed to shed rain and snow and not permit any moisture penetration. In addition, the free opening vent areas must be screened to protect against entry by insects, bats, rodents, squirrels, and similar invaders. A good size metal/steel roof louver to use is 12 × 18 inches.

Soffits, the flat painted surfaces under a roof or overhang, should be constructed of prepainted or vinyl coated sheets that are maintenance-free (Fig. 10-9). They're available not only with a smooth unbroken surface, but also with perforated or slotted surfaces that will encourage ventilation of the attic or roof space. Eave soffit vents allow circulation of air through an attic to prevent moisture condensation and its consequential damage to roof structure and insulation. However, venting the eaves alone is not sufficient. Because warm air cir-

Fig. 10-7. Gable ventilation.

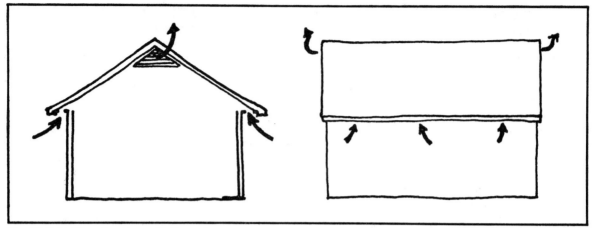

Fig. 10-8. Soffit and gable ventilation action.

culates upwards, you should install the roof or gable vents mentioned earlier. Have sufficient free opening areas near the top of the attic to match the collective opening space of the eave soffits.

FLASHING

Flashing is sheet metal or other material used to prevent the leakage or driving in of rain water and general moisture infiltration at joints near openings or where different materials or planes meet, such as around chimneys, vents, roof valleys, and stacks (Figs. 10-10, 10-11, 10-12, and 10-13). If a material projects horizontally from the surface of the house, as at window and door trims, or at the insulation around a foundation, flashing is required. It's also needed wherever roofs and walls join, as with a split level house or a 2-story having an attached garage (Fig. 10-14). Flashing is important at the juncture of a dormer's siding with a main roof, to prevent water from leaking through the roof system (Fig. 10-15).

Aluminum is the most common flashing material. It's produced in long rolls in several widths and is inexpensive, lightweight, and resistant to corrosion except in industrial areas and near the seacoasts. It has one drawback—a shiny appearance that must be painted. And aluminum does not paint well.

Galvanized steel and terne are also employed as flashing; but they must also be painted. Stainless steel, zinc alloy, and even lead have all been used in similar fashion.

An excellent, though expensive, choice for flashing is copper. It seems to last forever and requires almost no maintenance.

Fig. 10-9. Soffit under a roof overhang.

148

Fig. 10-10. Exterior flashings.

Fig. 10-11. Flashing around a chimney.

Fig. 10-12. Flashing around a vent pipe.

Fig. 10-13. Valley flashing.

Fig. 10-14. Flashings at roof and wall intersections.

Fig. 10-15. Flashing around a dormer.

Asphalt roofing material is sometimes used for valley flashing on roofs, and plumbing stacks are frequently flashed with special neoprene plastic collars. Vinyl materials in various colors are also on the market. Vinyl is an excellent choice and costs substantially less than copper.

INSTALLATION

No matter how good a roof's materials are, the roof won't be able to do what it's supposed to do if the installation is shoddy. Here are some things to watch out for:

When anchoring the felt paper and topping materials, power staple guns are the most economical way to go, *but* they don't do as sturdy a job as nails do. If your roofing is self-sealing asphalt shingles (shingles with glue underneath each tab that will stick to the shingle below it when baked in the sun), consider that it takes at least one and preferably two hot summers for them to "melt" together to form a strong bond. Up until that time, staples will not provide the holding power of wide-headed nails. High winds are more likely to blow stapled shingles from a roof.

Roofing nails having very sharp points and flat head diameters from 3/8 to 7/16 inch should be requested. Most modern installations are done with 11- or 12-gauge galvanized steel nails having rough shanks for superior holding power. Naturally, the length of the nails must be enough to allow for full penetration of the roof's sheathing.

When wood shakes and shingles are applied, the second or overlapping layers must be laid so the joints are at least 1 1/2 inches apart, and the adjacent shingles or shakes should have at least a 1/4-inch gap between themselves to allow for expansion during wet weather.

Make sure masonry and metal chimneys, skylights, and other obstructions are in place before the roofing begins. Otherwise the roofers will have to make an extra trip to complete the remaining shingles and flashing, at extra expense.

WATER DRAINAGE

Gutters and downspouts work together to collect runoff water from the roof and divert it away from the house so foundation seepage can be prevented (Figs. 10-16, 10-17, and 10-18). They're vital necessities in most cases, but can be troublesome to maintain—tree leaves, seeds, and twigs tend to collect in them and clog the downspouts, squirrels and chipmunks use them as freeways and store winter food in them, plus they can be damaged by ice that collects and hangs from their not-too-strong edges.

Here are some important points to consider when planning gutters and downspouts:

1. The least expensive gutters to buy are galvanized steel models, but they have to be painted before they can be secured to the fascia or rafter ends of a roof.

2. Another inexpensive option is to use unpainted aluminum. It's durable enough unless exposed to salt air near seacoasts or air laden with chemical contaminants. Even when no pollutants are present, however, this gutter should be painted for the sake of appearance.

3. Your other choices of gutter materials are much better: aluminum with factory baked-on enamel, and aluminum covered with a thin layer of vinyl. Vinyl gutters (solid vinyl) are also available. All three are in the long run durable, economical, and require little maintenance. Vinyl tends to be brittle in very cold weather, but it never requires refinishing because the color is integral to its form. Copper gutters are also available at substantially higher prices than the others.

4. Metals used in gutters and downspouts vary in thicknesses; 26-gauge galvanized steel is quite strong, but 28-gauge (a thinner metal) is more commonly used because it costs less. Aluminum is measured in a fraction of inches. The most frequently used thicknesses of aluminum are .025 and .032 inches.

5. Cleaning gutters and preventing downspouts from getting clogged can be easily done if you specify removable caps or screens. Then you can just pop off the caps or screens when necessary and flush the small accumulation of silt from the gutters with a garden hose.

6. Gutters should be mounted on the fascia boards (especially when the fascia is not made of or coated with vinyl) so the gutter backs are offset slightly, with an air space between the fascia so the fascia surface will not deteriorate from lack of ventilation.

7. Large houses with great expanses of roof require that both the gutters and the downspouts have sufficient capacity to handle expected volumes of rain water. Small houses up to 30 feet long are adequately serviced with standard 5-inch gutters and two downspouts. Standard building formulas state that a 4-inch gutter is the minimum size to be used for roofs that are no larger than 750 square feet. Roofs up to 1,400

Fig. 10-16. A roof drainage system.

square feet can be drained with 5-inch gutters, and larger roofs, with 6-inch material. Downspouts 3 inches in diameter will do if the roof is under 1,000 square feet, and 4 inches in diameter if the area is greater.

8. A roof plane will collect water during any rainfall, especially if there is wind. The higher the roof ridge, the more true roof area there is and the faster water will race into the gutters. If you plan a steep roof, try

Fig. 10-17. Metal gutter shapes.

Fig. 10-18. A gutter and downspout arrangement.

154

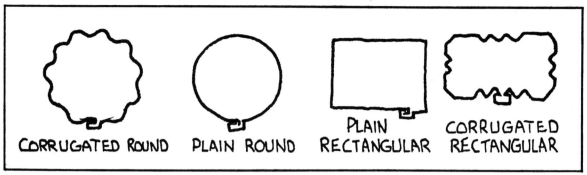

Fig. 10-19. Standard downspout shapes..

to be generous with gutter sizes regardless of the roof area size.

9. Your downspouts may be rectangular or round, and plain or corregated (Fig. 10-19). In cold-climate locations when there is a possibility of standing water freezing in the downspouts, the corrugated type

Fig. 10-20. A column.

is preferable because it can expand without damage.

10. The water discharge can flow from the downspouts into drainpipes that run to storm sewers, or it can be directed to a natural runoff area, perhaps to the street or a storage cistern or rain barrel—anything so it won't run down into the foundation or form a swamp on adjacent ground.

COLUMNS

Depending on the construction of your house, unsupported roof overhangs can protrude from the house a considerable distance. But beyond a certain point they must be supported around the outer edges by columns (Fig. 10-20).

Although wood columns, except in very small sizes, are hollow, they still have the strength to bear a substantial load. Trouble is, they're so intricately made, they cost a fortune.

Factory painted aluminum columns are the answer. They cost considerably less, are maintenance-free, and because they're hollow—made only of thin metal—they can be installed around a weight-bearing wood or steel post.

Chapter 11

Exterior Wall Finishing

The exterior wall covering is the single most dominant feature of a home's outer appearance; its color and texture are the first things noticed by anyone approaching the house. Good design calls for simple lines, common sense in the selection of materials, harmonious textures and colors, plus good proportions and scale. A hash of materials such as a bit of stone here, some brick over there, with shingles and clapboards and stucco all mixed above, will give the impression that the house is desperately trying to trick observers into liking *some* detail, and more often than not will ruin a dwelling's appearance.

Beyond its cosmetic nature, the exterior wall covering also acts as the final protective layer between a home's occupants and the great outdoors. Attached to the frame or masonry walls can be wood siding in various forms, brick, stone, stucco, aluminum and vinyl sidings, shingles of metal, asphalt, or plastic, and many other lesser-used siding materials.

MATERIAL REQUIREMENTS

When selecting the material for your home's siding, consider how the following characteristics stack up against those materials in the running.

■ Cost of materials

■ Cost of installation labor, ease of handling by size, weight, and shape

■ Resistance to natural weathering, chemical attack, and atmospheric pollution

■ Resistance to scratching and impact

■ Appearance of color and texture

■ Dimensional changes resulting from temperature and moisture

■ Resistance to moisture penetration

■ Combustibility

■ Sound insulation and absorption

■ Strength under conditions of compression, bending, shear, and tension to carry applied loads and resist the pressure of wind

■ Adaptability to future expansions and other modifications

■ Susceptibility to insect damage

ALUMINUM AND VINYL SIDINGS

Aluminum and vinyl sidings are available in different colors, forms, and textures for both vertical and horizontal installation. They can be used alone or in combination with each other to cover all exterior wood surfaces and eliminate wood painting and refinishing chores.

Aluminum Siding

Aluminum siding is a low-maintenance exterior wall covering that won't rust in the ordinary sense. Its baked-on finish lasts for 20 to 40 years, depending on the grade purchased. Aluminum siding won't rot, split, warp, or crack. It's manufactured in a wide variety of colors and shades from light pastel tints to whites to deep rich tones, and comes in numerous textures and finishes—some resembling wood siding. Two thicknesses, a thin (.19-inch) and thick (.24-inch) are available. The difference in price and insulating qualities are small, but if you're planning top-notch construction that calls for aluminum siding, opt for the thicker gauge because it's stiffer and holds up better to abuse. The width of the horizontal type of aluminum siding is a single panel of 8 inches, or "double-four" panels that are essentially single panels, each having a horizontal crease across the middle so it resembles two 4-inch-wide lengths of clapboard (Fig. 11-1). There are also clapboard-like panels with 5-inch exposures, 9-inch exposures, and two beveled edges to give the appearance of two strips of bevel-edge siding.

For vertical applications, aluminum siding is available in many of the same colors and finishes as the horizontal panels, so the two materials can be mixed and matched on the same dwelling. The vertical siding comes in 10-, 12-, and 16-inch-wide board and batten strips, plus in V-groove and other styles.

There are several drawbacks to aluminum siding you should know about. First, unless reinforced by being installed directly over a stiff backer board material, aluminum siding will dent when soundly struck by a baseball, rock, or other hard object flung by a neighbor's 10-year-old son. Secondly, the surface color can be scratched off, exposing the silvery bare aluminum beneath. And thirdly, bare aluminum exposed to industrial pollutants and sea-coast environments can gradually react to the air-borne chemicals in a negative way.

Vinyl Siding

Vinyl siding is another popular low-maintenance wall covering that's manufactured essentially in the same forms as aluminum siding. As with aluminum siding, vinyl can be backed with a polystyrene or other board reinforcement both to give the siding a strong base and an insulating R-value (especially when also backed with several layers of reflective aluminum building foil) of about 6.0. A major advantage to vinyl siding is that the color is molded throughout the entire thickness of the material, so a scratch will do little damage (Fig. 11-2). Neither will vinyl siding dent; it's resilient nature allows it to spring back into shape after all but the most violent blows.

The main drawbacks to vinyl are that it's inclined to buckle or ripple if not installed exactly correct, it's not as readily adaptable as aluminum is to cover exposed wood trim, and it comes in fewer colors—traditionally light shades.

Vinyl/Aluminum Siding

A superior siding to the previous two is a combination of vinyl and aluminum, referred to as vinyl/aluminum siding. It's actually an aluminum siding coated with vinyl. It offers all of the advantages of both materials. In particular, the vinyl coating holds the color, prevents fading, and limits damages from scratches. The aluminum's strength and stiffness helps resist rippling, buckling, and dents, especially when applied over a sturdy backer board. Whenever possible, this siding should be chosen instead of plain aluminum or vinyl grades, even though it carries a higher price tag. Some manufacturers offer this type of siding with up to a 40-year warranty.

No matter which type of aluminum or vinyl siding is selected, make sure the contractor correctly applies caulking around the doors, windows, and corners—wherever the siding forms a seam across its grain or meets with different building materials (Fig. 11-3). He

Fig. 11-1. "Double-four" aluminum siding.

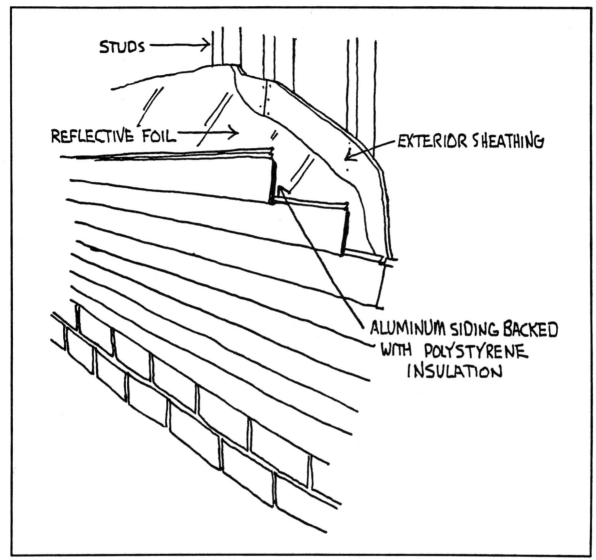

Fig. 11-2. Aluminum siding installed with backer board.

should also use aluminum nails for fastening the siding materials to exterior walls. Aluminum nails won't rust and form unsightly streaks.

MASONRY EXTERIOR WALL COVERINGS

Masonry exterior walls of brick and stone have always held a certain attraction for individuals who prefer the beauty, feel, and apparent strength of brick and stone construction. Masonry also enjoys an intangible prestige value that houses sheathed with wood, aluminum, or vinyl sidings seem to lack.

It's true that brick and stone sidings are more expensive than most other exterior wall coverings, mainly due to installation costs. To use brick or stone, a contractor must either move exterior walls inward 5 or more inches to allow space for the full masonry veneer so the specified outer wall dimensions can be retained, or he can keep the load-bearing foundation walls and exterior walls true to their specified dimensions by installing the brick veneer against the outside of these walls (Fig. 11-4). To permit the latter method, a separate outer foundation must be constructed to support the brick or

2. They hold onto their looks indefinitely, with little maintenance.

3. Brick and stone houses have historically held their value well, and have enjoyed good resale demand.

4. They're durable and have a reputation of permanence.

5. Due to their strength they don't "sway" as much in the wind as wood-sided homes, and will not develop as many interior plaster or drywall cracks and imperfections as in less rigid non-masonry dwellings.

6. Their resistance to fire is excellent.

Disadvantages:

1. They add considerable extra expense to the purchase price of a house.

2. It's difficult and expensive to make modifications to the exterior walls or additions to a structure. Removal of part of an old wall is expensive, and matching up the brick for the new walls can be difficult.

3. Brick and stone have low insulating values. Despite popular belief, even the thickest masonry offers exceptionally poor insulation properties, which is why masonry houses seem particularly cold and are hard to heat. Consequently, the proper amount of exterior wall insulation must be insisted upon regardless of the type of brick or stone used.

Brick

Brick makes a very attractive exterior, with numerous colors and textures available. All of one color can be used, or a mottled effect can be had by using many different shades or colors in the same surface. The best way to arrive at what you'd like is to take a drive through neighborhoods that have plenty of brick homes. A few color snapshots of what most appeals to you can be handed to your contractor. He'll be able to tell you where those bricks can be purchased or ordered.

It should be noted that most bricks are manufactured with holes in their centers. If the top of any bricks need be exposed, enough solid bricks without holes should be ordered along with the others. The same principle applies to textured bricks where typically only one side—the exposed side—is textured (Fig. 11-7). When bricks are needed to have a textured surface exposed on two or three or four surfaces, special bricks that are textured all around will have to be ordered.

Two or three days after the brickwork is completed, it should be cleaned with a muriatic acid solution by the mason crew.

Fig. 11-3. Caulking around a window.

stone walls (Fig. 11-5). In either case, there must also be a space between the masonry and interior wall surfaces, and this space should contain proper insulation and a vapor barrier (Fig. 11-6).

The first choice (of moving the walls inward to accommodate the thickness of bricks or stones) should require no change in the home's roof structure, particularly to the overhang and exterior trim. The second choice (of building out) might, however, require alterations in these areas to accommodate the wider dimensions of the outer limits to the exterior walls.

Advantages:

1. Brick and stone make beautiful, unique exterior wall coverings.

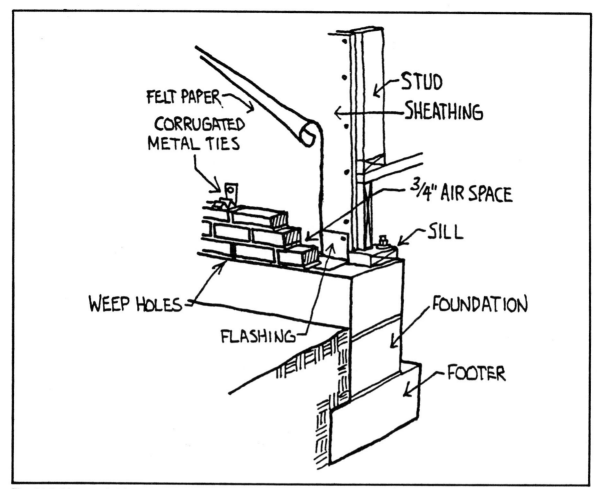

Fig. 11-4. Masonry-brick construction.

Stone

Stone also makes an attractive exterior that's durable and practically maintenance-free. It has most of the advantages and disadvantages of brick, except you wouldn't be faced with the problem of the holes in the bricks, and the textures not on all sides.

Stonework is usually more costly than brick. In general, stones cut with rectangular corners are more commonly used for covering exterior walls. Rubble or fieldstones having irregular shapes and no corners are seldom used anymore in ordinary house construction, but can be employed in a feature wall for dramatic effect to create a rustic appearance.

Openings In Masonry Exterior Wall Coverings

Due to the weight of brick and stone, door and win-

dow openings require special supports to hold the brick or stone securely in place above those openings. There are three common ways to provide such support: with steel lintels, curved brick arches, and flat brick arches.

The steel lintel is the simplest to install (Fig. 11-8). It consists of steel ''angle iron'' of appropriate length having flanges about 3 1/2 inches wide and a metal thickness of about 1/4 inch. It should overlap the top of the door or window opening by at least 8 inches on either side, so the weight of the brick or stone above the opening can be transferred to and distributed throughout the adjacent masonry structure. The curved brick arch is constructed of standard size and shape bricks or stones to span the opening (Fig. 11-9), and the flat brick arch is formed with specially cut bricks or stones (Fig. 11-10).

Fig. 11-5. Masonry-brick construction on supporting footer.

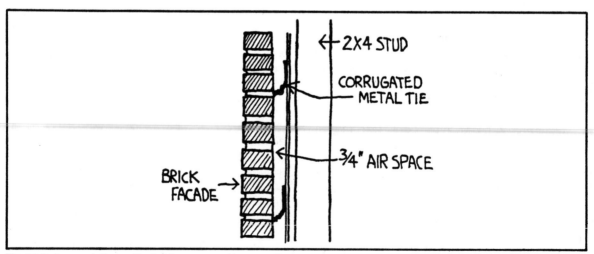

Fig. 11-6. A space between the brick veneer and frame wall.

162

Fig. 11-7. A textured brick.

STUD WALL

STEEL LINTEL

BRICK VENEER

Fig. 11-8. A steel lintel.

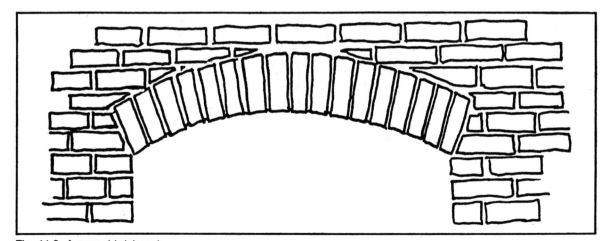

Fig. 11-9. A curved brick arch.

163

Fig. 11-10. A flat brick arch.

WOOD SHINGLES AND SHAKES

Wood shingles and shakes are usually made of cedar, sometimes redwood or cypress (Figs. 11-11 and 11-12).

Cedar shingles or shakes are used as siding when a homebuyer wants an eye-catching rustic appeal to his home, and a "warm" siding that's naturally resistant

Fig. 11-11. Wood shingle construction.

Fig. 11-12. Wood shake construction.

to decay and is an excellent insulation. Cedar has a golden brown color when new, that gradually darkens with age, and finally weathers into an attractive silver-gray, depending on the climate (the amount of sunlight and humidity) it's located in.

The difference between shingles and shakes is that both sides of shingles are sawn smooth, while shakes have at least one rough-textured side created by splitting it from the mother log. These are the same shingles and shakes that are also used for roofing.

While cedar, redwood, and cypress shingles and shakes can be installed in some climates without being coated with preservatives, weathering everywhere is best controlled by applying recommended weatherproofings every five or six years. Pressure-treated Southern yellow pine is also an acceptable substitute, when available.

The best wood shingles and shakes are free from knots and pockets of pitch. You can tell the difference between the best and lower grades by the wood grain—it should be regular and clear with few or no defects.

Drawbacks to wood shingles and shakes include their cost: they're expensive and time consuming to apply. Unless you opt for shingles and shakes prefabricated into 8-foot panels, each shingle or shake must be hand nailed into place, one at a time. They're also susceptible to fire.

Only non-rusting nails that provide sturdy holding power should be used to fasten wood shingles and shakes to the walls. Common nails won't hold well enough and will rust and streak the siding.

SOLID WOOD SIDING

Almost any type of wood can be used for solid plank siding, including such species as cedar, redwood, fir, cypress, pine, spruce, and hemlock. Redwood siding in particular is very durable. It resists deterioration from the weather and from insects. Unpainted redwood surfaces will darken season by season to a deep grayish brown.

Solid wood siding comes in many styles for horizontal and vertical applications, including beveled, dropped, and beaded planks for horizontal sidings, and V-groove, tongue-and-groove, board-and-batten, and channel vertical sidings (Figs. 11-13, 11-14, and 11-15).

Beveled horizontal wood siding is probably the most popular of all solid wood exterior wall coverings because it so nicely complements most styles of architecture. It consists of long boards, available in varying thicknesses and widths having beveled edges and tapered to exaggerate the deep, long horizontal shadow lines at the lower edges of the planks that help provide character to a dwelling's appearance. The individual boards are usually installed over a sheathing and building paper, nailed through them to the exterior wall studs. Corners are covered with either metal corner pieces or wood corner boards (Figs. 11-16 and 11-17). The thickness of the corner boards should be at least 1 inch to provide a substantial caulking base.

The old-fashioned clapboard siding that once covered (and still covers) many a home consists of wood planks of uniform thicknesses.

All solid wood sidings should be weatherproofed with water-repellant treatments, oils, varnishes, preservatives, or exterior paints and other coatings that will have to be periodically renewed as needed. A vapor barrier beneath the wood siding is also required to prevent condensation from within the home from causing paint to peel and wood to rot.

PLYWOOD SIDING

Plywood siding can be supplied in many varieties of wood and patterns at varying costs (Fig. 11-18). Check with your local suppliers to see samples. Only exterior types of plywood should be considered—those having their layers of veneer bonded together with a tough waterproof glue.

Plywood panels are manufactured in 4-foot widths and 8- to 12-foot lengths that, due to their size and ease of installation, help hold down labor costs. If you plan to use plywood siding, match the correct length panels

BEVELED SIDING   V-JOINT SIDING   SHIPLAP SIDING

Fig. 11-13. Solid wood siding.

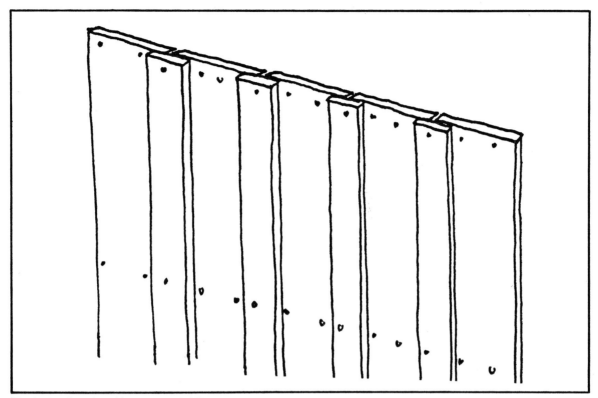

Fig. 11-14. Board and batten wood siding.

NAILING BLOCKS BETWEEN STUDS

SHEATHING

NAILING BLOCK

Fig. 11-15. Vertical application of wood paneling.

Fig. 11-16. Siding corner construction.

to the requirements of your home to have as few horizontal joints as possible, because they'll detract from the overall appearance and can be a source of water and moisture leaks if not correctly installed.

Because of its strength, plywood siding is sometimes applied directly to the wall studs, without the use of an underlayment sheathing.

HARDBOARD SIDING

Hardboard sidings are manufactured panels consisting mostly of wood products. They come in more finishes and textures than plywoods, but are not as strong.

On the positive side, factory-made hardboard sidings are free from natural defects. Their panels are stiffer, and less likely to warp or bend. Both the texture and depth of wood are presented in authentic-looking wood grains and grooves. The better hardboard or wood fiber sidings are over 50 percent denser than real wood planking, and won't crack, split, check, or delaminate. They can also be purchased primed—

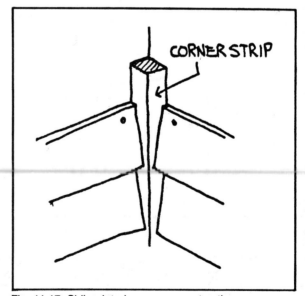

Fig. 11-17. Siding interior corner construction.

168

Fig. 11-18. Plywood panel vertical application.

ready for custom finishing in multiple lap sheets that are easy to install without having to nail narrow individual boards.

Remember that even factory-finished hardboard sidings will probably have to be refinished eventually.

STUCCO

Stucco is that plasterlike material so popular with English Tudor construction. It's an excellent exterior finish, having a long lifespan and needing very little maintenance. It's really a version of portland cement that's troweled on like plaster to either masonry or frame walls, with no seams or joints. It's usually made in white, but can be colored with any paint manufactured for application over masonry. There's also a wide selection of pigments available that can be added to the stucco mix. If deeper hues are desired, the stucco can be painted in the same fashion as concrete block.

Stucco can be applied as a finish coat to both existing houses and new construction. It can be finished to give a number of interesting textures to conform with traditional or modern architectural styling.

Three coats with a total thickness of about 3/4-inch are generally recommended. When applied over the sheathing of a wood-framed house, a layer of building paper is placed over the sheathing to protect it from moisture in the stucco. Next comes the application of metal lathing nailed to the exterior wall. The lathing should be self-furring or should be applied with self-furring nails. The most important point in the application of the stucco is that the scratch or first coat of material must be pushed through the metal lathing and behind it to form a solid layer between the lathing and studs or sheathing.

Fig. 11-19. Stucco application.

Where stucco is applied over large uninterrupted areas, control joints should be installed to permit expansion and contraction of the stucco material. Without these control joints the stucco will crack. As a general guide, control joints are planned for at least every 3 feet of travel.

The second coat is applied over the scratch coat (the first coat) after allowing sufficient time for the first coat to dry. And finally, after the second coat has dried, the third and finishing layer is applied with whatever pattern you have selected—smooth, stippled, swirled, or other (Fig. 11-19).

When applying stucco over masonry, the finish coat can be troweled directly to the block or concrete.

VAPOR BARRIERS

The living areas between the exterior walls should be sealed with appropriate material applied to the inside of the walls to prevent the movement of moisture from the living areas into the exterior wall insulation and outer wall coverings (Fig. 11-20).

Wall moisture originating from within a building can be an even more damaging factor than outside moisture penetration. Without proper vapor barriers, as the temperature increases within a home, inside water vapor is transmitted into the walls, where it condenses. This condensation results in a wetting of the structural materials and in a loss of the insulating qualities of the

Fig. 11-20. A vapor barrier view.

exterior walls. It also gives rise to such serious problems as the chemical, physical, or biological deterioration of the wall materials, such as the corrosion of metal, the spalling of brick, and the rotting of timbers.

Aluminum foil used as insulation on the outside of a wall, against the wall sheathing, should be perforated to negate condensation problems. If used without the perforations in cold-climate areas, wood frame damage could result.

Insulating batts in the exterior walls should have a vapor barrier backing such as treated Kraft paper or aluminum foil with the vapor barrier on or toward the living area side.

If blown insulation or unfaced insulation is chosen, aluminum foil-backed drywall can be used, or friction fit or other types of insulation can be applied after polyethylene sheet material is stapled or nailed to the interior of the wall studs and ceiling joists. The polyethylene film should not be less than 3 mils thick.

INSULATION

Underneath the sheathing and exterior wall covering, between the studs, one of the following types of insulation will generally be used:

■ Batts or blankets—these are prepared thicknesses of expanded glass fiber, mineral fiber, or organic fiber that are placed in the walls between the studs. They can be faced (having a vapor barrier on one side) or unfaced.

■ Blown or poured—this is composed of loose expanded mineral or organic fibers that are placed or blown into frame spaces. They're more useful for in-

Table 11-1. Insulation for Resistance Values of Siding Materials.

Material	Thicknesses in Inches	Resistance Rating
Air space	3/4 or wider	.91
Aluminum foil		2.44
(sheet type with		
3/4-inch air space		
Blanket insulation	3	11.10
Common brick	4	.80
Cinder block	8	1.73
Concrete	10	1.00
Concrete block	8	1.00
Gypsum board	1/2	.35
Insulation board	1	3.03
Plywood	3/8	.47
Roofing roll vapor barrier	1/8 to 1/4	.15
Sheathing and flooring	3/4	.92
Shingles		.17
Stone	16	1.28
Wood siding	3/4	.94
Window glass (single)		.10
Window glass (double)		1.44

sulating existing buildings when it would be impractical to completely remove the inner or outer wall sheathings.

■ Formed-in-place cellular plastic foam—this material is injected into framing spaces, where it solidifies. Like the blown varieties of insulation, the injected materials are also convenient to take care of existing structures so wall sheathings do not have to be removed to get at the wall insides.

If you choose to use vinyl, aluminum, or vinyl/aluminum siding, remember that backer boards or styrofoam or other materials can be used behind the siding panels to provide extra insulation as well as added strength.

Table 11-1 lists insulation values of various sidings.

Chapter 12

Stairs

Who can forget the dramatic confrontations in *Gone With the Wind* between Rhett Butler and Scarlet O'Hara on a huge, spectacular staircase?—the dominating structural feature of the deep-South estate of Tara. And what about the shrieking panic of Martin Balsam in Alfred Hitchcock's *Psycho*, as he tumbles backwards away from the murderer—down a stark wooden flight of stairs.

No doubt, staircases are custom-made for grand entrances and exits. At the same time, stairways can be convenient and dangerous, healthy and harmful, attractive and ugly, space-saving and space-stealing. They're less expensive and much more practical than elevators or escalators, and they enable us to make better use of small building sites by permitting several living levels to be positioned one atop another. Staircases can be made of wood, metal, stone, concrete, or any combination of construction materials having the strength to do the job.

A stairwell is the term for a shaft or opening through one or more floors of a house in which a staircase is constructed or placed. A completed stairway consists of the following (Figs. 12-1 and 12-2):

■ Stringers—diagonal or circular supports for the steps.

■ Treads—the horizontal upper surfaces of individual steps, the part your foot steps on.

■ Risers—the vertical pieces between the treads. Some basement stairways don't have risers, but are wide open between the wooden treads (Fig. 12-3).

■ Handrails.

■ Newel posts—the posts at the top or bottom of a flight of stairs that support a handrail, or the central upright pillar around which the steps of a winding staircase turn.

■ Balusters—any of the small posts that support the handrail of a railing. In olden times, these were frequently elaborate woodwork. Latter-day balusters, railings, and posts are more likely to be black wrought-iron, or simple wooden handrails attached to the sides of the stairwell walls.

GENERAL STANDARDS

1. Stairway angles—In staircase design and construction you'll find a rise and run similar to that of a

Fig. 12-1. An open main stairway.

roof's slope. The angle of a stairway is determined by the arrangement of the tread depths and riser heights.

2. Stairway treads and risers—A good stairway construction will find that the sum of two risers and a tread will fall between 24 and 25 inches. For safety's sake, all treads should be equal and all risers should be equal in any one flight.

3. Stairway widths—Main stairways should be at least 2 feet 8 inches wide, clear of the handrail(s). The overall width without the rail(s) should be at least 3 feet (Fig. 12-4). A basement stairway can be slightly narrower, with a minimum clearance of 2 feet 6 inches, and overall without the rail(s), 2 feet 10 inches.

4. Stairway landings—The minimum dimensions for a regular stairway landing is 2 feet 6 inches square, preferably 3 feet. For safety, landings must be level and free from intermediate steps between a main up flight and a main down flight.

Table 12-1 provides a checklist for interior stairways.

STAIRWAY TYPES

There are four basic types of staircases used in modern houses: straight stairways, L-shaped stairways hav-

174

Fig. 12-2. Staircase parts.

Fig. 12-3. A basement
stairway with open risers.

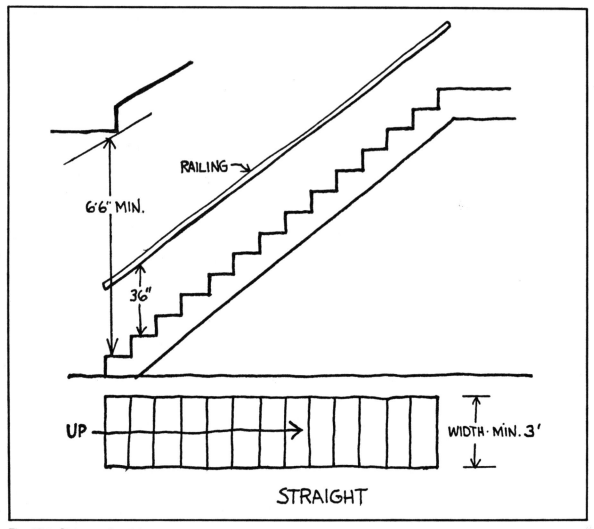

Fig. 12-4. Stairway dimensions.

Table 12-1. Checklist for Interior Stairways.

	Minimum	Best Quality
Vertical rise	7-1/2"	6-1/2"
Horizontal run	10"	12" to 14"
Tread width	3'	4'
Railing	Some Vibration	Firm
Baluster spacing	10"	6"
Number of landings	0	2
Natural lighting	Fair	Excellent
Artificial lighting	Fair	Excellent
First-floor foyer	Skimpy	Generous
Second-floor stair hall	Skimpy	Generous
Two-story-high walls	2	0

ing a landing or winders at the turn, U-shaped stairways having a landing or winders at the turn, and spiral staircases (Figs. 12-5 and 12-6).

Straight

Straight stairways are by far the stairways used in most house construction today.

Advantages:

1. They cost the least to build.
2. They're the easiest to carry items and materials on.

Disadvantages:

1. They're dangerous. If someone happens to trip at the top, a fall all the way to the bottom could result. Small children especially must be protected from them.
2. They're tiring to climb because there's no space to stop and catch your breath.
3. They're not very attractive.

L-Shaped and U-Shaped

Both of these stairway types were used frequently during past decades when individual carpentry efforts were more predominant than today's pre-built and space-saving methods.

Advantages:

1. They have landings to rest on to catch your breath.
2. The landings provide good spaces to hang decorations or art on.
3. They can be so attractive as to add character to a house.

Disadvantages:

1. They're more difficult to work into a house's floorplan.
2. They're more expensive to build.
3. They make it harder to carry large items such as bedroom furniture up and down. This is especially true of narrow U-shaped stairways.

Spiral

Only people thoroughly familiar with spiral staircases should plan them into their houses.

Advantages:

1. They save space. Spiral stairways can be installed where you cannot possibly fit conventional stairs.
2. Because they're rarely enclosed, they serve as attractive decorative units that generally become focal points of the rooms they're in.
3. They're good for outdoor use to provide access to second-story balconies, decks, and regular rooms.

Disadvantages:

1. They're dangerous. The treads are not full depth at both sides. If you fall from the top you can tumble a long way. Most people aren't used to them. Spiral stairways take a different gait to ascend to descend; visitors unaccustomed to them might find this a nuisance.
2. They're almost impossible to use for moving large items such as bedroom furniture between floors.

STAIRWAY STYLES

Regardless of their shape, stairways can be open on both sides, open on one side, or closed in by walls on both sides. The most attractive seem to have one open-type support or stringer. That way at least one wall provides a surface to hang artwork or plants, and the stairway still looks and feels spacious—even if it isn't.

As an added note, if you're planning to use stairwell walls as a gallery, arrange for blocking in between the studs while the wall is being erected. Then heavy items can be fastened to the blocking if need be, and not just into the weaker plaster or drywall.

WIDTHS AND HEADROOM

A staircase should be wide enough so two people can pass each other on the steps, and furniture can be transported up and down with a minimum of trouble. Consider 36 inches as a minimum width to be safe; 42 inches, if space permits, is ideal. Extra inches are especially handy if a staircase is closed, or makes a turn and involves winders as in L-shaped and U-shaped types. If wider, they're a lot easier to maneuver large pieces of furniture on.

The stairs should also be plenty deep for a good step. Twelve to 14 inches deep for individual treads is both the maximum and ideal range. The headroom between any part of the tread on any individual step and the nearest vertical obstruction should not be less than 6 feet 6 inches (Fig. 12-7).

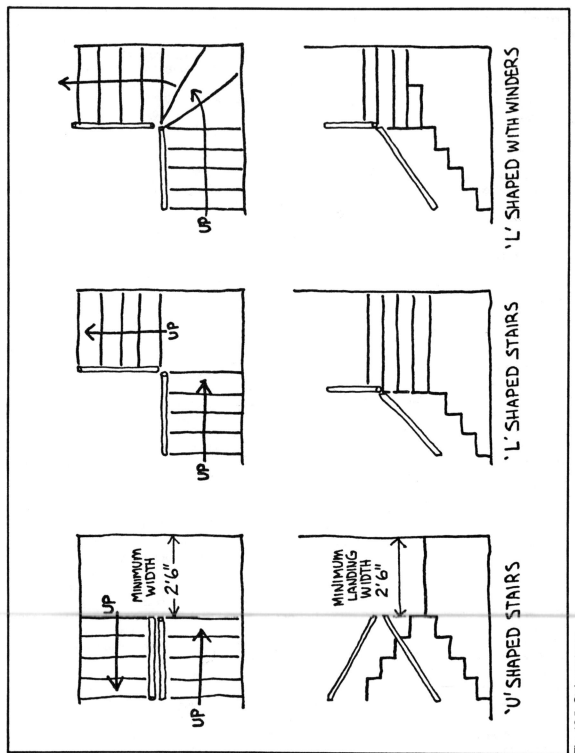

'L' SHAPED WITH WINDERS

'L' SHAPED STAIRS

UP

UP

MINIMUM WIDTH 2'6"

UP

UP

MINIMUM LANDING WIDTH 2'6"

'U' SHAPED STAIRS

Fig. 12-5. Stairway types.

Fig. 12-6. A spiral staircase.

RAILINGS

Stairway railings can be fun to think about. Should you have a flashy brass rail? Or maybe an intricately-carved wooden one? Then again, maybe it would be better to put up a half wall and top it off with a nice modern slab of sanded-smooth oak with a great, round finial (ornamental post top) at the bottom to keep youngsters from using the rail as a slide.

Stairway railings can add a lot to the safety and decor of your home, so keep the following in mind:

■ All stairways should be equipped with permanent and substantial handrails 36 inches in height from the center of each tread. Narrow stairs can get along with one rail, but wide stairways should have two.

■ The railings should be continuous from floor to floor, even when there are landings.

■ All handrails should have rounded corners and a surface smooth and free of splinters.

■ Since you're going to need *some* kind of railing(s) anyway, you might as well use railing that adds to the attractiveness of your home.

FOLDING AND EXTENSION STAIRWAYS

Folding or extension stairways are widely used for necessary and convenient access to attics, finished living quarters, or out-of-the-way closet storage spaces (Fig. 12-8). The use of these stairways also saves the floor space of the room or area below the device, which allows for freer planning.

Most folding or sliding extension stairways come completely assembled for installation in a prepared opening, and are attached to a ceiling door so you just have to reach up and pull them down. The two most

6'6" MIN. HEADROOM

Fig. 12-7. A stairway's minimum headroom.

Fig. 12-8. A pull-down extension staircase.

popular types are rigid extension units that slide up and down parallel with their pull-down door, and three-section units held together with hinges. When not in use, the hinged sections fold up into a compact bundle that's stored on top of the closed door.

Although these setups provide somewhat less sturdy means of access than access gained by permanently fixed stairs, the folding and extension models are not used as frequently and are the better choice for top level spaces.

The folding units fit into homes with floor-to-ceiling heights of 7 feet 6 inches to 8 feet 9 inches. Rigid extension stairways are available in many more sizes ranging from 7 feet 6 inches to over 16 feet.

All folding and extension staircases that fold or slide up into unheated areas such as attics must not be overlooked when it comes to insulation and air infiltration. The door of a pull-down stairway should be hinged on one side of the frame and equipped with latches so when closed the door can be snugly pulled against a rubber seal to prevent drafts.

Insulation to match the other parts of the attic floor should then be placed on top of the door and around the sides of the frame. Caulking compound can be applied around the seam at the juncture of the door frame and subfloor.

EXTERIOR STAIRWELLS

Exterior stairwalls that lead from basements and lower living levels directly to the outdoors face special problems that are solved by construction specifications different than those used for inside stairways (Fig. 12-9).

1. Exterior stairwell sidewalls are normally formed with poured concrete or concrete blocks.

2. The steps are practically always made of poured concrete.

180

Fig. 12-9. Outdoor basement stairs.

3. The entrance doors, single or double, can be wood, but should preferably be steel with a wood center or core. These composition steel/wood doors hold up very well against moisture in all climates.

4. The bottom of an outside stairwell should be lower than the basement to prevent a direct flow of water into the house. A 6-inch sill is standard, and when possible, have a stairwell drain installed to convey excess water into an appropriate drainage area.

Chapter 13

Windows

Imagine a house without windows and you're likely to conjure up some prison-like dwelling or subterranean earthen home constructed into the side of a hill. Windows, large and small, perform many important functions in the typical home. First of all, they provide natural lighting. Secondly, they admit fresh air for ventilation and allow oxygen-depleted used air to be expelled. They also provide access for passive solar-heating sunrays, and openings in the house's outer shell for air conditioning units. From within the house they enable the occupants to attain a visual continuity with the outdoors, and provide exits to those same outdoors in case of emergencies.

GENERAL CONSIDERATIONS

When planning for and selecting windows for your house, keep the following points in mind for each potential window candidate:

- Insulating and anti-air infiltration properties.
- Ease of operation.
- Necessary maintenance.
- Easy or difficult to clean.
- Its style and how it will fit in with your overall exterior scheme.

- Price—a cheap window could cause a loss of any initial savings through consequential increases in heating, cooling, and maintenance costs.

DRAWBACKS

Like any other feature in a house, windows can also have their drawbacks:

1. Large expanses of glass increase heat loss during periods of cold temperatures and become a source of unwanted, and at times uncontrollable, solar heat gain during warmer months.

2. In the summer, windows not only let the hot sun in to make the inside of a house uncomfortably warm, but they also permit sunlight to fade the color from carpeting, paneling, furniture upholstery, and practically anything else.

3. In addition to providing views that might not always be pretty, windows can turn a house into a goldfish bowl by enabling outsiders and strangers to see into the interior living areas.

4. Windows require a fair amount of washing—a chore no one enjoys. Every now and then a child's softball smashes through one, or a limb from a nearby

tree gets blown too close, and they must be replaced.

5. Windows are the first surfaces in a house to fog up when the interior humidity rises. They can stream with condensation, which might ruin the finish on the sills.

6. Windows can admit annoying neighborhood noises.

7. When not secured properly, windows provide encouragement to burglars and intruders.

8. Windows become ugly black mirrors from inside when you turn on the lights at night.

9. Windows require expensive curtains, draperies, and blinds that also need periodic cleaning.

10. If not carefully and tastefully selected and located, instead of improving the appearance of a house they can actually detract from it.

WINDOW TYPES

When selecting from the eleven basic window types used on today's modern houses, there's no reason why you can't employ two or three types of them as long as they look nice together. At the same time, there's no reason why you can't combine one type of movable window with a fixed window, or even with another type of movable window in the same opening.

Casement Windows

A casement window is a miniature version of a hinged door, except it's opened and closed with a crank or lever mounted on the inside of the window sill (Fig. 13-1). Thus you don't have to disturb a screen or additional storm sashes which might be mounted on the inside. A latch locks each sash tightly.

Advantages:

1. Because of their method of operation, casement windows are ideal for installations behind counters and hard-to-reach or difficult-to-move furniture. Wherever you can't stand right next to the window for leverage, or can't reach the entire window due to its high placement, you won't be able to comfortably position a double-hung window and should opt for the casement.

2. They offer excellent sealing against air infiltration.

Fig. 13-1. Casement windows.

3. Depending on the particular design, interior and exterior surfaces can be cleaned from inside the house.

4. Since the entire sash opens, a casement window admits 100 percent of an available breeze.

5. Even during storms a casement window can be cracked open an inch or two without letting in water.

Disadvantages:

1. You have to carefully consider where you locate casement windows in relation to outdoors activities. They shouldn't open out onto terraces, porches, decks, or sidewalks where people can bump into them.

2. Because the screens or storm windows are fastened to the inside, they hinder fast exits through the window openings if needed during a fire or other emergency.

Double-Hung Windows

These are the most common windows found in construction today. The double-hung window has two sashes or panes that slide up and down in channels or tracks called stiles. As a rule, the sashes are the same size; but in some cases, the bottom sash is taller than the top sash. The two sections are held in place by either springs or friction. If by friction, It's a sign of a good, tight fit (Fig. 13-2).

Advantages:

1. Because they're held firmly in place, double-hung windows rarely warp or sag.

2. Unless they're painted shut, double-hung windows are simple to open and close as long as you can stand directly adjacent to them for good leverage.

3. Relatively little air leakage occurs around their edges, even when the sashes are not weatherstripped.

4. They can be cleaned from the inside if the sash is removable. The older models had pulleys and weights suspended within the walls to help the windows open and close—they had to be cleaned from the outside. Some modern double-hung window sashes can be popped out of their stiles, to the inside, for convenient cleaning and replacement.

Fig. 13-2. Double-and-single-hung window operation.

Disadvantages:

1. If you will have to reach over a counter or piece of furniture to open and close them, double-hung windows will be difficult to operate and shouldn't be used.

2. These windows can never open to more than half of their total area.

3. Even when they're open only an inch or two, double-hung windows are likely to admit hard-driven rain.

4. If the sashes aren't removable, the only way to clean them is from the outside—with a ladder when they're located on second-story levels.

Single-Hung Windows

Not as popular as their double-hung cousins. The single-hung window looks exactly like the double-hung unit but differs in that the single-hung window's top sash is fixed. It can't be moved. Only the bottom half of the window can be screened.

Advantages:

1. Because half of this window is sealed shut, there's somewhat less maintenance, and less chance of air leakage.

2. The cost of a single-hung window is less than that of a similar double-hung model.

Disadvantages:

1. Ventilation is limited to the bottom part of the window opening.

2. Washing a single-hung window can be a problem unless the lower sash is removable.

Awning Windows

Awning windows are hinged similarly to casement windows, but along their top edge so they swing out and up when you turn a crank or lever or simply give them a push (Fig. 13-3). Some units are made with special hardware that provides pivot action—the top of the sash moves down as you push the bottom outward. The screens and storm windows are installed from the inside.

Advantages:

1. They can be opened wide enough so you can get almost 100 percent of possible ventilation—even during a rainstorm—without letting in water.

2. They can be used as clerestory windows,

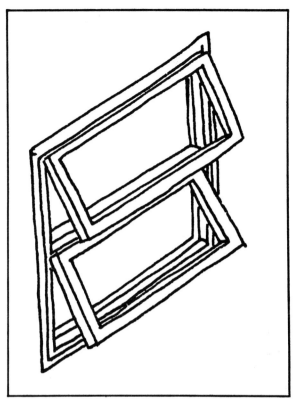

Fig. 13-3. An awning window.

placed high in walls to provide natural light and ventilation while assuring privacy and leaving a maximum amount of wall space for furniture placement.

3. They offer an excellent seal against air infiltration.

Disadvantages:

1. Awning windows shouldn't be installed overlooking porches, decks, terrace, or sidewalks because someone might run into their open projecting sashes.

2. Because they're slanting, open sashes are so exposed that they become dirty in short order and require more frequent washing than any other type of window.

3. As with casement windows, having storm windows and screens mounted on the inside can hinder a quick escape.

Hopper Windows

Hopper windows are the reverse models of awning windows. Hinged or pivoted at their bottom, they open in-

185

Fig. 13-4. A hopper window.

ward and downward from their top so that the entering air flows upward (Fig. 13-4). Operated by a lock handle at the top of the sash, they're most commonly used in basements and clerestories. Screens and storm windows are installed on the outside.

Advantages:

1. Hopper windows provide almost 100 percent of possible ventilation.

2. Both their inner and outer surfaces can be easily washed from the inside.

Disadvantages:

1. Hopper windows interfere with draperies and curtains and are impossible to darken with shades when opened.

2. Because they stick out inside a room or hallway they can cause traffic problems in living areas of a home.

3. Because of their unusual open position they can be difficult to exit from in case of an emergency.

Horizontal Sliding Windows

In effect, a horizontal sliding window is a double-hung or single-hung window laid on its side (Fig. 13-5). With some units, both sashes slide from side to side in a channel; in others, only one sash (usually the right sash) slides. In still others made with three panels of glass, the two outer sashes slide to the center over a fixed sash that is twice the width of each sliding sash.

Advantages:

1. As with double-hung windows, because they're held firmly in place they won't warp or sag.

2. They're easy to open and close as long as you're standing next to them.

3. There's relatively little air leakage around the edges, even if they're not weatherstripped.

4. Washing is usually easy because many sashes can be removed from the inside of the house.

5. Horizontal sliding windows are a good choice in locations where you want an operating sash with a large expanse of glass and a minimum number of framing members to obstruct the view. No other operating unit fills this requirement as well.

Disadvantages:

1. Horizontal sliding windows cannot be opened to access more than 50 percent of their possible ventilation space.

2. They will admit driving rain into the house.

3. If you must reach over a counter or piece of furniture to open and close them, horizontal sliders are difficult to operate and should not be used.

Fixed Windows

Fixed windows are panes of glass mounted in frames that are installed directly into a wall. They can't be opened and closed, and can be ordered in a variety of sizes, shapes and glass types.

Advantages:

1. They're the most weathertight windows available.

2. Because they can't be opened they don't require screens or hardware.

3. They're less expensive than other windows.

Disadvantages:

1. They can provide no ventilation.

2. They're impossible to exit from in case of an emergency.

3. They must be cleaned from the outside.

186

Fig. 13-5. Horizontal sliding windows.

Fig. 13-6. Fixed windows.

Fig. 13-7. A bay window.

Bay Windows

A bay window consists of three adjacent windows or sections of windows in a series. Two side sections are angled back from each side of a straight center window or section of windows (Fig. 3-7). An entire bay window unit can be made from combinations of windows such as casement or double-hung windows on the sides and a fixed center section.

Bay windows are ideal when you want to increase both the real and apparent size of a room. They'll also add a graceful note to an otherwise rather severe facade.

Advantages:

1. They open up a 180-degree view.
2. They not only increase ventilation but also enable you to scoop in breezes traveling parallel to the house walls.
3. They form a delightful niche for sitting or dining near. They're often featured in designer rooms and houses.
4. They provide an excellent place for growing houseplants.

Disadvantages:

1. At least the center part of a bay window must be cleaned from the outside.
2. Because this arrangement protrudes out from a wall it's not as energy efficient as other window installations.

Bow Window

The bow window is a close relative of the bay window, and they're often confused with each other (Fig. 13-8). The bow window is gently curved rather than angled, and is considered a more graceful feature when used in a house's living area. The bow window receives its name from the arrangement of a series of windows that arc out from the house's exterior walls.

Because of their curved shape, bow windows are necessarily made up of relatively narrow sashes or of many small fixed panes. When sashes are used, they're generally the casement and occasionally the double-hung type. With casement windows, either all of the sashes can open, or only the two end panels, or only the end panels and every other intermediate panel. When double-hung windows make up a bow, only the end panels open. Bow windows are also available in models in which none of the sashes open, with fixed large or small-pane styles.

Advantages:

1. They open up a 180-degree view.
2. They not only increase ventilation, but they enable you to scoop in breezes that run parallel to the house walls.
3. They form the same kind of delightful niche found with bay windows, for dining near, growing plants in, and sitting in a kitchen, dining room, or living area.

Disadvantages:

1. Because they protrude out they are not as energy efficient as windows closer to the walls.
2. These windows are usually narrow or small, so there's little opportunity to escape through them in case of an emergency.
3. Some of their sections are normally fixed, and cleaning will have to be done from the outside.

Jalousie Windows

A jalousie window is made up of a series of narrow horizontal panes or glass slats that open outward with a crank.

Advantages:

1. They can be opened far enough to gain virtually 100 percent of available ventilation.
2. They can be opened during rainy weather without admitting water into the room.
3. The air flow from jalousie windows can be adjusted in any amount or vertical direction.
4. They're easy to open over a counter or furniture.

Disadvantages:

1. They possess many small glass sections that need to be cleaned.
2. They're difficult to wash from the inside.
3. They're very poor for preventing air infiltration.

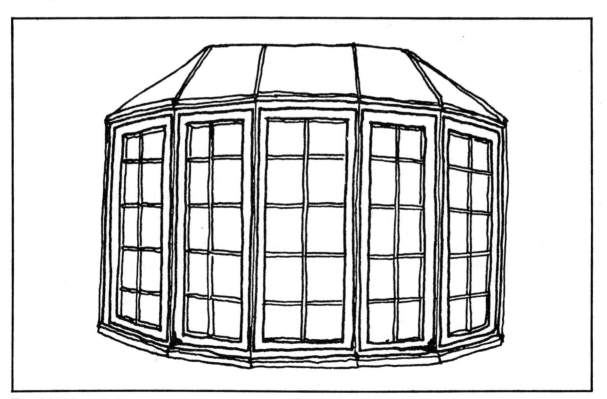

Fig. 13-8. A bow window.

Fig. 13-9. A jalousie window.

4. Jalousie windows are impossible to exit from in case of an emergency.

Skylight Windows

Skylight windows do two things better than other windows can: they admit more natural light and distribute the light more evenly. They also make rooms look larger and can help equalize the light in a room that might have windows on only one side. Frequently the skylight chosen is a fixed window unable to be opened. Fixed skylights come in many shapes—dome, smooth, low profile curb, pyramid, and ridge (Fig. 13-10). The same windows also are available in models that can be opened and closed.

Another kind of roof skylight window is a single sash that pivots at the sides about a third of the way down from the top. When opened, the bottom swings out and the top swings in. It can be held open at any angle—even completely reversed for easy washing from the inside.

In a third type of roof window, the sash is hinged at the top and raises like an awning window to give an unobstructed view while keeping out rain. The outside of the more progressive models is washed with a special tool provided by the manufacturer.

Advantages:

1. They permit installation of smaller and fewer windows and simplify interior decoration and furniture placement by allowing more uninterrupted wall space in a room.

2. They can be used to illuminate rooms and areas which share no outside walls for regular windows, such as inside bathrooms and halls.

3. They supply privacy without sacrificing any natural lighting.

4. Skylight windows can provide more flexibility in the planning of a house because you needn't have to include sidewall windows in every room.

Disadvantages:

1. Because heat rises, skylight windows tend to be less efficient than a plain windowless roof or ceiling in cold weather, especially during times then the sun is not shining. Because of this, consider only skylights that have a minimum of double-glazed glass (not all of them do).

2. They can be difficult to clean. The fixed types must be cleaned from above. Leaves, dust, and even ice buildup during winter can pose real problems.

3. Provisions need to be made for draining off water—either condensated moisture or water that collects from rain and melting snow.

4. If not properly located and slanted, skylight windows can create irritating glare and will actually mar a dwelling's appearance.

5. On hot, sunny days when solar heat is not desired, skylights will raise a room's temperature, making your air conditioner work overtime.

TYPES OF GLASS

There are three types of glass panels typically used to construct modern windows: plate glass, tempered glass, and insulating glass panes.

Plate Glass

This is the standard glass normally used in house windows. It can be plain or tinted to reduce glare, and can be doubled up to sandwich in a thin air space for insulating qualities. When struck or put under stress it shatters into pieces that are usually very sharp. It's an

Fig. 13-10. Skylight windows.

excellent material for windows, though, because it can be manufactured free of flaws and distortions.

Tempered Glass

Tempered glass panes are three to five times stronger than those made of ordinary plate glass. Tempered glass should be used in doors and in glass panels adjacent to doors and other areas where it is likely that people, especially children, might run or fall against. In fact, many building codes require the use of tempered glass in wall areas that are within 4 feet of any door. When struck or put under stress, tempered glass develops hairline cracks or breaks into many small, rounded pieces instead of into the sharp, ragged pieces that come from standard plate glass. Auto windshields are a common example of tempered glass.

Insulating Glass

This glass exists in two forms. One is a special double or triple sheet of glass (double or triple pane/double or triple glazed) separated by an air space(s) with the glass edges welded or formed together to make an air-tight center spaces(s) much like the liner for a thermos bottle. The other consists of two or three sheets of glass held in the frame with an insulating air space in between. Double-pane insulating glass used in most stock windows is made of two panes of sheet glass having a 1/4 to 1/2 inch of air space between them. Triple-pane glass windows consist of three panes of glass with approximately 1/4 inch of air spaces between each pair of panes (Fig. 13-11).

In this day and age it no longer makes sense to install windows having only a single thickness of glass. Instead, all exterior windows located in living areas or heated areas of a house should be double- or triple-paned. The additional expense is well worth it; you should save enough in fuel and energy costs to more than make up for the initial investment within a few years and have a more comfortable home in the meantime.

More advantages to using double- or triple-pane windows as opposed to a combination of single-pane sashes and storm windows are that the multi-panel windows offer a better appearance, are easier to clean, provide increased soundproofing, a permanent installation, and even a better overall price.

In all multi-pane windows, specify that the thickness of the sash is at least 1 3/8 inches.

TYPES OF WINDOW FRAMES

There are three primary types of materials used in window frame construction: wood, metal, and plastic.

DOUBLE PANE

TRIPLE PANE

Fig. 13-11. Insulating glass—double- and triple-pane windows.

EXTERIOR RIGID VINYL COVERING

Fig. 13-12. Vinyl-covered window construction cutaway.

Wood Frames

Wood frames are the most handsome window frames available, and are often preferred for this reason alone. They can be purchased with their exterior and side surfaces (the parts that abut the tracks or stiles and face the outdoors) covered with a layer of tough vinyl or a layer of aluminum with a coat of factory-applied, baked-on paint. The only part of such a window frame that retains it wood surface is the interior section—what you can see that faces the inside of a room. That inside surface is the most important part of a window frame from an appearance point of view. It can be stained, varnished, or painted to match or complement the rest of a room's decor. The choice of colors for either the vinyl-

or aluminum-clad surfaces are usually limited to white, brown, or bronze.

Because no staining, varnishing, or refinishing is ever required to the non-wood surfaces, maintenance there is extremely infrequent. And the wood part of the frame, the part you can see from inside the room, will rarely need attention either, because it isn't exposed to the elements. Vinyl-clad or aluminum-clad factory-finished wood frames are excellent choices (Fig. 13-12).

Advantages:

1. A primary advantage of wood-framed windows is their appearance. They look less clinical than aluminum or steel sashes.

193

2. Condensation is not much of a problem with wood frames as it is with steel and aluminum ones.

3. Wood frames can be successfully used with all architectural styles.

4. If you prefer multi-pane windows, it's much easier to find what you want in wood than in aluminum or steel.

5. Wood frames are poor heat conductors, so they provide effective insulating qualities.

6. Provided you select the wood frames having vinyl- or aluminum-clad exteriors, the outer and inside-track surfaces will be practically maintenance-free.

7. The interior surfaces of wood sashes can be painted to match any color scheme or can be stained and finished in natural tones to match inside wood baseboard and other trim.

Disadvantages:

1. If the exterior surfaces are not vinyl- or aluminum-clad, they'll need refinishing, especially if no awnings or substantial overhangs are in place to shield the windows from rain, sun, and snow. If uncovered wood frames are not treated to resist decay and moisture absorption, they'll rot in short order.

Metal Frames

Metal window frames are available in aluminum and steel models, in natural finishes or various selections of anodized coloring. Aluminum windows became popular because people thought aluminum models suffered from none of the problems associated with wood—that they wouldn't rot, swell, contract, warp, or need refinishing. This, however, proved not altogether true. Near seacoasts and corrosive industrial locales, aluminum frames can corrode so badly that they need to be painted for protection. For these reasons it's advisable to purchase aluminum or steel windows that are protected by factory-applied finishes.

Advantages:

1. They cost less than wood frames.

2. They lend themselves nicely to places that receive rugged use, such as basement windows and garage windows.

3. They're easy to operate.

4. They won't warp.

Disadvantages:

1. Because aluminum and steel are excellent conductors, they permit high heat loss from within a house through the window frames.

2. Also due to their conductivity, they cause excessive moisture condensation on the interior portion of the frames when significant temperature differences exist between the inside and outside of the windows. If you insist on aluminum or steel-framed windows, make sure you select a brand that has a thermal break—where an insulating material separates the interior and exterior sections of the frame. The most efficient and practical insulating materials include plastic, urethane, epoxy, and vinyl.

3. Aluminum and steel frames often provide a looser fit than that of wood frames, allowing for more air infiltration.

Plastic Window Frames

An available option on some window types is plastic frames. Plastic frames are lightweight and corrosion-free, but not very popular.

Advantages:

1. Painting is never needed unless you decide to change their color.

2. They're easy to operate.

3. They're not as expensive as wood frames.

4. Plastic is not a good conductor; it doesn't have the condensation or heat transfer problems of aluminum or steel.

Disadvantages:

1. The main drawback is their lack of strength. Wood, aluminum, and steel frames are much stronger. Plastic frames are more likely to break, especially during cold weather when plastic turns brittle.

2. Plastic windows don't seal as well as wood-framed windows, and they permit more air infiltration.

WINDOW SIZE AND ALIGNMENT

Make certain that the size of some windows, particularly in bedrooms, are large enough to escape from in case of a fire. As mentioned previously, all sleeping areas should have at least one easy-to-open window having an opening of not less than 5 square feet, with a minimum clear width of 22 inches and a sill height not more than 48 inches above the floor.

Also be aware that the size and placement of windows will limit where you can comfortably arrange furniture. Major items such as desks, sofas, bureaus, dresser, china cabinets and buffets all normally require wall space. However, if window sills are high enough, some furniture can be placed beneath them. Many pieces of furniture are only 30 to 32 inches high.

Windows should be aligned in a pleasant manner, especially when viewed from the outside. That typically means that windows across each living level conform to one long horizontal line. Small windows should line up with the top or bottom halves of large windows. Then for maximum effect, have the tops of exterior doors line up with the window tops on the first level. When possible, arrange vertical window placements directly above one another (Fig. 13-13).

WINDOW LOCATION

No matter where you locate your windows, all ex-

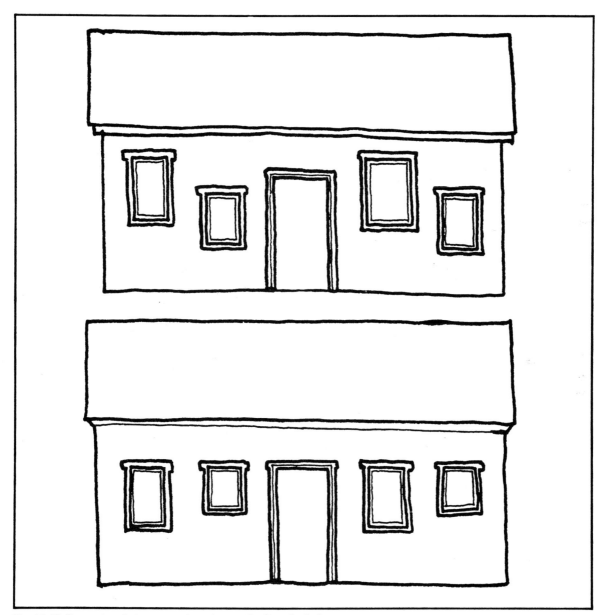

Fig. 13-13. Comparison between poorly and correctly aligned windows.

posures will be able to provide a sufficient amount of natural light (Fig. 13-14). Southern exposures, with their high sun angle, offer the most light and the best opportunity to control and use sunshine to good advantage. East and west exposures are the most difficult to control, having low-angled rays, with the west being particularly troublesome. Northern exposures, lacking direct sunlight, are the easiest of the four to control. For the typical home, windows on the north side of the house are discouraged. However, when windows are required due to an unavoidable or advantageous orientation or positioning of the dwelling—such as a master bedroom overlooking a scenic view—a northern exposure can be well suited to the dwelling.

Other points to remember:

■ Horizontal window openings are especially useful for controlling light from southern exposures.
■ Vertical window openings are most useful for controlling light from eastern and western exposures.
■ Windows located high in a room offer the most illumination and the deepest penetration by natural light.
■ Clerestories and skylights offer good possibilities for lighting interior spaces in a home.
■ Windows can be located to illuminate parts of a house or room where specific tasks will be under-taken. Of course, artificial lighting should also be arranged for times of insufficient daylight or during evenings.

WINDOW VENTILATION

In addition to providing natural light, the second major function of windows is to provide ventilation to get rid of stale air in the home (Fig. 13-15).

Large openings allow for the best natural ventilation when arranged to encourage a cross-current of air. Openings should be oriented to pick up prevailing summer breezes.

Good air flow occurs when the inlets and outlets are approximately the same size. Better air flow can be attained by having a larger ratio of outlet to inlet area. A combination of openings can direct air flow as desired; openings placed lower in the wall surfaces result in better cooling than those placed higher in a room.

Exterior features such as overhangs, porches, fences, garages, shrubs, and trees can be used to block or encourage air flow.

If you plan to install central air conditioning in the house you won't need side windows for cross ventilation. By not putting them in you'll save their installation costs. plus money on energy costs as well.

Fig. 13-14. Good window location.

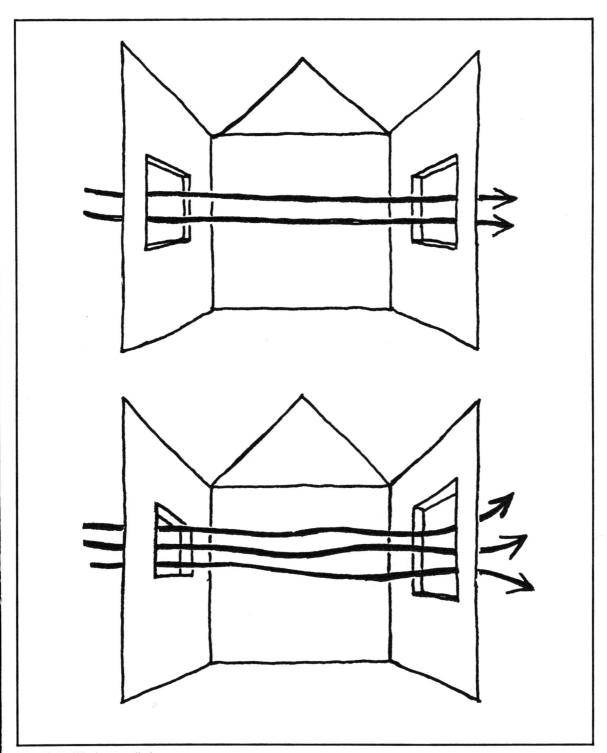

Fig. 13-15. Window ventilation.

STORM WINDOWS

Storm windows cut down on conductive heat losses and gains, depending on the season, by impeding air infiltration and by providing an air space between inner and outer panes of glass (Fig. 13-16).

On most types of windows (double-hung, single-hung, awning, sliding, and jalousie) conventional storm windows are put up in the fall and taken down in the spring. But they can also be used to keep cool, conditioned air in a home during summers in warm-climate locations. On casement windows, storm windows are frequently fastened to the outside of the sash.

Combination aluminum-framed storm windows and screens are used on double-hung, single-hung, and sliding windows. In cold-climate areas they have generally replaced conventional storm windows because combination storm window/screens can be left in place year round. In winter they keep out the cold, and during summer, the insects.

Two kinds of combination storm windows are available: double-track and triple-track. Both include an upper panel of glass, a lower panel of glass, and a single panel of screen. In the double-track arrangement, both glass panels are fitted into separate channels or tracks so they can move up and down independently of one another. In summer, when ventilation is required, the lower panel of glass slides to the top of its track and the screen panel is inserted into the lower half of the track. During winter the screen is removed entirely and stored.

The triple-track storm window arrangement eliminates the necessity for putting in and taking out the

CHANNEL FOR EACH OPERATING PANEL

SCREEN

STORM PANELS

WEEPS

TRIPLE TRACK

Fig. 13-16. A storm window arrangement.

screen because each of its two glass panels and the screen all fit into their own tracks. You can convert from a storm window to a screen by simply sliding the lower glass panel up and pushing its replacement screen down. In winter the screen can conveniently be stored in the top part of the frame, within its own track. Although definitely more expensive, this setup is highly desirable if it fits into your budget. You gain the flexibility to have screens at a moment's notice, or storm windows, and it sure beats having to lug and store screens someplace where they could get damaged during the off season.

When planning for storm windows, consider that caulking is needed around permanently installed units, at the outer edges of their frames. Two weep holes must be kept open at the bottom edges of each storm window to provide drainage and some ventilation, or wooden sills might rot. To prevent corrosion and maintain a good appearance, select only storm windows that have frames finished in baked enamel or that are coated with a layer of maintenance-free vinyl or similar material.

SCREENS

About the only choice to be made when purchasing individual screens is whether you want aluminum or fiberglass mesh. Both are very durable. They're available in several colors and are suitable for installation in aluminum and wood frames. Aluminum screencloth has greater resistance to impact, but can't be straightened when dealt a sharp blow or raked by a dog's nails. Fiberglass tends to "belly out" readily, but can be pulled flat again with little trouble. Most screens installed in today's homes are made of the aluminum mesh (Figs. 13-17 and 13-18).

SHUTTERS

The main purpose of outside shutters is to beautify the exterior of a house, to harmoniously balance out the appearance of the windows. Maybe years ago they would prevent an Indian's arrow or a marauder's bullet or grizzly bear's paw from entering a window, but they don't anymore. In fact, most of them no longer even close.

You can save money by using ornamental shutters made of plastic or aluminum. They're less expensive than wood and don't require costly operating hardware. Plastic and aluminum shutters are both made in louvered and raised-panel designs—usually in black, brown, green, or white (Fig. 13-19). Aluminum units are

Fig. 13-17. A screen in place.

the stronger of the two, but because the finish is only baked on, and they're exposed to the rain, sun, and snow throughout the year, they'll eventually need repainting. Vinyl-covered shutters, on the other hand, are integrally colored so they don't have to be touched up even if damaged.

Shutters are typically installed by fastening them to window casings or adjacent wall surfaces with screws. But that also means there's no easy way to clean out wasp's nests, bats, or debris that can accumulate behind the shutters.

Wood shutters present an additional problem: if they're not adequately treated with wood preservative, they'll rot. They require frequent stripping, sanding, and painting or staining, which can be quite a time-consuming process, especially if the shutters are lou-

Fig. 13-18. Screens.

Fig. 13-19. Shutters.

LOUVERED   RAISED PANEL

vered. Like the other site-painted items, painted wood shutters won't clean as easily as those having factory finishes. On the positive side, wood shutters can be custom-finished and refinished to match any decorating scheme.

DIAMOND AND RECTANGULAR PANES

If you'd like to have windows with either small diamond—or rectangular-shaped panes, you could invest a small fortune into obtaining them. Alternatively, you could buy single-pane windows and turn them into multi-pane ones with devices called snap-in muntins. The muntins are made of plastic or wood, painted or coated to match the window frames. They snap onto the back of the window, on the inside, depending on the style of window. You can remove them quickly when you wash the windows, then snap them back into the sash when the glass is clean (Fig. 3-20).

Seen from the inside of a house, muntin inserts are difficult to distinguish from permanent muntins. From the outside, although you might be able to tell that they

don't project through the glass, the insets impart an appearance and effect of the real thing at a much lower cost.

EXTERIOR CONTROL OF SUNLIGHT

Regular double- or triple-pane windows, even though they're great for controlling the loss or gain of heat through conductivity, cannot prevent sunrays from entering and heating a room. This is good in winter, and bad during summer. There are tinted window coatings that act like tinted glass when applied (as in a car windshield), or to a certain extent, like sunglasses. For the sake of appearance, however, few people want to darken their windows to anything beyond clear.

Wide roof overhangs, like wide-brimmed hats, will effectively shade windows facing south. The same overhangs will also admit winter sun that strikes at a much lower angle. Overhangs, though, are not very effective at shielding east and west windows, because the morning and afternoon sun rises and sets at low angles all year round. Beyond installing canvas, plastic, or aluminum awnings, deciduous trees can be the answer (Fig. 13-21). Deciduous shade trees planted on the south, west, and east sides of a home will provide much

Fig. 13-20. Snap-in muntins.

Fig. 13-21. An aluminum awning.

appreciated shade during summer, and because they lose their leaves toward the end of each autumn, sunshine will be let through when you need it the most—during winter.

INTERNAL CONTROL OF SUNLIGHT

In addition to conventional curtains and draperies there are two other effective ways to control the sun internally. The first is with the use of roll-up shades that come in many colors, patterns, and textures. The least expensive ones are made of a thin layer of vinyl. Vinyl-coated cotton shades are better. Fiberglass laminated to vinyl results in the most durable and nicest-looking shades available. While most of these materials are translucent or semi-translucent, some are designed to give complete darkness. Also on the market are shades that are coated on the back to prevent and reflect the incoming sunrays. The alternatives to fabric shades are those constructed of slender strips of wood, metal, or semi-rigid vinyl.

A second method of regulating available sunshine is by using venetian blinds (Fig. 13-22). Venetian blinds of horizontal wood, metal, or plastic strips or louvers are made in two basic styles. The standard blind features 2-inch-wide slats held in place by wide cloth tapes. The mini-blind has 1-inch-wide slats held by slender cords. Simply because it's made of heavier material, the standard blind is more rugged, flaps less in a breeze, and can be used to darken rooms more effec-

Fig. 13-22. A venetian blind.

Fig. 13-23. Vertical blinds.

tively. But the mini-blind is more attractive in every way, and when it's opened wide the slats are almost invisible. These and other size units come in a great range of colors and dimensions up to a maximum of about 100 square feet.

Vertical blinds are, in effect, venetian blinds turned on their side (Fig. 13-23). They take the place of draperies and conventional blinds or shades on windows of above-average size, and are particularly suited to unusually tall or wide windows. They're ideal for glass sliding doors because they can be set to shut out glare without barring the view, and won't catch in the doors like draperies will.

HEAT LOSS THROUGH WINDOWS

The loss of heat through window openings can be reduced by:

■ Adding additional layers of glass (storm sashes or panels).

■ Sealing cracks around glass, sashes, and window frames to prevent air infiltration.

■ Making sure that weatherstripping and thresholds seal the door edges tightly.

■ Providing movable shutters or closures on the exterior to trap an additional layer of air next to the glass.

■ Installing similar closures or heavy drapes on the inside to trap an additional layer of air next to the glass.

■ Providing shelters for window openings in the form of overhangs, baffles, recesses, or plantings.

Chapter 14

Doors

There's absolutely no doubt about it. You can't have a house without doors. Even primitive men had them—hides draped from poles or vines. The Egyptians used woven reed mats that rolled up and down, and early Britons employed huge stones that pivoted in a circular fashion. Doors as we know them came into vogue during the Middle Ages—sturdy wooden models held together by strips of wrought iron or tightly fitted dowels.

A door provides a lot more than a simple entrance or exitway in a house. A door is a moving part. It lets in air, and seals out weather, dirt, and noise. It takes up space, gets in the way, batters walls and furniture, and gets battered in turn.

Doors protect our privacy and belongings. They keep out the heat and cold, and allow ventilation of a closed-in space, even when shut. They provide an access for natural light and bring the outdoors inside. They'll hold warmth-giving heat indoors during winter and cool air in the summer. They'll even pull teeth in a pinch.

When considering the entrance/exits that your house will have, here are some questions to ask yourself about each door:

■ Will it operate easily and reliably?

■ Will it close securely?

■ Will it permit easy passage of people and objects?

■ Will it interfere with the use of space on either side of the door?

■ Will it effectively close off whatever is supposed to be closed off?

■ Will it retard the spread of fire?

■ Will it minimize the transmission of sound?

■ Will it permit you to see through to the other side?

■ Can you hang things on its back side?

■ Can it cause injury if someone walks into it?

DOOR TYPES

There are five main types of doors that a house can contain: exterior, interior, storm, patio/garden, and basement.

Exterior Doors

In addition to providing privacy and security, exterior

doors can serve as effective weather barriers and sound reducers (Fig. 14-1). They're about 1 3/4 inch thick, between 2 feet 8 inches and 3 feet wide, and at least 6 feet 8 inches high. There should always be a secure layer of weatherstripping around an exterior door's edges to ensure a tight weatherseal.

A house's front door deserves extra attention because it's the part of a home visitors see first. The main entrance can make an impression that adds considerably to a dwelling's appearance, and even to the home's value and saleability. Because of the special importance of front doors, they're constructed differently in some respects from just "plain" exterior doors. Details such as door caps, exterior moldings and panels, windows or lights in or to the side of the door, reeded or fluted pilaster trim at the jambs, all help make the front door something to be approached by its own sidewalk, lit up from the outside at night, flanked by landscaped shrubbery beds, and protected by a roof overhang.

Home security should not be forgotten in connection with any exterior doors. Security means a combination of the proper door locks and hardware plus the ability to see who is calling before the door is opened. One way that front entrance visibility can be accomplished is by the installation of sidelights. Sidelights are narrow glass panes or panels that run the height of an entrance door, placed either at one or both sides of the door for added beauty and natural light. Two drawbacks are their poor insulation value and their susceptability to intruders who, if the sidelights are not positioned correctly in relation to the door's locks, can break the glass then reach inside to unlock the door. Sidelights vary in size, but are normally less than 1 foot wide. They're available in many finishes and textures.

Interior Doors

Interior doors exist mainly for privacy and noise reduction. To be effective for each, they must be well fitted. However, if the house heating system depends on a free flow of air from room to room, interior doors should be undercut at least 1/2 inch above the finished floors to permit air passage. Although this doesn't apply to

Fig. 14-1. Exterior doors.

Fig. 14-2. A louvered door.

Fig. 14-3. A storm door.

areas having their own air supply and return outlets, remember that door fit is still important. You don't want doors to work too tightly, especially over carpeting, because large amounts of friction will be created when the doors are opened and closed.

Interior doors are available, in addition to flush and paneled models, in full louvered, top and bottom louvered, or partially louvered with either a top or bottom of paneling (Fig. 14-2). Louvered doors, while more expensive than flush or paneled doors, are particularly useful in locations requiring a free flow of air—namely closets or rooms containing mechanical equipment that must "breathe" to operate safely and efficiently, such as water heaters and certain heating units and furnaces.

Storm Doors

Storm doors for all outside entrance/exits are a must in cold-climate locations (Fig. 14-3). They should be designed for easy changeovers from screens in summer to glass inserts during winter. A storm door is nothing to skimp on: good styling on the front entrance storm door can add a lot to the appearance of a house, and can also cover up a nondescript front door.

When selecting your storm doors look for the following:

■ The main frames and frames for the glass and screen inserts should be strong. If you can easily bend or flex the frames, they're too weak to make an adequate door.

■ A functional design will typically enable you to remove the glass and screen inserts from the inside of the house, in a simple manner.

■ Look for weathertightness to prevent the entrance of water, cold air, dust, and insects.

■ Make sure storm doors have easy-to-work locking mechanisms.

Patio/Garden Doors

Sliding glass doors and French doors will bring the outdoors inside and will provide convenient access to a patio, deck, or garden. They'll make whatever room they're part of appear larger than it really is, and they'll supply a flood of natural air and light when their screens are in play. Practically any house built today will use one of these doors somewhere within its walls (Fig. 14-4).

Factors to consider when selecting patio/garden doors are thermal insulation quality, weathertightness,

Fig. 14-4. A glass sliding patio door.

a secure locking system, and a nice appearance.

Sliding glass doors are generally the bypass type. All units include at least one fixed and one sliding panel. Some 3-panel models are available having a center panel that slides open in one direction. For larger wall spans, 4-panel models can be installed having two center panels opening in opposite directions to make a convenient access to large decks or patios.

French doors generally have a single panel that swings while the other panel remains stationary (Fig. 14-5).

Basement Doors

Outside basement doors allow you to transport such items as screens, storm windows, and garden tools inside and out without lugging them through the living areas. As mentioned earlier, basement doors are ideal for lower-level laundry rooms, and will provide children with a means of getting to the back and side yards without having to traipse through the rest of the house. A basement door should be at least 36 inches wide to accommodate large appliances.

Although basement doors should be able to effectively withstand the weather and provide security against theft by themselves, attractive basement stairwell covers can be purchased in ready-to-install packages for use in all types of houses. They're typically steel double door covers that enclose and protect the outer stairwells leading to basement doors.

DOOR STYLES

There are six basic styles of doors used throughout to-

Fig. 14-5. French doors.

day's modern house: hinged, bifold, sliding, pocket, folding, and cafe.

Hinged Doors

A hinged door is essentially a simple rigid panel that swings open and closed on hinges. It's the most common type of door for openings 18 to 36 inches wide. Hinged doors come in several architectural styles: flush, contemporary, Colonial (six-panel) and glass (Fig. 14-6).

Advantages:

1. When closed, hinged doors seal very well, curbing energy loss and limiting the amount of sound transmission.
2. They're initially one of the least costly type of doors to purchase and hang.
3. They require a minimum amount of maintenance and cleaning.

4. They're great for providing (on their back sides) space for door-hung shoe racks, necktie racks, belt racks, and for fastening all-purpose hooks to.

Disadvantages:

1. They take up precious room to swing in. Consequently, they can't be used where there's an obstacle in the way of their swing.

Bifold Doors

Bifold doors are similar to those used in telephone booths, except house bifolds open outward (Fig. 14-7). They're used indoors only, because there is no way to seal the cracks around their edges. The most common bifold door consists of two fairly narrow vertical panels that are hinged together. One panel pivots next to the door jamb; the other glides in an overhead track. To

SOLID FULL LENGTH - 3 HINGES

FLUSH   CONTEMPORARY   COLONIAL WITH GLASS

Fig. 14-6. Hinged doors.

open a bifold door, you either shove the track-mounted panel toward the opposite door jamb or pull the knobs fastened to the panels near the hinged edges. Both methods of opening force the panels to fold together back-to-back at right angles to the doorway opening.

For small openings, a single bifold door with two narrow panels is adequate. For larger openings, the door is made with wider panels or a double bifold door is employed having four panels hinged together. These doors are designed to be operated from one side only and are best suited for closets. They're useful in providing wide door opening coverage to shallow closets, making the most of the available square footage.

Advantages:

1. When both sections of a bifold door are completely opened you have an almost clear view through the doorway. Or you can have access to one-half of the same closet without disturbing the door covering the other half.

2. Bifold doors are ideal in cases where there is little room to swing a regular door outwards.

3. They allow the maximum opening to a closet with the minimum extension into the room. Extending the height of the doors a full 8 feet to the ceiling eliminates headers and permits full-width shelves and access to the upper area which would otherwise be lost.

4. In walk-in closets, space used for the "walking in" cannot be used for storage. In shallow closets with bifold doors, however, the users do not enter the closets—so almost all of the space within is available for storage.

5. Bifold doors come in styles that include solid or louvered panels. The louvered doors have a beautiful appearance that alone makes them popular with many homeowners.

Disadvantages:

1. Bifold doors can come untracked or malfunc-

Fig. 14-7. Folding doors.

tion easier than simple hinged-hung doors can. They won't take the abuse that other types of doors can take.

2. Bifold doors are somewhat more expensive than their counterparts.

3. Bifold doors cannot be used where a tight weather seal is required, nor where the door must be operated from two sides.

4. Bifold louvered doors are time consuming to clean and difficult to repaint or resurface because of the many individual slats.

Sliding Doors

Sliding doors consist of usually two and sometimes three door panels that slide by each other (Fig. 14-8). These units hang from and move along double or triple tracks installed against the underside of conventional head

Fig. 14-8. Sliding doors.

jambs. To prevent operating problems, sliding doors made of glass should be of top quality construction with sturdy tracks, double or insulated shatterproof glass, and mohair or other stripping laid along the door's tubing edges for maximum insulation.

Advantages:

1. They can be used wherever doors are needed, but door swing space or projections are not permitted.
2. Sliding doors are easily maintained and cleaned.
3. Glass sliding doors are wonderful for expanding a view and for making rooms seem a lot larger than they really are.

Disadvantages:

1. Sliding doors give access to only half (or when three doors are involved, to one-third) of an opening at once.
2. These doors are sensitive to any settling that might occur in a house. The sliding doors can stick against the bottom floor guide, which causes the doors to ride up and either jump or damage the hanging tracks and trolleys.
3. Sometimes the screen inserts to glass sliding doors can be a nuisance if they're not fitted exactly because they tend to pop out of their tracks.
4. A cheaply made closet sliding door, when handled gently, will do the job, but a cheaply made glass slider that isn't weathertight and fitted correctly will let in a lot of cold or hot air and can be difficult to operate.

Pocket Doors

A pocket door slides in and out of a pocket built into the wall framing (Fig. 14-9). These space-savers can have silent-action rollers and rubber stopping bumpers inside the pockets so the doors will work without clatter.

Advantages:

1. Pocket doors are useful in locations where there is little room to swing a standard hinged door out of the way due to interference with traffic or the operation of other doors.
2. When open, pocket doors are completely out of the way (disappeared into a wall). They take up no floor or wall space and they don't obstruct the door opening even a fraction of an inch.

Disadvantages:

1. Recessed sliding door openings and surrounding walls must be specially constructed. Pocket doors can malfunction if not carefully installed with quality materials. Normally, in order to provide a pocket, the size of the wall studding is substantially reduced, producing a wall frame where warpage can occur more readily. Wall warpage can interfere with the operation of the pocket door by causing the door to bind against the side of the pocket. To minimize the probability of wall warpage, it's best to start out with 2-x-6-inch studs in walls that will be used to place door pockets.
2. A wall that has a door pocket built into it should not contain electric wiring in that section of wall. You'll have to place electrical outlets and the light switch on the other side of the doorway.
3. Pocket doors are not structurally strong. They are best used in a wall that will have some cabinetry on one side to give the wall stability. Avoid locating them in a wall that will be tiled: the wall will give when leaned upon—enough to crack the grout and loosen the tiles.

Folding Doors

Folding doors are made of many thin, narrow vertical strips or creases that fold back-to-back into a compact bundle when the doors are pushed open. The simplest folding models have very small strips that are tied together with cords. Most, however, have wooden or metal slats about 4 inches wide that are hinged together with vinyl fabric. In all cases, these doors hang from and run in a track. They open and close between the door opening's side jambs.

The main applications for folding doors are in closets, laundry niches, and some storage pantry-like areas in the garage, basement, or hallways. They can also be used to divide large open spaces into two smaller rooms.

Advantages:

1. They don't swing out or protrude from doorway openings. Their accordian sections fold up into a small bundle when opened.

Disadvantages:

1. Folding doors give scant protection against fire and noise.
2. They don't operate very smoothly. They're not

Fig. 14-9. A pocket door.

made as sturdily as other doors, and cannot stand heavy use well.

3. A main objection of homeowners is that folding doors don't *look* like doors. Rather, they resemble stiff draperies. Folding doors are generally neither strong nor secure and they have a tendency to wave in a strong draft.

Cafe Doors

These are the doors you've probably seen a thousand gunslingers walk through, into television and theater saloons. They're the double swinging doors on the short side—about 30 to 60 inches long—installed on opposite sides of an opening roughly midway between the top and bottom of a passageway (Fig. 14-11). They're attached to the opening's walls or jambs with gravity pivot hinges that enable the doors to be pushed open in either direction, only to be closed automatically once a person passes through them. Cafe doors are available in both louvered and paneled designs.

212

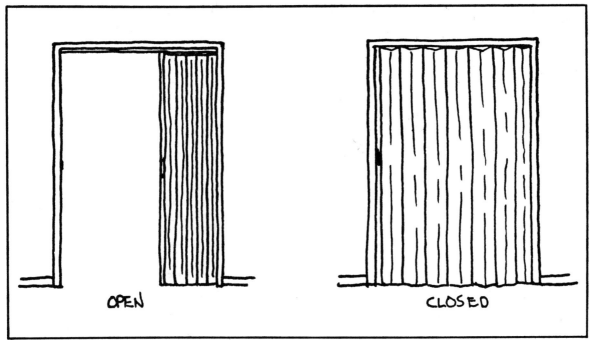

Fig. 14-10. Folding doors.

Because you can push your way through them without using your hands, these doors come in handy in places where you're likely to have your hands full, such as between a kitchen and dining room, where the cook is likely to shoulder his or her way from the kitchen carrying a Christmas turkey. Cafe doors don't provide much privacy from noise or odors (then again, who wants to hide the delicious smells that loft from a kitchen stove?), but they will block the sight of stacks of dirty dishes in the kitchen sink from guests in the dining room.

Advantages:

1. They're inexpensive and easy to install.
2. They're simple to use.
3. They add a unique and attractive character to the interior of a home.

Disadvantages:

1. They provide little protection from fire and noise.
2. Because people are constantly pushing against them, they're difficult to keep clean. If louvered, they're time consuming to wash or resurface.
3. Children can be injured running in and out of them.

DOOR OPENINGS

Door openings are permanent features of a house that should be well thought out in advance. Here are some pointers to consider:

Fig. 14-11. Cafe doors.

213

1. Door opening heights should be either 6 feet 8 inches or 7 feet.

2. A front door should be at least 36 inches wide. If less, then the house should have another outside door at least 36 inches wide so you'll have the ability to move large items in and out. The recommended thickness of a 36-inch exterior door is at least 1 3/4 inches.

3. Secondary outside door openings to a basement, kitchen, laundry, or garage can be 32 inches wide, with thicknesses of 1 3/8 inches.

4. If you happen to position a bedroom door at a right angle to the end of a hallway, make sure the hallway is extra wide, and the door opening is 36 inches, or you'll never be able to maneuver large pieces of furniture such as dressers and headboards into the room.

5. Consider each individual passageway if you want interior doors to open inwards or outwards, as safety and convenience dictate. Basement stairway doors should always open away from the stairs.

6. When planning patio/garden openings, figure that they consist of sliding and stationary glass panels with widths of at least 36 inches per panel.

HINGED DOORS

Because hinged doors are used in every house in some fashion, exterior or interior, they deserve additional comment:

1. They are easier to open and close than all other doors except swinging cafe doors.

2. They permit you to use their backsides for storage space, a major advantage in closets.

3. Their operation is noiseless except when slammed.

4. When properly hung, exterior models close tightly and stop drafts, dirt, and insects from penetrating around the edges.

5. Solid full-length doors should be hung with three hinges to prevent saging. Two hinges are sufficient for hollow doors.

6. Interior doors should swing into the rooms they close off from hallways. Otherwise they interfere with hallway traffic.

7. Doors on hall closets must obviously swing into the hall, however, and it's often advisable to swing a kitchen door into a hall so it doesn't create traffic problems within the kitchen work areas.

8. Whenever possible, bathroom doors should swing in; but if the bathroom is too cramped, the door can be hung to swing outwards.

9. Doors between adjoining rooms other than bathrooms can be swung whichever way will cause the least inconvenience.

10. Doors at the head of stairways must swing away from the stairs.

11. Whether a door should be hinged on the right or left depends on which position will interfere less with furniture placements and passage through the doorway.,

12. Ideally, storm doors should be placed so they can swing back into a corner out of the way.

13. Every exterior hinged door (and other types for that matter)—front, back, or side—should have a step down to the outside, a sill forming the bottom part of the frame and entrance. This 8-inch step down is to prevent water from entering the house during heavy rains or snows.

PET DOORS

In warm-climate regions you might consider, if you own a dog or cat, installing a pet door next to one of your side or rear entrances. Several companies manufacture ready-to-install units that are two-way, self-closing, silent, chew proof, energy efficient (fitted with a weather seal), and lockable.

Advantages:

1. You needn't be present to let your pet in and out during the day or night.

2. You needn't worry about heating or cooling the great outdoors by leaving a front or rear door propped open while you peer into the darkness and whistle for your pet to come home.

Disadvantage:

1. What might happen if your neighbor installs the same type of door, for his pet?

DOOR CONSTRUCTION

The construction of interior and exterior doors are substantially different from one another (Fig. 14-12).

Interior Doors

Interior doors can be built out of wood, plastic, metal, or any combination thereof. A solid wood door is the best and most expensive interior door available. It's made by sandwiching a wooden core between two sheets of high-quality hardwood veneer such as birch,

Fig. 14-12. Door construction.

oak, mahogany, or pine. Lighter-duty hollow-core doors can have the same expensive veneer faces, but sheets of sturdy wood-grain plastic also make a practical, easy-to-clean surface. The lowest-priced hollow-core models are frequently covered with less durable wood composition board.

When selecting door veneers, keep in mind that birch and oak doors hold up better than mahogany or pine. Oak and birch are harder woods that can resist greater impacts.

Exterior Doors

Exterior doors can be categorized as wood, steel, and composition patio doors.

Exterior Wood Doors. There are two basic styles of exterior wood doors: flush and paneled. Flush doors are simple flat-surfaced doors that can be constructed with particleboard cores between two outer surfaces of a durable wood such as a high-grade fir, or a solid lumber core between the two outer faces. The latter are much stronger—a better choice, and the most expensive. Paneled doors have decorative designs cut into them. Both flush and paneled models can be made with various amounts and shapes of glass inserts.

Advantages:

1. They're beautiful, particularly when stained to show off their grain.

2. They lend themselves to an infinite variety of designs, and can even be customized by a manufacturer to your specifications.

3. They can be sawed and planed to fit existing door openings or new openings that are not carefully made, or openings that later warp because of settling.

4. Solid-core wooden doors have good insulating qualities and are relatively soundproof.

Disadvantages:

1. If not taken care of they may expand, warp, crack, shrink, and cause sealing problems.

2. They must be periodically refinished.

3. Wood doors are combustible.

Exterior Steel Doors. Most steel doors are really part steel, and part wood: steel sheets fabricated around a wooden frame in which urethane foam or another insulation material is sandwiched to provide outstanding protection against cold and heat.

Steel doors are more practical than doors made of wood. They're designed to prevent the principal cause of wood door failures—warpage that results in improper closure and air infiltration. Newer door frames provide a thermal break between interior parts and exterior surfaces, eliminating winter condensation inside. Other features that apply to many steel door models are adjustable thermal-break thresholds, magnetic weatherstripping for uniform sealing, and fire-resistant qualities that result in excellent safety ratings.

Exterior steel doors come in many styles and finishes and are commonly fitted with glass inserts and peepholes for viewing. The ones with magnetic gaskets along their edges that grip like those on a refrigerator door give the tightest seal against the weather. They're usually prehung so the frame and hinges are included with the door as a complete unit. The cost of steel doors is quite competitive with those made of wood.

Make sure that any steel door you buy is predrilled by the supplier to accept the door hardware (handles, knobs, and locks) you desire. If you don't, it could cost you extra to have it done at the job site because some carpenters don't have the right tools to drill this type of door.

Advantages:

1. They provide an excellent seal against air infiltration.

2. They're well insulated.

3. They're difficult to force open.

4. They're not affected by moisture.

5. They're fire resistant.

6. Plastic decorator panels are available for some steel door models to provide elaborate and varied finishes.

Disadvantages:

1. They come in standard sizes and are not easily adjusted to slightly uneven openings or odd-sized openings.

2. In most steel doors the panel embossing is more shallow and not as handsome as that on a wood door.

3. They just don't have the wood grain beauty of wood doors.

Composite Patio Doors. Patio doors are typically sliding glass doors framed with steel, aluminum, or wood, or French doors that provide access to a patio, porch, deck, or garden. The wood frames are frequently clad in vinyl to eliminate the need to repaint or refinish the wood.

Double- or triple-pane glass that has a good insulating value should be used in these doors, and weatherstripping must be included along the door edges. Design improvements have separated the inner and outer faces so the frames will not easily transmit heat or cold.

Advantages:

1. Glass allows a view of the outside surroundings.

2. They require very little maintenance.

3. They're great if lots of light and sunshine are desired.

Disadvantages:

1. They're easy to force open if proper precautions are not taken.

2. Poorly constructed ones will sweat.

3. When constructed without thermal-type double or better glass, they have poor insulating value.

4. Even with the better patio doors, the wear and tear caused by the regular door movement across the sealing material is prone to leaks in the sealing that will allow air and moisture infiltration.

Chapter 15

Garages

The garage is not often given the attention it deserves by the typical homeowner. A brief analysis of the situation reveals no sound reason to skimp when it comes to this important part of a house. To be sure, many owners don't realize that a garage costs considerably less per square foot than the rest of the house: it doesn't have as much electrical, plumbing, lighting, floor or interior wall finishing costs. Consequently, instead of being stingy with your garage money, you're better off to build one a little larger than you initially want. It won't cost much more (Fig. 15-1).

When you do plan your garage, here are some ideas to keep in mind:

1. Few garages built today are constructed apart from the house. The conveniences of initial construction, maintenance, and homeowner traffic patterns of an attached garage far outweigh the disadvantages.

2. However, an attached garage requires safety features not necessary in detached models. In some locations the ceiling and wall connecting the garage and house must be plaster, masonry, or some other fire-retardant material, and the door from the garage to house must be solid wood or covered with sheet metal.

3. One disadvantage with attached garages is

that carbon monoxide fumes can gather within them and seep into the house's living quarters. Carbon monoxide fumes are heavier than air and can travel into a basement or lower living level if the garage is on that same level or higher. To prevent this from happening, most codes require at least a 4-inch stepdown from the house to garage. A curb can also be a partial remedy. In any case, if you plan a house with an attached garage, realize that there will always be a potential danger with exhaust fumes if a car is allowed to idle in the garage with the garage door closed.

4. Request a minimum of two electrical wall outlets in the garage, as well as an outlet or outlets centered in front of the overhead garage door(s) so you can put up automatic door openers at a later date if desired.

5. Besides having overhead lights installed in the main level of the garage, if there's a garage attic in your plan, specify a light fixture there also.

6. Provide adequate ventilation in any garage attic.

7. If you want to heat the garage, have the outer walls as well as those of adjoining living spaces insulated.

8. Proper flashing should be installed between

Fig. 15-1. A view of a garage interior.

the garage roof and the wall of the house, or vice versa, to prevent water seepage there.

9. Have a water faucet installed somewhere inside the garage. The best place is on a common wall with the living quarters of the home.

10. If the back of a fireplace chimney will protrude into the garage, make sure you take this into consideration when planning the garage's length.

11. To keep out the elements, soil, and energy-wasting summer and winter temperatures, make certain that the garage car doors will seal tightly against the floor and sides of their tracks when closed.

12. For appearance's sake, specify if you desire the trim around the car doors to have "cut corners" at the top. This can add an interesting flair to a garage for a minimum cost.

13. If you're having a 2-car garage, decide if you'd prefer two separate car doors or a single larger model. Two doors might look more attractive, but they require two separate door openers and need a support post in the center, which reduces the opening clearance that might be wanted for maneuvering a boat trailer in. Such a center post between two doors can also become an expensive obstacle for inexperienced drivers to crunch into.

GARAGE DOORS

Modern car doors on garages either swing or roll up out of the way. Old-fashioned sliding bypass types and hinged garage doors are much less practical. The typical single garage car door is 7 feet high and 9 feet wide,

218

while a double garage car door is the same height and at least twice as wide (Fig. 15-2).

Swing-Up Doors

These are less expensive than roll-up doors. They can be made on the building site from the same materials used for the house's siding. Often, though, swing-up doors consist of thin sheets of exterior plywood that are easy to work with and light enough for the owners to open with little bother. The drawbacks to this type of door are that it's so lightly constructed it often has problems operating on its tracks, it can warp easily, it's not very energy efficient, and it's not as attractive as roll-up doors (Fig. 15-3).

Roll-Up Doors

Roll-up doors are the most popular and practical garage car doors available. They come in wood, steel, aluminum, fiberglass, plastic, masonite, and composite models. Insulated garage doors made of plastic, aluminum, or steel having fiberglass insulation inside are terrific for saving energy. The plastic doors are also maintenance-free, with the door's color an integral part of its makeup so it never needs resurfacing. Any roll-up door should be trimmed on the bottom with an astragal (rubber) strip to ensure a tight seal with the floor. Specify the type of roll-up door propelled by torsion springs for operating safety and ease. All roll-up doors can be efficiently connected to automatic door openers.

Automatic Door Openers

It's nice to come home late at night in a thundering rainstorm or blizzard and push a tiny button from within the car so your garage door raises open for you, like magic. For the sake of convenience and safety, automatic garage door openers are relatively inexpensive and reliable contraptions.

Fig. 15-2. Garage doors.

Fig. 15-3. Swing-type garage door.

Make sure that the opener you select will reverse itself automatically if it encounters any obstacles (such as a small child) while closing toward the garage floor. Also arrange for the control buttons to be positioned high enough so children can't reach them. This is especially important for the buttons near the inside garage man door to the house. Children will sometimes push a control button to close the door then attempt to dash through the opening before the door is fully lowered.

Man Doors

A man door entrance to the garage should be included to permit access into the garage from outdoors without having to raise the garage overhead door(s). The best man doors are steel-covered. They're strong, burglar-resistant, and fare well against the weather. As with all other outer man doors, garage entrances should be insulated with weatherstripping. Depending on your lot, if the back of the garage faces the back yard you might consider having a second man door installed there.

GARAGE ATTIC DOORS

If your garage will have a usable attic, specify a set of pull-down steps as described in the chapter on stairs. Folding or roll-down type stairways make accessing the attic quick and easy. At the same time, with attic space in the garage quite high off the garage concrete floor, these types of stairways are far safer than using conventional step ladders.

crete and by making the floor a lot easier to clean.

GARAGE WINDOWS

A garage needs ample outside light and ventilation during the hot summer months. Two windows, one at the front (or several if included in your overhead car doors), and the other at the back of the garage are usually enough. Try to keep one side of your garage either clear to be lined with pegboards for storage or to be kept open-studded (if insulation is not installed and covered) for hanging garden tools and other items.

GARAGE FLOORS

The following points should be considered to ensure a trouble-free garage floor:

1. It should have a 4-inch gravel base covered by a sheet of polyethylene or similar moisture barrier.

2. If no garage floor drains will be installed, the garage floor should be poured so it begins about 1/2 inch above the driveway's surface then slopes up to about 2 inches toward the back of the garage.

3. If possible, include floor drains, one for each car space. Naturally the floor should be sloped toward the drains to prevent any standing water.

4. To support the edges of the garage floor where the fill around the footer might settle, specify 6- x -6-inch reinforcement wire to be used throughout the entire floor.

5. The concrete should be poured at least 4 inches thick.

6. The garage (and basement) floor should be steel-troweled to achieve a smooth finish. This is a time-consuming task that requires going over the surface with a trowel many times during an 8-hour period or until the surface hardens.

7. The part of the garage floor that sticks out past the garage door—normally about 6 inches, should taper away from the door down toward the driveway, ending 1/2- to 1-inch higher than the driveway surface. By tapering it down toward the driveway, rain water and melting snow will not drain into the garage. Keeping the floor slightly elevated prevents water from backing up into the garage from the driveway.

8. Before parking cars on a new garage floor, the floor should be cleaned with a solution of muric acid, then coated with two or three applications of a clear sealer. The sealer improves the floor's appearance by preventing oils, grease, and dirt from staining the con-

GARAGE STORAGE

In years past, there was a time when garages fit the automobiles for which they were built like a glove. Now garages are erected not only to hold cars, but to shelter a wide variety of other work, maintenance, and leisure-time items.

Consider the following ideas when planning your garage storage space:

1. Provide enough space to store lawn mowers, snow blowers, bicycles, and other bulky objects along a side wall, and a place to keep even larger items such as garden tractors or small boats along the back wall.

2. No matter what size garage you plan, many cubic feet of relatively inactive storage space can be salvaged by building one or more shelves 24 to 30 inches deep around the two sides and the back of the garage about 6 feet above the floor, attached by brackets fixed to the walls.

3. A huge rack for storing screens, oars, water skis, and other long articles can be hung from the ceiling over each car bay. If built over a "walking area" at the sides or in front of where a car is stored, the rack should clear the floor by at least 6 1/2 feet, but where it hangs over the car it can be dropped to within 5 feet of the floor because it isn't over a highly traveled pedestrian zone.

4. The more you can keep items elevated off the floor, the easier and faster you'll be able to clean out the garage.

5. You'll find it handy to leave one wall with the framing studs and planks exposed—not covered with plaster or drywall board. This permits large sheets of pegboard to be attached directly to the studs and will provide an air space to arrange hanging pegs into the pegboard itself. Pegboard is ideal for hanging all sorts of hand tools, garden tools, hoses, and many other items on. If pegboard will be secured to finished walls, an air space must be left behind it using 2- x -2-inch studs to establish a gap.

6. Heavy garden tools and items such as hand mowers, empty lawn rollers, and long wooden extension ladders can be hung from 4-inch boards nailed across open studs when steel brackets are fastened to the top edges of the 4-inch boards.

7. The garage attic—even when trusses are running through it—can be used to store off-season out-

doors equipment. It's smart to request that a plywood floor be laid down, at least on the main part of the attic where headroom is high enough so you can walk around with ease.

GARAGE POSITIONING

This subject is covered in the chapter on house orientation and positioning, but in a nutshell: the garage must fit on the lot without encroaching on an adjacent site. It should also conform to the slope of your lot. For instance, when a garage is located on a downward side-sloping lot, instead of completely lowering the garage to meet the natural lot line or placing load after load of fill dirt to raise the lot level, it's best to reach a compromise between the two solutions.

Position a garage to reduce energy loads. In cold-climate areas garages should be located on a northern part of the house, if possible. In hot-climate areas, a garage is best suited on the east or west side of the home to help reduce heat gain.

Chapter 16

Fireplaces

The first fireplaces were logical extensions of an open campfire—functional features that were designed to provide heat, illumination, and a way to cook foods in early dwellings. They were originally a necessity in all but the warmest climates, and supplied homeowners with basic needs for centuries.

Then, rather quickly, other mechanisms were invented that more efficiently heated our houses, illuminated our rooms, and cooked our foods. Consequently, fireplaces lost some of their functional value and became more decorative options than necessities. They still, however, remained a part of our houses for the atmosphere they created and the warmth that homeowners appreciated on cold winter evenings. But once the energy crunch hit, fireplaces were accused of being the energy-wasters they really were, and fell further out of favor.

Not to be upstaged, hundreds of large and small companies set out to modify the standard fireplace into units which, through various add-on tubes, gadgets, and heat exchangers, would burn fuel much more efficiently.

So today's fireplaces can be divided into two main types: those used primarily to heat a room, area, or entire house, and those used primarily as decoration or novelty. Even with continuing concerns for energy usage, the latter models—the fireplaces installed for decoration and atmosphere—are still the most popular. They're included in modern homes because of what they add to a room, in ways that excite our senses of sight, sound, and smell. They're a bit of the past to which we still cling.

Any fireplace should at least do five things:

■ Permit combustion of fuel.

■ Exhaust the by-products of combustion from the house.

■ Deliver as much heat as possible into the house.

■ Function safely.

■ Be located so people sitting in all areas of a room can enjoy it. In a long, narrow room, for example, it's better to plan a fireplace on one of the long side walls than on a short end wall. Because people tend to cluster around a fireplace, if it's located at one of the short walls in an end of a long room, the opposite end of the room (without a fireplace) would likely go completely unused.

Advantages:

1. They can provide additional heat.

2. They can provide *all* the necessary heat in mild climate locations.

3. They enhance the appearance and comfort of any room in any house in any climate.

4. They provide a traditionally romantic and reassuring atmosphere.

5. They can burn as fuel certain combustible materials that would otherwise be wasted, such as coke, briquets, scrap lumber, and even tightly rolled-up newspaper.

Disadvantages:

1. The major problem with the typical fireplace is that it gulps substantially more air than needed to complete combustion. Much of this "consumed" air is simply drawn from the room up and out the chimney without ever contributing to the combustion process. In modern homes the room air used for fireplace combustion has *already* been heated by the primary heating system, and most of it is lost up the chimney. Therefore, rather than assisting the primary system in its heating function, the fireplace interferes with its operation and usually increases the workload of the primary system. Without pulling punches, the fireplace is one of the least efficient ways known of heating a house. As much as 90 percent of the heat produced by a wood fire can dissipate up and out of the chimney.

2. A fireplace heats by radiation only. Much of the air that comes into contact with the hottest surface of the fireplace ends up outdoors.

3. Some warm air from the house will go up the chimney even when a fire is not lit and the damper is closed.

4. Because brick and stone are poor insulators, a fireplace and chimney can create a thermal opening in the wall of a dwelling if located somewhere in the building's outer shell.

5. A fireplace adds to the cost of a new house.

6. When positioned within a building, a fireplace occupies valuable floor and wall space.

7. When positioned in an exterior wall a fireplace can occupy scarce land.

8. A fireplace requires periodic maintenance and occasionally the services of professional chimney cleaners.

9. Fireplaces can be messy. Storage space for wood and other fuel is needed, and so are places to keep dirty fireplace tools such as pokers, tongs, and shovels.

10. Fireplaces can be dangerous when incorrectly built, installed, or used. They can be hazardous when small children have free reign throughout the house.

TYPES OF FIREPLACES

With the development of modern manufacturing processes, materials other than traditional stone and brick have been made available for fireplace construction. Today there are numerous fireplace designs that come ready-made or can be custom-built to suit any application.

Masonry Fireplaces

These are the original models that are still popular today. They consist of a stone or brick exterior having a lining of firebrick—a brick that can stand high temperatures encountered when wood or other fuels are burned. Of course, built into the masonry are various operating accessories such as grates to hold the wood while it's being burned and dampers to regulate the amount of air flow through the chimney passages (Fig. 16-1).

There are several basic masonry fireplace designs, including fireplaces with a single opening, constructed against a single wall (the most popular); fireplaces constructed into a wall that divides two rooms so the fireplace has two openings, one per each of the back-to-back rooms; fireplaces built into an outside corner, open to two sides; and even circular fireplaces of brick and stone having sheet metal chimneys suspended from the ceiling that flare out over the round firebox like an inverted funnel.

Because these masonry units must be built from scratch on the building site, the labor construction costs of masonry fireplaces are high. Once a masonry fireplace is up, however, its maintenance expenses and efforts are minimal. The beauty, durability, and reputation of its brick or stonework can enhance the overall appearance of a house's interior and exterior, and will increase the home's saleability and value.

Stone generally costs more than brick and requires a higher degree of skill on the mason's part. If you opt for an all masonry fireplace, make sure that whoever will be putting it up has had experience completing others as well. There are many tricky steps to masonry

Fig. 16-1. An all-masonry fireplace.

fireplace construction, and unless they're all done exactly right, they can cause serious structural and safety defects that might appear years later. All-masonry fireplaces, because of their weight, require at least an 8-inch-thick concrete footer for adequate support.

Masonry And Steel Box Fireplaces

This type is similar to the all-masonry fireplace except that the firebox, or where the fuel is burned, is prefabricated of steel instead of being constructed from firebrick. The rest of the fireplace is all masonry. Masonry and steel box fireplaces are less costly than all-masonry units, and they still retain a handsome appearance and low-maintenance characteristics (Fig. 16-2).

Circulating Fireplaces

A typical circulating fireplace consists of a specially designed prefabricated steel shell with firebox and damper. It greatly simplifies construction because all a mason has to do is build the foundation and hearth, set the prefabricated fireplace shell on the hearth, and lay finishing bricks or stones up around the shell. The shape and approximate exterior dimensions of the fire-

place are already determined.

The prefabricated shell more than makes up for its restricted originality with increased efficiency. The real advantage of a circulating fireplace is that it heats not only by radiation (as a conventional fireplace does) but also by circulating heated air directly into a room or adjacent area. As a result it provides almost twice the heat output of a conventional fireplace.

The walls of most prefabricated circulating fireplaces consist of two layers of steel separated by an air space. Cool air enters the air space through grilles installed at the floorline next to the fireplace. As air passes through the bottom grilles and moves between the hot layers of steel it gets heated and then expelled through other grilles installed near the top of the firebox or high in the wall. To encourage constant air movement and an even amount of heat, fireplace manufacturers offer air inlets containing electric fans. Fans can also be used with the warm-air outlets placed in adjacent rooms. The chimney may be of conventional masonry construction or made of prefabricated steel.

Prefabricated Built-In Fireplaces

This type of fireplace not only reduces construction

costs but makes it possible to install a traditional design almost anywhere in an existing house. The prefab built-in fireplace consists of a steel firebox complete with hearth, damper, and a prefabricated chimney as well. Prefabricated built-in fireplaces are made with either one opening in the front, or with one front and one side opening. Most are constructed with zero clearances, meaning that they can actually touch any building or framing material around the bottom, sides, and back, without becoming a safety hazard. Because of this, they can be constructed right into a wall at floor level or above, or they can project out from a wall. In any case, you can finish up to them with wood paneling or any other wall surfacing material (Fig. 16-3).

Another advantage to the prefabs is that they don't weigh much, so there's no need to place them on an elaborate and expensive masonry foundation. Instead, they can be set directly on the subfloor or finished floor, and a fire-resistant forehearth can be integrated into the finished flooring directly in front of the fireplace's main opening.

Freestanding Fireplaces (See Fig. 16-4)

These are the least expensive fireplaces and the easiest to install. All it takes to erect one is to set it in position and run a flue through the roof or the nearest exterior wall. Freestanding units are usually made of steel or cast iron and come in a variety of designs and shapes—round, square, hexagonal, egg, even triangu-

Fig. 16-2. A masonry and steel box fireplace with circulator vents.

Fig. 16-3. A prefabricated fireplace.

lar. They're typically finished in bright porcelains or enamels of red, green, yellow, black, white, and other tones. While some units stand on the floor, others hang on the face of a wall or are suspended from the ceiling on chains. They're modern-looking but give off plenty of old-fashioned heat because the sheet metal chimney, in addition to the firebox, also gives off radiant heat directly into the room (Fig. 16-4).

These fireplaces can be placed right against unpainted masonry walls, but if a wall is constructed or surfaced with combustible materials, a clearance of several inches is needed behind them and even a greater safety margin is required on the sides.

Woodburning Stove Fireplaces

To someone looking for more heat than glamour, a wood- or coal-burning stove is a good choice. Made of either steel or cast iron, freestanding stoves can pro-vide over 90 percent of the available heat from wood or coal combustion. And because they're not con-structed at the building site, they'll definitely keep in-stallation costs low. Another plus is that many models have doors that can be swung open so all the pleas-ures of an open fire can be enjoyed (Figs. 16-5 and 16-6).

When strapped for funds, a homebuyer can elect to install a freestanding woodburning stove after the house is constructed, without much trouble.

Like freestanding fireplaces, stoves must be set out from combustible walls, and although most of them stand on legs, they should be centered on a non-combustible pad of some kind.

FIREPLACE DESIGN

Fireplace efficiency depends largely on proper design (Fig. 16-7). Here are some important considerations to

NON-COMBUSTIBLE MATERIAL

Fig. 16-4. A freestanding fireplace.

NON COMBUSTIBLE MATERIAL

Fig. 16-5. A Franklin wood stove.

FLUE

FIREBOX DOOR

DRAFT REGULATOR

NONCOMBUSTIBLE FLOOR

Fig. 16-6. A wood stove.

review if you're thinking about including one or more fireplaces in your house:

1. A common but inefficient way to include a fireplace in a home is to have it built so it employs a chimney at a side or end of a house where three sides (the back and two sides) of the chimney are exposed to the weather (Fig. 16-8). As mentioned in the chapter on insulation, brick and stone are of limited insulating value. They'll lose much more heat to the outdoors than they'll save for the indoors.

2. Don't settle for a fireplace with a large throat and a flue without an adjustable damper. An adjustable damper will limit the amount of warm air lost up the chimney to only that necessary to remove the smoke.

3. The best place for a fireplace to be located is away from outside walls, so most of the heat stored in its brick, stone, or metal parts is eventually delivered into the house.

4. A low chimney is a dangerous design feature.

5. The most efficient fireplaces follow established guideines, mathematical relationships between the fire-

box, opening, throat diameter, and flue diameter dimensions. A good ratio of front opening to cross sectional area of the flue is seven to one. If the front opening is proportionally too large, the draft will be poor. A cheap way to decrease the acceptable size of the opening would be to raise the hearth, but because a fireplace requires a sizable draft, savings on energy expenses in this manner would be insignificant—and very often negative. You don't have to actually *know* what all the technical points are before you plan a fireplace, but you should be able to throw out a few ideas so your builder thinks that he's dealing with someone who understands and will accept only letter-perfect work.

Even though fireplace dimensions vary tremendously, as a general rule, the height of the opening should be 2/3 to 3/4 of the width, and the depth of the firebox should be 1/2 to 2/3 of the opening height. For openings up to 6 feet wide, the height should usually not exceed 3 1/2 feet; 4 feet is a good maximum for fireplaces over 6 feet wide. The average fireplace has an opening 30 to 40 inches wide and 30 inches high. The depth is between 16 and 18 inches. A typical fireplace facing will be 6 feet wide all the way from floor to ceiling.

6. A fireplace should be more or less in scale with the room it will be located in. As a rule of thumb, there should be 5 square inches of fireplace opening for every square foot of floor area.

Fig. 16-7. Fireplace design.

Fig. 16-8. A three-sided fireplace chimney exposure.

7. Each fireplace should have its own flue, and it's not a sound idea to vent any other heating unit, such as a furnace, into a fireplace flue.

8. Glass doors will eliminate lazy drafts and reduce smoking, plus they're good safety features to have.

9. If natural gas is available in your area, it's smart to include a gas starter in a fireplace, for convenience.

WHERE TO LOCATE THE FIREPLACE

Keep in mind that once you decide where a typical fire-

place will be positioned, and once it's installed, you won't want to move it to somewhere else in the house.

Whenever possible, outside wall locations should be avoided. Why heat the outdoors? The heat loss from a placement along an outside wall is nearly 25 percent. A chimney that's exposed to the weather along its entire length on one or more sides is bound to cool off quickly when the fire is low. Then, when the fire is rekindled, the products of combustion must try to force their way out of a chimney filled with dense, cooled gases. In contrast, heat that escapes through the chimney walls from an inside wall fireplace will help warm the house or at the very least the garage (Fig. 16-9).

By locating a fireplace on any inside wall, especially the wall separating the garage from the living areas, the part of the fireplace and chimney facing the garage does not need a facing of finished brick or stone. Thus, by locating a fireplace in the way most favorable for energy savings, material and installation costs can also be lowered.

There should be ample room around a fireplace for furniture. But the furniture placement shouldn't interfere with traffic in and through the room, and it shouldn't be situated closer than 6 feet away from the fireplace front opening. Chairs and sofas positioned closer than 6 feet will likely make occupants uncomfortably warm no matter what the outdoors temperatures are.

FIREPLACE SUPPORT

A masonry chimney is usually the heaviest part of a house. It must rest on a solid foundation to prevent settlement. Concrete footers are recommended. They must be designed to distribute the load over an area wide enough to avoid exceeding the safe load-bearing capacity of the soil. They should extend at least 12 inches beyond the chimney on all sides, and be at least 8 inches thick for single-story houses and 12 inches thick for two-story houses with basements. If there is no basement, the footers for an exterior wall chimney should be poured on solid ground below the frostline.

CHIMNEYS

Stone and brick fireplace chimneys can be domineering features no matter if they're located at the exterior of a home, or the interior.

Chimney Walls

Walls of chimneys with lined flues, not more than 30 feet high, should be at least 4 inches thick if made with brick or reinforced concrete, and at least 12 inches thick if made of stone. Flue lining is recommended, especially for brick chimneys, but it can be omitted if the chimney walls are made of reinforced concrete at least 6 inches thick or of unreinforced concrete or brick at least 8 inches thick. A minimum thickness of 8 inches is recommended for the outside wall of a chimney exposed to the weather.

Brick chimneys that extend up through a roof might sway enough in heavy winds to open up mortar joints at the roofline. Openings in the flue at that point are dan-

Fig. 16-9. A fireplace chimney through a garage.

gerous because sparks from the flue could start fires in the woodwork or roofing. A good practice is to make the upper chimney walls 8 inches thick by starting to offset the bricks at least 6 inches below the underside of roof joists or rafters.

Chimneys can contain more than one flue. Building codes generally require a separate flue for each fireplace, furnace, or boiler. If a chimney contains three or more lined flues, each group of two flues must be separated from the other flue or groups of two flues by buck divisions or "wythes" at least 3 3/4 inches thick.

Neither the chimney nor the fireplace should touch any wood or flammable materials in the house structure. A 2-inch space is required at all points except behind the firebox, where it must be increased to 4 inches. The space between the chimney and floor members should be firestopped with similar noncombustible materials.

Chimney Height

Proper chimney height on any house depends on the shape of the house's roof as well as the positions and sizes of surrounding trees, buildings, and even hills. The chimney should extend or rise at least 3 feet above flat roofs and at least 2 feet above a roof ridge or raised part of a roof within 10 feet of the chimney. A chimney hood should be provided if a chimney cannot be built high enough above a ridge to prevent trouble from wind eddies caused by breezes being deflected from the roof or nearby trees. The open ends of a chimney hood should be parallel to the ridge (Fig. 16-10).

Chimney Lining

Although a chimney can be lined with brick, there's less chance of soot accumulating inside the flue if a clay tile lining at least 5/8 inch thick is used instead. The joints between such tiles must be completely filled with mortar and finished smooth on the inside. The lining is naturally surrounded with at least one sturdy course of 4-inch common brick, with stone, or with concrete blocks.

Fig. 16-10. Chimney heights.

233

Chimney Mortar

Brickwork around chimney flues and fireplaces should be laid with cement mortar. It's more resistant to the action of heat and flue gases than lime mortar is. An example of a good mortar to use in bonding flue linings and all chimney masonry except firebrick consists of one part Portland cement, one part hydrated lime (or slaked lime putty), and six parts clean sand, measured by volume. Firebrick should be laid with fire clay. Mortar will not stand up to intense heat as well as fire clay will.

This is an example of why it pays to have only experienced masons work on fireplaces. It's a shame, but they're a disappearing breed, the fireplace specialists—going the same way as plasterers and stone masons. Even some of the masons who claim to know what they're doing, do not. They leave behind a trail of fireplaces that smoke, draw poorly, and have flues that are constantly accumulating thick deposits of creosote. That's why, when putting up a masonry fireplace, it's best to select someone (or make sure the contractor chooses someone) who is a proven professional.

Chimney Damper

The fireplace and chimney are large heat gobblers, particularly when they're not in use. Because of this, a fireplace chimney should be equipped with a metal damper that can be closed when the fireplace is not in service, to prevent a continual draft. The damper should be the kind that opens and closes by increments, so air flow can be regulated when the fireplace is needed. A fireplace damper is typically about 36 inches wide.

Chimney Cleaning

All wood fires produce some creosote, and flues should be inspected periodically to check the amount of creosote present inside them. Creosote can build to thick deposits that become extremely flammable, and can result in dangerous house fires.

When the flue's insides are coated with uneven surfaces of black gummy-looking material and you can't discern the tile and mortar joints, it's time for a cleaning. The term "chimney sweep" isn't really accurate. Creosote is so hard and gummy that it can't be swept loose. Rather, it has to be chipped away from the masonry with a blade. In the meantime, care must be taken not to knock our mortar joints or damage the flue lining. Because it's such an infrequent and important task, and specialized tools are needed, it's best to hand the job over to a commercial chimney sweep or cleaning outfit.

HEARTHS

The fireplace hearth is really the floor of the fireplace: the brick, stone, or concrete pad beneath and in front of the firebox. Practically any non-combustible material can be used, including brick, concrete, adobe, terrazzo, quarry tile, marble, mosaic tile, slate, fieldstone, and even bronze or copper if laid over a non-combustible base.

Because the hearth cannot safely rest on the floor or wood framing, it should either be cantilevered out from the foundation or the foundation should be extended under it. It must be supported from the ground up, beginning with a concrete pad 12 inches thick or more typically part of the entire fireplace footer, and then brought up to the subfloor level with concrete block.

A hearth should extend at least 20 to 24 inches into the room and at least 6 to 8 inches on both sides of the fireplace opening. It can be flush with the floor so that sweepings an be brushed into the fireplace, or it can be raised. Raising the hearth to various heights and extending its length is a common practice, especially in contemporary designs. This can create a natural seat for people to rest on and warm up for brief periods after a cold winter evening of skiing or tobogganing. When done on a smaller scale, though, raised hearths can also present a hazard for tripping over.

COVERS AND SCREENS

Suitable screens should be placed in front of all fireplace openings to minimize the dangers from sparks and exploding embers, and to keep young children from playing with and in the flames. Some wire mesh screens tend to be messy looking, especially when they're not permanently attached to the sides of the fireplace opening. It's better to go with ones that are suspended across the opening, and can be removed when the fire needs tending. Glass screens or doors are even better because they're ideal for safety, are attractive, and they can be closed at night to reduce heat loss that would otherwise go up the chimney during cold weather. They can also be used to prevent any smoke from being blown back down the chimney into a room by a strong fluke downdraft.

MANTELS

There's no strict rule dictating that every built-in fireplace must have a mantel. Many do not; quite a few contemporary designs carry the wall covering material—usually brick or stone—right up to the edges of the fireplace opening. Most traditional fireplaces, however, still have mantels. Mantels can be made of practically any sturdy material, from a slab of rough-hewn oak, to granite, marble, slate, and even concrete. Some consist of elaborately carved wood.

All wood mantels must be set back from the fireplace opening edges to keep the wood from catching fire. The Federal Housing Administration requires a minimum clearance of 3 1/2 inches, but most fireplaces look better if this is increased to 7 or 8 inches. Mantel shelves and other wood accessories projecting more than 1 1/2 inches beyond the fireplace breast should be installed at least 12 inches above the opening.

ASH PITS

A fireplace ash pit or soot pocket is formed in the hollow space within the foundation walls and is connected with the fireplace by a small metal door called an ash dump. Some fireplaces have them and others don't. It depends on the fireplace design and how often the owners will be burning wood. With an ash pit, to get rid of ashes and soot you simply open the ash dump door and scrape or shovel the ashes into the pit. Then they're removed through a tight-fitting metal cleanout door (about 10 by 12 inches) located in the foundation wall of the basement. In houses without basements the pit takes the form of a metal bucket that is lifted out through the hearth when full.

In recent years, placement of the ash pit has started to be located on the outside of the chimney. This arrangement is workable in houses with or without basements, providing access to the ash pit from the outside, where the ashes are ultimately disposed of anyway.

Chapter 17

Plumbing and Electric

The plumbing and electrical systems in any dwelling can be considered the actual lifelines of the house. Without them practically all modern conveniences would be impossible. Due to their importance, both systems are strictly regulated by local and national codes, and both are included in ambitious inspection programs required to ensure the occupant's safety.

PLUMBING

The entire plumbing system of a house can be broken into five basic categories: water supply, pipe types, fixtures, water heaters, and drainage.

Water Supply

Your water source can be either public, private, or a combination of both. Public water systems are the most worry-free, from a homeowner's point of view. Large water pipes called mains deliver the water directly to your house in practically all urban locations, ensuring adequate water pressure and supply. Large private water systems often do the same thing for subdivisions beyond the reaches of public water mains. These latter full-blown water systems, both public and private, have strict rules and codes to follow that guarantee proper

hookups to residential and other dwellings (Fig. 17-1).

Beyond the big water supply systems, things get personal. Since the first settling of America, private wells or springs have provided rural farms and homesteads with their water. Wells are not as regulated as large water systems, and if you need one it's best to consult local professionals familiar with the conditions and special situations that exist in your area. There are certain constants, however, you should be aware of when putting in a well as your main source of water supply:

1. Locate the well as close to the house as possible, yet as far away from any septic disposal system as practical—preferably uphill from it. You certainly don't want to contaminate your water with sewage leachate. Also consider the possibility of sewage systems that future neighbors might install to the sides or back of your property, and keep in mind that presently uncontaminated rivers, streams, lakes, ponds, and swamps might not always remain so. Try to position the well at least 50 feet away from any of them.

2. When the well is completed, have it tested for water flow, purity, taste, and even color. If any irritating characteristics persist, such as "hard" water (having higher than usual concentrations of dissolved solids),

Fig. 17-1. A water supply system.

a metallic taste, or a cloudy or off-color appearance, determine if those qualities can be effectively handled with a water softener/treatment unit that attaches to the well.

3. Protect the pipe that transports water from the well to the house from freezing temperatures. This can be achieved by burying the pipe below the frost line, lagging or wrapping it with insulation, or attaching electrically heated wire to the pipe (heat tape). It won't matter how good the rest of your plumbing is if the supply line freezes in mid winter. Frozen pipes are expensive to thaw and can cause a lot of damage if they burst.

Pipe Types

Cast-Iron Pipe. Cast-iron pipe might be heavy and awkward to handle, but it's so strong and durable it's been in general use for drain, sewer, and vent lines for years. Compared with plastic pipe, cast iron has two major advantages. First, it delivers its contents with much less noise. If you can't avoid running a waste pipe through a wall adjacent to a living or dining room, cast-iron pipe will prevent many uncomfortable moments you otherwise would be spending listening to water and waste gurgle through the line. Secondly, cast-iron pipe is more durable, and able to withstand the rigors of the "Roto-Rooter" and other mechanical and chemical pipe-cleaning equipment.

Galvanized Steel Pipe. Galvanized steel pipe is what was most commonly used for transporting water before copper lines gained widespread acceptance. It's available in various lengths and diameters, and has a rust-resistant coating on its inside and outer surfaces. Connections are usually made by cutting threads into the pipe, and because this is done after the protective coating is applied, any exposed threads will rust.

Because of its strength and wall thickness, it has a long life expectancy even when buried, and is frequently used to supply natural gas as well as water. It's still a popular choice whenever a "rough" plumbing line is called for. The principal drawback is the pipe's roughness, which contributes to water-flow friction and the collection of mineral deposits and sediments that over the years tend to reduce the inside diameter of the pipe in the same way that cholesterol can accumulate and restrict a person's arteries.

Copper Pipe. Copper makes an excellent pipe. It comes in soft, flexible, and hard varieties, has a very long life, and—if not spliced into galvanized steel pipe—is generally not affected by corrosion. It's the Number

1 choice to supply hot and cold water throughout today's modern houses. Copper lines are somewhat more expensive when compared with other piping, and so is the labor needed to install them. Individual pieces are connected by soldering them to copper unions, T's, elbows, and other fittings. A major advantage to copper pipe is that it can be bent or curved rather easily, a characteristic not available with cast iron, galvanized steel, or most plastics. However, due to the nature of copper pipes, they require special pressure chambers to prevent "water hammer"—a shuddering, rapping noise created by a sudden turning off (or on) of the water. Like most other pipes, copper can be ruptured by water freezing inside.

Plastic Pipe. Plastic pipe is substantially less expensive than copper, galvanized steel, or cast iron. Plastic pipe is also simple to install, and keeps labor costs low. There's no melting of solder with torches, and no measuring exactly and threading the ends of galvanized steel stock. There's only convenient plastic fittings, easy-to-cut plastic pipe, a can of plastic cement, and a brush. Almost anyone can do it.

There are other advantages to plastic pipe. It's extremely lightweight and easy to support. Plastic is chemically inert and unaffected by corrosive materials. The smooth inside surface of plastic pipe aids the movement of materials through the lines.

A few major disadvantages, though, eliminate the use of plastic lines from several parts of a house. It's not as strong as cast iron, galvanized steel, or even copper, plus it doesn't have the bending capability of copper. Because of its tendency to crack under heavy loads or stress, it shouldn't be used beneath or within a concrete slab where long, strong lengths of pipe having no splices are needed. As mentioned before, to the unwary, the noise that water makes when running through plastic lines can be a serious source of irritation if used above or near first floor living areas, especially in multi-level houses. Plastic pipe will also burst when water freezes inside of it.

A good-natured battle wages between the proponents of cast iron, galvanized steel, copper, and plastic for dominance over the entire house plumbing material question. The most sensible approach, and one favored by many residential plumbing experts, is to employ each kind of material where its strong characteristics can be used to best advantage. This means cast iron and steel for drainage pipes and for piping that's embedded in ground or concrete, copper for water supply, and plastic for venting.

Plumbing Fixtures

Plumbing fixtures include the water heater, sinks, tubs, showers, toilets, and even outside faucets. The time and effort you spend planning their locations will be the most important part of their installation.

A major consideration is where to position the water heater, and how to keep as much of the plumbing as possible concentrated in one area of the house. Hot water will cool off if it must stand in sizable lengths of pipe or travel long distances between supply and outlets.

For ease of installation and savings on materials, bathroom fixtures should line up along one wall, and, depending on the floorplan, the other side of such a "wet wall" full of plumbing could perhaps accommodate the plumbing support systems of a kitchen, a laundry, or another bathroom. Compact fixture placement will keep the plumbing clean and simple.

Another benefit of compact plumbing is that certain fixtures can share vents. Whenever possible, locate the toilet between the tub and sink, so the vent from the toilet can perform extra duty. Rookie bathroom planners sometimes ask builders to put the toilet on one wall, and the bathtub on the opposite wall. That causes unnecessary piping to be installed at an extra cost.

To save yourself from headaches later on, make certain every fixture can be turned off, preferably via a local or easy-to-get-at water shutoff valve, in case it's necessary to isolate the fixture so it can be serviced or repaired.

Don't ignore the economy that can be realized through careful planning of the house's entire plumbing system.

Drainage

Sufficient capacity and pitch, tight sealing, proper venting, and provisions for cleanouts are all important to a good, effective drainage system. Used water and the waste it carries must be disposed of, both for convenience and for the sake of your health. Waste creates unpleasant and potentially harmful gases that must be expelled. A drainage system therefore has two functions: to transport water and solid wastes to a sanitary sewer or septic tank, and to dispel noxious gases into air you won't be breathing.

If everything has been well planned, the waste water and materials will flow by gravity alone. Consequently, all drainage parts must be pitched or sloped downhill. These parts must be connected with special fittings and be large enough in diameter and smooth

enough inside, to prevent accumulation of solids at any point. This also means the fewer drainage parts, the better. It might not sound very exciting, but it's true: the entire plumbing installation is generally planned around the drainage system—to drain as many fixtures as possible with the same main pipe, and, the fewer main pipes, the better. This setup makes venting easier, too.

There are three topics important enough to discuss by themselves when it comes to drainage: the plumbing lines that drain the house's internal plumbing fixtures, the private, individual sewage system or septic tank, and the ventilation of toilets.

The Fixture Drainage Lines. All drainage lines must be pitched toward the sewer pipe. Tests have proven that a 1/4-inch pitch per foot of run will allow practically all waste to move freely, even though it doesn't seem like much of a slope. At the other extreme, a pitch of 45 degrees or more (1 foot of pitch per foot of run) provides a much stronger flow. If a straight run can't be accomplished with a sizable pitch of 1/2-inch per foot maximum, the largest part of the run should be made using the latter pitch and the remainder sloped at 45 degrees.

The slope and size of the trenches required to be excavated for the disposal lines are also important, and depend on the type of soil and the contour of the land. Ideally, the excavator will dig all the drainage trenches at the same time he is completing the foundation preparations.

Septic Systems. Because drainage systems that malfunction can affect the health of the entire community, there are usually stringent regulations governing private or individual sewage systems.

The best all-around individual sewage system is the septic tank (Fig. 17-2). The septic tank system consists of a house drain, a septic tank, an outlet sewer, a distribution box, and a disposal or leachate field.

The location of a septic tank and leachate field on a building site will be partly determined by the lay of the land, since the drainage must be downhill and away from the water supply. Also remember that you need access to the tank when it must be serviced or repaired.

The soil pipe is the main drain from the house to the septic tank. It's usually made from a sturdy 5-inch cast-iron pipe that's pitched at an ideal slope.

The septic tank will most frequently be tar-coated steel or concrete (Fig. 17-3). Concrete costs more but lasts longer. When raw sewage enters the septic tank, bacterial action breaks down the solids into liquids and gases that drain off through an outlet sewer into a box

Fig. 17-2. A septic tank system.

Fig. 17-3. A septic tank cutaway.

that distributes them into the disposal field. This field is an area beneath the surface of the ground that "absorbs" and "cleans" the waste through natural processes of decay. The size of a septic tank is important because a tank that's too small will accumulate sludge that can back up into the house.

If you will need a septic system, be careful when you select your building site. Building sites seem to be getting smaller all the time. Once wide-open land is being built up. The space required for safe waste disposal is dwindling, and in some close-quartered neighborhoods, nonexistent.

Toilet Venting. Gases and odors that cannot break through the toilet's trap water are instead vented from the house through piping that runs up into the attic then out past the roof. It's best to specify that the vent pipe or pipes should be installed near the rear of the house so they're not visible from the street (Figs. 17-4 and 17-5).

Water Heaters

Most water heaters are fired by either natural gas, oil, or electricity. A standard tank of 40 or 50 gallons is usually sufficient for a family of four living in a house having 2 1/2 bathrooms, but larger tanks are available for heavy water users.

A number of points regarding water tanks should be kept in mind:

1. Install them near the highest hot-water demand areas in order to reduce wasted energy from running the water long distances. If your house will have two equal demand areas, try to center the water heater between them. At the same time try to keep the tank near the furnace so both can be vented through the same chimney.

2. If you locate the water heater in an unheated area, put an additional layer of insulation on the sides and, unless natural gas is used as the power source, on the top as well.

3. A water heater tank must have a drain near the bottom so you can easily drain the tank when needed, and can periodically flush out sediments that collect at the tank's bottom.

4. When locating a gas water heater in a garage, keep it raised off the floor so when the pilot ignites the gas burners it doesn't cause any floor-hugging gasoline fumes from a lawn mower or auto to explode.

Plumbing Inspection

It's likely that the plumbing in a new house will be put together satisfactorily from a technical viewpoint because local codes usually specify that licensed plumbers must perform the work, and all of that work must then be inspected and certified. But no matter what the situation is with your house, it's still up to you to determine if the plumbing is installed in a fashion that you feel is adequate. Here are some guidelines to help you evaluate your plumbing system:

1. All piping must be well supported. Nothing should just hang there, wobbly or vibrating when liquids and waste materials pass through.

2. Pipes should run close to inside wall edges, as near to the ceiling as practical. When pipes must be run through the studding of exterior walls, they should be kept toward the interior edge of the studs to allow as much room as possible for insulation that is arranged between the pipe and exterior part of the wall.

3. Check to see if the plumber installs the correct number of shutoff valves and cleanouts as called for in the prints.

4. Before the walls and floors are closed up by the contractor, the rough plumbing must be inspected by the municipal inspector. You should also make a mental note (or drawings) of what's getting covered up.

5. The regular plumbing lines should be water tested and inspected for leaks, and if there's a gas line, it must be subjected to pressure leak tests as provided by the building code.

Plumbing Checklist

Few things in a house can be as irritating, as destructive, and as expensive to repair as plumbing. When it's working, it's taken for granted. When it's not, look out. Keep yours out of trouble by identifying and following the applicable items in this checklist. And also be aware of the choices you can make in the selection of various fixtures.

1. Plumbing Item Specs

—Sewer pipe from street to house: preferably cast iron or steel.
—Sewer pipe under the house: perferably cast iron.
—Exposed sewer and vent pipes: plastic, cast iron, or copper.
—Cleanouts in enough locations.
—Cold and hot water supply pipes: preferably copper.

Fig. 17-4. A drain/waste/vent plumbing system.

Fig. 17-5. Toilet venting.

—Natural gas piping: preferably "black" steel pipe.

—Toilet: specify type, manufacturer, model, and color.

—Bathtub: specify fiberglass, steel, or cast iron, manufacturer, model, and color.

—Molded shower/tub unit: specify type, manufacturer, model, and color.

—Bathroom vanity top/sink: specify porcelain or cultured marble or other, type, manufacturer, model, and color.

—Kitchen sink: specify stainless steel or enamel, type, single-, double- or triple-basin, manufacturer, model, and color.

—Faucets: specify single- or double-handle type, chrome, brass, or other, manufacturer, model, and color.

—Shower heads: specify type, manufacturer, and model.

—Garbage disposal: specify manufacturer and model.

—Dishwasher: specify manufacturer, model, and color.

—Clothes washer and dryer: specify type, manufacturer, model, and color.

—Laundry tub: specify manufacturer, model, and color.

—Water heater: specify electric, natural gas, or oil heat source, gallon capacity, and manufacturer.

—Water softener: specify manufacturer and gallon capacity.

—Water wells and pumps: specify size and type.

—Septic tank: specify size and type.

2. Plumbing Minimum Drain Specs

—Toilet: 3 inches
—Shower stall: 2 inches
—Clothes washer: 2 inches
—Laundry tray: 1 1/2 inches
—Tub and shower: 1 1/2 inches
—Bathroom sink: 1 1/2 inches
—Kitchen sink: 1 1/2 inches
—Dishwasher: 1 1/2 inches

3. Miscellaneous Plumbing Points

—You should be able to conveniently turn off the water or gas supply pipes to each plumbing fixture in the house with shutoff valves (Fig. 17-6).

—Include enough sill cocks. A sill cock is nothing but an outside cold water tap for lawn watering, car washing, and other outdoors use (Fig. 17-7). A typical

Fig. 17-6. Sink shutoff valves.

Fig. 17-7. A sill cock.

dwelling should have at least two of them, and very large houses as many as four—so you needn't rely on excessively long and unwieldy garden hoses. Have the sill cocks staggered around the house with one near the driveway. In cold-weather locations where freezing occurs, sill cocks should be of the freeze-proof variety.

—Locate a sill cock or water faucet in the garage, preferably on an inside wall bordering the heated part of the house.

—A neat way to arrange the plumbing needed by the clothes washing machine and laundry tub is to have the plumber install an in-the-wall-plumbing box (Fig. 17-8). This box will dress up the installation and provide protection to any plaster or dry wall that might be placed behind your laundry area. It's a good method to achieve the plumbing compactness mentioned earlier. All of the faucet, supply, and drain piping is conveniently tied together.

—Insist that single-unit bathtubs or shower stalls be installed while the house is being framed for a more custom built-in look.

—Vinyl, ceramic tile, and wood flooring should be installed in rooms that will receive plumbing fixtures. You'll have a much neater looking floor if these materials are laid down first. If carpet is the selected floor covering, the carpeting underlayment should be installed before the plumbing comes through, then the carpet itself can be laid after the fixtures have been set.

—To prevent cluttering up the appearance of the front of the house, have waste vents exit through the roof at the rear of the house.

ELECTRIC

If there is one system in a house that's the most dangerous to fool with or skimp on, it's the electrical. Like

Fig. 17-8. An in-the-wall plumbing box.

the plumbing in a house, the electrical system is taken for granted until something goes wrong. To properly understand your prints and drawings, familiarize yourself with the symbols on Table 17-1.

Appliance Wiring Checklist

When you think of it, there aren't many modern home conveniences that would work without electricity. How could we possibly get along without:

—Telephones
—Doorbells
—Refrigerators
—Freezers
—Furnace blowers

Table 17-1. Basic Electrical Symbols.

1.	$\ominus$	Is a duplex wall outlet. This supplies your power to lamps, vacuum cleaners, and other household devices and appliances.
2.	$	Means a single pole switch. This provides on and off control to outlets from one position.
3.	$_2$	Signifies a double pole switch. This provides on and off control from two locations. For example, it lets you turn off lights from either of two entrances to a room.
4.	S_3	Is a three-way switch.
5.	S_4	Is a four-way switch.
6.	$\textcircled{S}$	Is a pull-switch in the ceiling.
7.	S_{cb}	Is a circuit breaker.
8.	$\ominus_{wp}$	Means a weatherproof outlet.
9.	$-\bigcirc-$	Denotes a ceiling light fixture.
10.	$\boxed{TV}$	Is a television antenna outlet.

—Thermostat controls
—Cooking ranges
—Water heaters
—Televisions
—Video players and games
—Attic fans
—Garage door openers
—Dishwashers
—Clothes washers and dryers

and to a lesser extent:

—Hi-fi speaker systems
—Intercoms
—FM antenna hookups
—Burglar alarms
—Smoke alarms
—Garbage disposals or compactors
—Electric grills
—Sun lamps
—Cable television hookups

Clothes dryers and cooking ranges require special 240-volt receptacles. These receptacles differ slightly but significantly, so make sure the dryer and range receptacles will fit your appliances.

Electric water heaters are wired directly to the house's electric cable. Electric ranges are also often without cords and are wired directly to the main cable.

Electrical Outlets and Switches

The rule for electrical outlets is one duplex outlet for every 12 linear feet of wall, because lamps and household appliances usually have 6-foot-long cords. When a doorway comes between, the outlets around it should be located closer than 12 feet apart or you might have to use extension cords in that part of the room. Kitchen outlets are best located above the counter top to handle appliances safely. A duplex wall outlet for every 4 linear feet of kitchen counter space will do nicely.

Here are some other guidelines:

1. There should be light switches at every entrance to every room, and, for safety's sake, at the top and bottom of stairs and at garage and basement doors.

2. Insist on "intermediate" or "specification" grade outlets and wiring. There are also "competitive" grades of cheaper quality for only a minor difference in cost. Specification grade, the best of all, is usually

recommended for heavy-duty commercial installments, but is worth the extra cost if you want top quality, particularly in the kitchen and in all switches.

3. Wall switches should control overhead lighting. If a room has more than one entrance, install double or triple pole switches so these lights can be turned off and on from each doorway.

4. If a room has no overhead lighting, such as a living room or family room, wire at least one duplex outlet to a wall switch so a lamp that's plugged into the outlet can be turned on and off at the switch on a wall near the room's entrance.

5. Include at least two outside electrical outlets, and for a large, sprawling ranch home, up to four. They should be equipped with ground fault circuit interrupters for protection from the electrical hazards posed by wet lawns and driveways, as well as a weatherproof cap that covers each receptacle.

6. Install a ceiling outlet in the garage for an automatic garage door opener even if you're not planning to use an opener at first. If you decide to add such a convenience later, you can do it without having to spend a hefty price for a service call.

7. Be certain to plan several receptacles in the garage for power tools, an extra freezer, a vacuum cleaner, and other items.

8. All ceiling outlets should be securely anchored to hold the weight of any fixtures you might attach there.

9. There should be at least one duplex safety outlet above each bathroom vanity top for hair dryers, electric shavers, and other personal care appliances. Such outlets will prevent serious shock if you mistakingly drop or touch an electric appliance to water in the sink.

Lights And Lamp Wiring

1. Wire for overhead lights in the dining area, kitchen, kitchen sink and counter, laundry room, bedrooms, bathrooms, hallways, garage, and basement.

2. Avoid putting recessed lights in ceilings with unheated space above. They can't be properly insulated and will leak air badly.

3. Include recessed lights in all closets, with wall switches mounted outside the closet near the closet door (Fig. 17-9). Even shallow closets having bi-fold doors will need lighting on occasions.

4. Consider some outside spot or floodlights to provide illumination for general use of the yard and for security reasons. Ideally, some yard lighting should be

Fig. 17-9. Closet recessed lighting.

controlled from the master bedroom as an additional security measure.

5. Make sure there's wiring for lighting at the front and back doors and wherever someone can enter the house, with convenient switching at those locations.

Lighting should be wired near the main electric service panel so you can see the circuit breakers and their labels without using a flashlight.

Electrical Inspection

Here are some wiring features to watch for as the house nears the plaster/drywall stage, and how the electrical inspection proceeds:

1. See that in exterior walls, cables installed horizontally run along the top of the bottom plate. This provides the least interference with wall insulation. Most electric cable is sufficiently flexible to comply.

2. Make sure electric wiring around bathtubs is run below the top of the tubs to eliminate any chance of making contact with electricity by accident when long screws are used to fasten shelving, soap dishes, or towel racks to the wall.

3. Wires should not be placed too close to any plumbing to avoid damage from the plumber's propane torch.

4. After all the rough wiring is completed—known as the "roughing out"—and checked over, the electrician will contact the Underwriters Laboratory inspector who will inspect everything and, if satisfied, will issue a certificate and notify the utility company. The utility company will then send a representative to connect power to the house. At this point, at least one outlet will be in service.

5. The electrician can now wire all the outlet box receptacles as soon as the walls are insulated and covered. To finish his work, the electrician must know how you plan to complete the walls so that he can set the boxes to project the right amount through the finished walls.

Special Electrical Features

Remember to give some consideration to the following features:

1. Noiseless, no-click light switches. There are also silent touch-button switches that require only slight finger pressure. There are flat-plate models that are set flush against the wall. Slight pressure on the top turns a light on, and pressure on the bottom turns it off.

2. Ivory, colored, metallic, and other decorator switchplates are available instead of the standard plastic brown, off-white, or tan models that might clash with your decor. Electrical and lighting suppliers carry them in a wide variety of models and materials. These unique switchplates can give your house an interesting finishing touch overlooked in most dwellings today.

3. Dimmer controls enable you to adjust lighting intensity up or down according to your needs. Illumination in a room can be dimmed to a candlelight glow for a dinner party, kept subdued for television watching, or turned up brightly for reading. Dimmers are used chiefly for living and dining rooms, but can also be employed in a bathroom night light arrangement.

4. You can have built-in automatic switches that turn on the closet lights when the doors are opened, and then turn them off again when the doors are closed, similar to the refrigerator interior lighting setup.

5. Remote-control lighting is surprisingly simple to install. It can permit you to turn indoor and outdoor lights on or off from a central location such as the kitchen or master bedroom. A remote kitchen switch can control front door or garage lights, for example. A control panel next to your bed can eliminate that final tour of

Fig. 17-10. Snow-melting wire.

your house and grounds every night to turn out all lights.

6. Electrical snow-melting panels are available for sideways and driveways, and electrical snow-melting wire strung along your roof gutters will prevent the accumulation of dangerous snowdrifts and icicles on the roof slopes and gutters (Fig. 17-10).

7. No-shock outlets can be installed to prevent children (who frequently insist on jabbing hairpins or anything metallic into receptacles) from receiving shocks.

8. As already mentioned, safety-type grounded outlets can be added protection against shocks and help prevent the possibility of electrocution in a bathroom.

9. Even if you don't initially plan to have a security alarm system, the wiring for one can be easily installed within the walls at the time of construction. The system can then be inexpensively completed at a later date, if desired.

Chapter 18

Lighting

A long, long time ago, before Man discovered how to use fire, when it got dark, it got dark. There were no switches to flip on and no battery-powered flashlights to illuminate his pathways. Once Man mastered fire, he not only increased his comfort, safety, and gastronomical enjoyment, but the light he gained supplied him more time to constructively work with.

The first lamp could easily have been a lit pool of fat drippings that shimmered in a shallow depression near the glowing embers of a cave campfire.

Then clay and stone lamps came onto the scene— little more than saucer or cup-shaped receptacles for oil, grease, or tallow. They were either open or covered, with a carrying handle on one side and on the other a small trough or gutter in which a wick rested. This simple lamp persisted for about 10,000 years, resisting change even into the eighteenth and nineteenth centuries, when people were still using dangerous iron lamps that burned disagreeable-smelling whale oil.

Throughout the ages, light has always been something sacred to Man. It has been a symbol of religion, of ideas, of knowledge and understanding. It has enabled Man to take advantage of his most important sense: sight.

Lighting in these days, however, often takes a back seat. People will frequently buy lights because they fancy the fixtures, not the effects those features will provide. When you think of it, that's kind of silly. It's like buying food for only its looks, or for how it might conveniently fit into a refrigerator or freezer, with no regard to taste or nutritional value.

There's often little advance planning for lighting. While homeowners pride themselves in knowing about the last little detail about stereo systems, kitchen ranges, or video players, the lights that illuminate their private worlds are completely taken for granted. That's too bad, because precise lighting can create a variety of moods within a room, and can help exaggerate strong points while downplaying shortcomings. Lighting can increase efficiency in work areas. Consider how important lighting is to theater and dance. What show, what film, what live performance have you ever seen that didn't include strategic lighting as a major part of its overall effect? Consider museums and art galleries, and how they can make their showcase works stand out by merely lighting them properly.

The same principles hold true for lighting in home use, in the average household. Unfortunately, most people are afraid to make advance plans they feel might be too radical and difficult to change. Nothing radical is needed, however, to lend a home a pleasant lighting scheme. It doesn't take many built-in lights to add a distinctive touch and an interesting atmosphere.

TYPES OF LIGHT SOURCES

There are only two major kinds of electric lamps or "bulbs" as most people call them: incandescent-filament lights, and fluorescent lights.

Incandescent-Filament Lighting

The first practical incandescent-filament lamp was perfected by Thomas Edison in the late 1800s. An incandescent lamp produces light when its filament (usually made of tungsten) is heated by an electric current to a regulated temperature, at which point it glows and emits the amount of light it's engineered for (Fig. 18-1). The quality of light from incandescent-filament bulbs is warm in color and gives a friendly, homelike feeling to interiors. Beneath incandescent lights colors such as oranges, reds, and browns are enhanced, while the cool colors such as blues and greens are subdued.

These bulbs provide a source of light that can be focused or directed over a restricted area if desired, and because most incandescent household bulbs have the same size base, the lighting from fixtures or lamps can be increased or decreased within limits by changing to bulbs of different wattages. Most types of incan-

Fig. 18-1. Incandescent-filament light.

descent lights are less expensive than fluorescent types. And the incandescent lighting fixtures themselves are generally less expensive to purchase because they require no ballast or starter—a part of a light bulb that provides a high voltage for starting the bulb and a low voltage for running it.

Fluorescent Lighting

Fluorescent bulbs operate by an entirely different principle. They're typically long and tubular in design, either straight, circular—or angular in the case of neon signs used for advertisements. They're coated on the inside of the glass with phosphors, substances that give off light when subjected to ultraviolet radiation generated by a low-pressure electric charge. The low-pressure electric charge is regulated by the starter or ballast that forces the charge through a mercury vapor inside the bulb (Fig. 18-2).

Fluorescent bulbs offer greater lighting efficiencies, up to three or four times as much light per watt of electricity as incandescent bulbs have. They'll also last from seven to ten times longer, largely due to their cooler operating temperature range. Hot-burning incandescent lights, by their very nature, burn themselves out more quickly.

Fluorescent bulbs provide lines of light and are excellent choices in work areas where light coming from several angles is needed to effectively eliminate bothersome shadows. These lines of light are also frequently used over mirrors, kitchen work surfaces, in window valences, corners and covers, and other architectural features. Due to their long lifespans, fluorescent blubs are favorites in places that can't be conveniently reached for bulb changing. Circular fluorescent tubes have traditionally been used in high kitchen ceilings, for example.

When considering their use, remember that fluorescent bulbs aren't as quiet as incandescent lights. Most fluorescent models produce faint humming sounds, which admittedly are noticeable only if a room is library quiet or if you're listening for the noise.

HOME LIGHTING USES

Few homes will rely on one or the other type of bulb exclusively. Most dwellings employ a combination of incandescent and fluorescent fixtures. What follow are three different ways in which both types of lights can be used in modern dwellings.

General or Background Lighting

General or background lighting is a low level of illumi-

Fig. 18-2. Fluorescent light.

nation required for general living activities. It's the level of lighting provided by ceiling fixtures such as overhead lights in a bedroom, or from lighted valances (shield affixed to the wall that directs light from a source behind it upwards, downwards, or both ways), coves, cornices (shield affixed to the ceiling that directs light from a source behind it downwards), and wall lighting, or portable lights in groups of three or more that cast relatively low but adequate levels of light throughout a room.

Local or Task Lighting

Local or task lighting is the light you want focused on relatively small areas or limited areas used for specific activities such as reading, playing cards, typing at a computer keyboard, playing billiards, sewing, painting, shaving, cooking, and eating. These are the lamps and fixtures people are most familiar with, including floor, desk, and table lamps, and other lighting arrangements such as recessed ceiling downlights in a bathroom to illuminate a bathtub or hand basin.

Accent or Decorative Lighting

Accent or decorative lighting comes from fixtures planned to create certain moods and atmospheres, plus fixtures designed and positioned to emphasize artwork, plants, and any other items important to a person's lifestyle or interior decor. Many different kinds of fixtures and blubs can be used in creative ways here, limited only by the owner's imagination. Spotlights, floodlights, and all sorts of decorative lights, and even candles, will create a host of effects.

While planning your home you have to decide how much of each of the lighting types you need. Then look through designer books, magazines, and lighting sales literature for ideas, and visit actual houses that pay attention to lighting schemes, if possible. It doesn't take many well-placed lighting built-ins for a house to be considered unique, and such a home will gain a special quality not found in the vast majority of dwellings that treat lighting merely as a necessary after thought, not a part of the house to be carefully planned for maximum effect.

With general lighting, the light rays are usually diffused over a wide span. This diffusion is best accomplished by translucent glass or plastic. In a room in which the light is completely diffused, it will be coming from all sides, either directly or indirectly (bouncing off other surfaces) including the floor and ceiling, so no shadows are cast.

Dimmers can be useful to increase control over a variety of lights by taking charge of the voltage or current applied to the bulb or tube, thus giving a full range of light intensities.

LIGHTING TERMS

Lighting terms are constantly being tossed about by various experts and publications. Here are a few that you should be familiar with:

■ Semi-Direct—light is directed 60 to 85 percent downward.

■ Semi-Indirect—light is directed 60 to 85 percent upward.

■ Recessed—lighting fixtures affixed flush to a ceiling, wall, or other surface (Fig. 18-3).

Fig. 18-3. Recessed lighting.

■ Valance—light sources shielded by a panel attached parallel to a wall, usually employed across the top of a window. Valance lighting provides illumination both downwards and upwards, unless the valance possesses a top (Fig. 18-4).

■ Indirect—a system in which over 85 percent of

Fig. 18-4. Valance lighting.

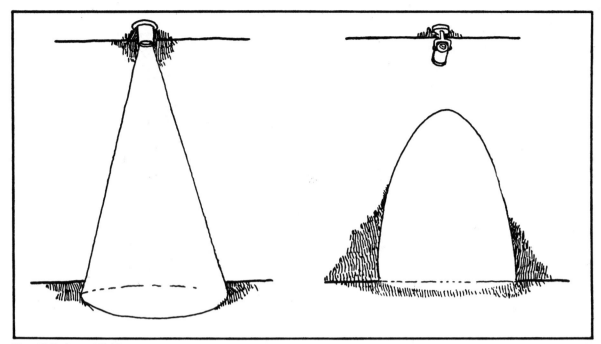

Fig. 18-5. Accent lights.

all the light is cast upwards toward the ceiling.

■ Direct—over 85 percent of all the light is cast downwards toward the floor.

■ Accent—directional lighting to emphasize **a** particular area or object (Figs. 18-5 and 18-6).

■ Cornice—light sources shielded from direct

Fig. 18-6. An accent light.

Fig. 18-7. Cornice lighting.

Fig. 18-8. Cove lighting.

254

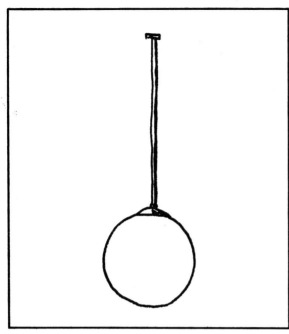

Fig. 18-9. General diffuse lighting.

view by a panel of wood, metal, plaster, or diffusing plastic or glass parallel to the wall and attached to the ceiling, and casting light over the wall through direct or downward lighting (Fig. 18-7).

■ Cove—light sources shielded by a ledge and casting light over the ceiling and upper wall, usually through indirect or upward lighting (Fig. 18-8).

■ General Diffuse—almost an equal amount of light produced in all directions, such as the light emitted from a suspended globe (Fig. 18-9).

■ Luminous—a lighting system consisting of a false ceiling of diffusing material with light sources mounted above it (Fig. 18-10).

■ Track—one electrical outlet supplying a number of separate fittings that can be positioned anywhere along a length of electrified track. It's very versatile for ceilings or walls, with vertical or horizontal tracks (Fig. 18-11).

■ Wallwashers—when installed about 3 feet away from a wall, wallwashers will light up the wall evenly from top to bottom or bottom to top, without spilling or wasting light away from the wall into the room. Angled closer to a wall of paintings or art groupings, wallwashers will

Fig. 18-10. Luminous lighting.

Fig. 18-11. Track lighting.

splash light onto varying surfaces, leaving interesting shadows contrasted in between.

■ Downlights—round or square metal canisters that can be recessed into a ceiling, semi-recessed, or ceiling mounted to cast pools of light on the floor or on any surface below them. They can be a spotlight, a floodlight, or an ordinary bulb. A spotlight will throw a concentrated circle of light. A floodlight will cast a wider, cone-shaped light. An ordinary bulb will provide soft, diffuse, all-over lighting.

■ Uplights—fixtures put on the floor, behind sofas, plants, or other appointments, under glass shelves, and in corners where they will lend a beautiful dramatic accent, bouncing reflected light off ceilings and into the room, creating moods that could hardly be imagined possible by day.

Chapter 19

Heating and Cooling

You could plan the most attractive and utilitarian home design imaginable, locate it on a site with a breathtaking view, and furnish it with the finest appointments, yet unless the temperature and humidity within such a dwelling are maintained properly, the house would be unpleasant to live in. Heating and cooling systems are probably the hardest working components of any home, especially in homes located in climates having temperature extremes.

For discussion purposes, heating systems can be broken down into two topics: types of fuel, and types of heat production and delivery systems.

TYPES OF FUEL FOR HEATING

There are six primary types of fuel available for home heating use: electricity, fuel oil, gas, coal, wood, and solar.

Electricity

Electricity is readily available almost anywhere.

Advantages:

1. It's a clean fuel that leaves no residue to contaminate the house or the atmosphere.

2. It requires no chimney to vent exhaust fumes and smoke.

3. It's a fuel that can be depended on in the future because it can be produced from almost any other type of fuel—gas, oil, coal, nuclear, solar, wind, and even geothermal.

Disadvantages:

1. Its fuel rate charge depends on the local utility company.

2. Breaks in the main supply line could leave you and your neighbors temporarily without supply.

Fuel Oil

Like electricity, fuel oil enjoys a comprehensive, efficient distribution that makes it readily available practically everywhere.

Advantages:

1. If the oil fuel burner is routinely serviced and adjusted for maximum efficiency, fuel oil is a relatively clean source of combustion.

2. The supply is stored on the property. Once it's in place there's no worry about it being cut off.

Disadvantages:

1. It must be stored in a tank that's usually located underground. The tank requires certain safety and fireproofing precautions, all of which combine to make the initial setup cost fairly high.

2. Money is tied up in the inventory stored.

3. The use of fuel oil requires a chimney.

4. If burned in an out-of-tuned unit, fuel oil will give off dirty emissions that will soil the interior of a home.

5. Fuel oil is not a renewable resource.

Gas

Gas comes in two forms: natural and liquid. Natural gas is not available in certain areas. Liquid or bottled gas can be purchased almost anywhere.

Advantages:

1. Gas is a clean fuel.

2. Natural gas requires no on-site storage.

3. The liquid type can be maintained on the property.

Disadvantages:

1. The natural gas supply and rate depend on the local utility company.

2. A liquid gas supply requires a storage tank.

3. Liquid gas ties up money in inventory.

4. The use of gas requires a chimney.

5. Gas is not a renewable resource.

Coal

Coal is an old fashioned fuel that, while remaining a popular fuel supply for the generation of electricity, is rarely used any more to heat individual homes. It's not readily available in all areas.

Advantages:

1. Its supply can be maintained on the property.

Disadvantages:

1. If you've ever read Victor Hugo's *Germinal*, then you know what a dirty, filthy substance coal is— even *before* it's burned.

2. It's dirty when burned, too, unless you invest in a few million dollars worth of emission controls.

3. Coal leaves large amounts of ashes after combustion occurs.

4. Naturally, the use of coal requires a chimney.

5. It's a hot-burning fuel that possesses above-average risk of starting house fires.

6. Its supply requires substantial storage capacity, as well as money tied up in inventory.

7. It's not a renewable resource.

Wood

The availability of wood depends on the local supply of hardwoods such as oak, ash, maple, cherry, and elm.

Advantages:

1. The supply can be maintained and even grown on the property.

2. If you have the equipment and time you can procure the supply yourself.

3. It's a renewable resource.

Disadvantages:

1. Substantial labor is required to place or have the hardwood put in storage, to move it to a stove or fireplace, and to remove the ashes after combustion.

2. If handling and preparing the wood yourself, there's danger involved with using chain saws, hand axes and mauls, and power splitters.

3. The inventory takes up a lot of room and ties up money.

4. The use of wood requires a chimney.

5. There's some risk of house fires.

6. Wood can bring dirt, insects, and fungi into the house.

7. The combustion of unseasoned wood can cause smoke to taint interior furnishings.

Solar

Solar energy depends on the amount of sunshine available.

Advantages:

1. It's clean.

2. There's no charge for sunshine.

3. It should be readily available for the next few billion years.

4. It lessens the reliance on endangered and imported fuels.

Disadvantages:

1. Unless substantial expense is incurred, not much long range solar energy can be stored up. Even at best, it's minimal when compared with the storage capacity of other fuels.

2. How often have you known our weatherman to be 100 percent accurate? There's an inability to predict with assurance the amount of sunlight that will occur during any one season.

3. Back-up systems are needed.

4. It's relatively expensive to plan and install a solar system.

TYPES OF HEAT PRODUCTION AND DELIVERY SYSTEMS

When it comes to selecting a heating system there are three major types to consider: forced-air, hot-water, and electrical. Beyond those are lesser used heat pumps and solar units.

Forced-Air Heating Units

These systems burn fuel to create a supply of warm air that's blown (or forced) by a fan through a duct system to various parts of the house. During its operation, this type of furnace heats air that's drawn from the rooms and then sends it back through ducts and registers out again into the rooms (Figs. 19-1 and 19-2). The sheet-metal enclosures that route the air back to

Fig. 19-1. Forced warm air heating system.

Fig. 19-2. Wall register with damper.

the furnace are known as the return ducts, and those supplying the air are supply ducts. Before the air comes back into the furnace from the return ducts it goes through an air filter where most of the fugitive air-borne dirt and dust is filtered out to protect the furnace from fouling its working parts.

Advantages:

1. Forced-air systems are very versatile. Air can be heated, cooled, humidified, dehumidified, filtered, and circulated, all through the same distribution system.

2. A forced-air system is quick to respond manually or automatically via the thermostat to temperature changes.

3. There's no freezing of pipes to worry about.

4. Forced-air units can operate with a pilotless ignition that uses an electrical spark generated only at the moment the fuel flow begins.

5. A forced-air unit requires simple, low-frequency maintenance. The filters need occasional cleaning or changing, the electric motors and a few moving parts need a few drops of oil periodically, and the blower belts need intermittent inspections. Only once every few years will a typical furnace *need* tuning up by a professional. In units that are heavily run, though, such tune-ups will usually save more in energy than they cost to have done.

6. The installation price of forced-air heat is fairly low.

7. The distribution system requires less floor or wall space when compared with radiators for water or steam heat or electric baseboard heat.

8. The forced-air system is adaptable to all six main types of fuel supplies.

9. Forced-air systems can be designed with more than one blower or with special devices so the heated air can be directed to particular locations, with each having its own thermostat. The ducts can also have fins inside them, or movable baffles that can be used to regulate the airflow to various rooms.

Disadvantages:

1. Some of the by-products of combustion—the unburned gas, oil, soot, or other contaminants—can come out through the registers. A poorly-maintained, dirty forced-air furnace, for instance, will inevitably transport small amounts of air-borne soot through the supply ducts and onto walls, floors, and furnishings. This process is likely to be so gradual and insidious that the graying of white walls, for instance, might only be revealed by washing a small area near one of the registers or by removing an old picture from a wall.

2. Another disadvantage of forced-air heating is heat cycling. Because the blower operates on an off-and-on basis, the room temperature varies near the thermostat setting from an over-warm to under-warm condition and then back again. There's a constant over-compensating going on.

3. If a blower motor is too large, or a duct system has not been properly designed, the movement of air in a room can be uncomfortable.

4. The sounds of the blower unit can be conveyed all over the house through the ductwork.

COPPER TUBING WITH FINS

Fig. 19-3. Hot water radiator.

Hot-Water Heating Units

With this type of system, water is heated in a boiler by a gas, oil, or other burner and then pumped throughout the house via a piping network equipped with radiators that ultimately distribute heat into the rooms (Fig. 19-3). Two basic types of radiators are used. One is the familiar under-the-window style modernized with an attractive grilled cover (Fig. 19-4), and the other is the low-profile baseboard radiator (Fig. 19-5). The disadvantages of the former are that they jut into a room and

Fig. 19-4. Radiator.

Fig. 19-5. Cast-iron baseboard radiator.

interfere with furniture placements and drapes. If recessed into an exterior wall, the available space remaining for insulation becomes inadequate. The baseboard system gives more outside wall coverage and efficiency. For these reasons it's generally the favored of the two. The best baseboard models are the long, low kind about 6 to 10 inches high and up to 10 feet long, and longer. They're often made with copper or aluminum heating fins. The most elite, quiet, and expensive baseboard units are the cast-iron makes.

Advantages:

1. Hot-water heat is a clean, effective, and fairly quiet way to heat a home.

2. Hot-water heating systems are adaptable to many of the more-popular solar heating arrangements.

3. Hot-water heating can be zoned to allow you to maintain different temperatures throughout the various parts of a house.

4. Hot-water heat is a more even heat than the heat supplied by forced-air systems.

Disadvantages:

1. Humidifiers, dehumidifiers, air cleaners, and air conditioners must be added as separate systems, thus incurring extra expense. There's also no ductwork already in place to provide for the movement of air needed by such add-ons.

2. Hot-water units can be difficult to keep clean. The insides of the pipes can develop mineral deposits that reduce heating efficiencies.

3. Some hot-water systems tend to rattle and clink when the heat goes on and off.

4. If a pipe breaks, a whole floor could get soaked.

5. The actual production of heat is not spontaneous; it takes time for the process to start up and become operational. The complete cycling of the water takes time initially when compared with the immediacy of forced-air heat.

6. This type of heating system is generally more expensive than forced-air units to purchase and have installed.

7. Having to keep furniture away from the radiators can be annoying.

8. If a home with hot-water heat will be idle during periods of freezing temperatures, there's the possibility of frozen and broken pipes, or else the entire system might have to be winterized.

Electrical Heating Units

Another popular but less used heating system than forced-air and hot-water, is electrical. Straight electrical heating systems use electricity directly as a heat source. In essence, electricity is converted to heat when it moves through conductors that resist the flow of current. The conductors or heating elements become hot and give off heat. The heat is then typically distributed via baseboard heaters that come in a wide range of sizes (Fig. 19-6), with different output ratings so they can be used for general heating or for supplementary purposes. It's also possible to install small electrical heating panels in a wall.

Advantages:

1. An electric heating system is one of the simplest and least expensive heating systems to install. There's no ductwork, plumbing, or expensive furnace or boiler needed.

2. It's an unobtrusive and practically silent heat.

3. Depending on the circumstances—where and how it's used—electric heat can be very cost efficient. Each room can have its own control, permitting variations in the amount of heat provided. If these controls

are judiciously used, the cost of operation can be quite low.

4. Maintenance is practically non-existent. There's no furnace or boiler to service, repair, or replace.

5. Electric heating units offer the cleanest heat available.

6. Because there's no fuel burned, there's no need for a chimney.

7. Electric heat is adaptable to wind and other locally powered electrical sources.

Disadvantages:

1. Electric heat is a very dry heat. Humidification is usually necessary during times of low humidity.

2. Sometimes electrical heating element surfaces get hot enough to pose a danger to young children.

3. The practical justification for electric heat can depend heavily on what the local electric rates are in the area: how they compare with other fuels economically. And historically, they haven't fared very well.

The following two heating systems are not commonly used, but you should at least be aware of how they operate.

Heat Pump Units

In areas having mild winter weather this system can make sense. It employs available cool outdoor air to chill refrigeration coils. The liquid in the coils compresses and in the process heat is created and given off into the house. In summer the entire process can be reversed and used to cool the home. Electricity is the fuel used to create the mechanical energy that then converts unusable heat to usable heat.

In nontechnical terms, a heat pump can be described as an air conditioner that runs two ways. During warm weather it removes heat from the inside air and pumps it outside. In cold weather the unit takes heat from outside air and brings it inside (there is, believe it or not, heat in winter air). The efficiency of a heat pump falls off as the outside temperature drops, so the most practical design includes conventional electrical resistance elements that come into play automatically should outside temperatures fall to about 15 degrees Fahrenheit or below.

Fig. 19-6. Electric baseboard heating unit.

Solar Units

A typical solar heating system that uses the sun for an energy source includes four major components:

■ Collectors to harvest sunlight and convert radiant energy to thermal energy.

■ A storage medium that can hold enough heat to last through a night or through periods of cloudy weather.

■ A distribution system to convey the heat to points of use.

■ A back-up system that will take over when stored solar heat is exhausted because the sun has not been cooperating.

The collector is often a heat-absorbing metal plate with an array of tubing that can be an integral part of the plate or merely bonded to it. Usually this assembly is in a black frame with a layer of insulation at the back and a glass or plastic-covered air space on the surface that should face the sun.

The heat transfer can be made through liquid or air that passes through the collector, picks up heat, transports it to storage and later distributes it through the house.

COOLING SYSTEMS

Because all of the same ductwork will be shared, most forced-air heating units are readily adaptable at a modest surcharge to include mechanical air conditioning. The heat pump automatically includes it because the same equipment is used for both heating and cooling operations. If your primary heating system is not one requiring ducts, and you still want central air conditioning, it will have to be a separate system.

There are a number of important points that shouldn't be overlooked when you're considering the installation of air conditioning:

1. When comparing bids from air conditioning contractors, look at the seasonal energy efficiency ratings (SEERs). That's how the various units are compared. Ten and above is excellent. Eight and nine are good, and six and seven, acceptable.

2. Don't oversize a system in hopes of getting quick cooling. If oversizing is much greater than 15 percent, a fast cooldown of the air will occur, but without efficient moisture removal. The result will be cold, clammy, very uncomfortable air.

3. If possible, locate outdoor compressors away

from decks, patios, bedroom windows, and dryer vents. Also avoid interior corners that tend to accentuate the compressor noise. Try instead for a shaded area out of the limelight. That way the noise won't bother anyone and no direct sunlight will unnecessarily increase the compressor coil's workload.

4. If you're not starting out with air conditioning and you plan to add it later, arrange for a large enough heating equipment room so the air conditioning coil, air cleaner, and humidifier can be comfortably installed.

5. Provide a means of draining condensation water from the coil to a nearby fixture.

6. If you plan on expanding your home at a later date, make certain the size of your air conditioner will accommodate it.

7. If you're not going to have a forced-air system with ductwork, consider having the proper ductwork for air conditioning installed when the house is being constructed even if you're going to wait to have the air conditioning put in.

DUCTS

Two materials are generally used for making the ducts for forced-air heating and independent air conditioning systems: fiberglass and galvanized sheet metal.

Fiberglass duct board makes a quiet system. It's usually fabricated in a shop and assembled at the job site with staples and duct tape. This type of construction doesn't give the rigidity nor the ruggedness of galvanized sheet metal. The fiberglass costs about the same as the galvanized wrapped with insulation.

Galvanized sheet metal is also made up in a shop then brought to the job site for final installation. Sheet metal ducts are put together using stronger fasteners, including rivets. By itself, a sheet metal duct tends to be noisy. The noise can be substantially reduced, however, by lining the duct interiors with insulation. The insulation brings another benefit by providing thermal support along the ductwork and prevents heat loss from the ducts wherever it's installed. In the interest of economy, costs may be held down by lining only the sections that produce the most noise. Listed below are the areas that need to be lined:

■ The first 6 to 10 feet of return duct from the return air grille toward the furnace.

■ Supply ducts from the furnace through the first "T".

■ All corners.

■ Ducts mounted outside the heated parts of the

house. These sections should have at least 2 inches of insulation applied.

■ All of the open seams in ductwork need to be sealed with reinforced duct tape.

Whenever practical, run the ducts within heated areas. This will substantially increase the amount of heat retained within the ducts, and will place more heat where you want it at a lower cost. If the ducts run through heated areas, even what little heat that escapes the ducts will not be wasted.

If there's a basement, make sure a few heat outlets are placed there to cut down on dampness. They can simply be adjustable louvers installed in one of the main heat supply ducts.

REGISTERS

The purpose of heating and cooling outlets or registers is to supply the heat or cooled air where it's most

needed, usually near the exterior walls and windows (Fig. 19-7).

Standard practice finds at least one outlet located at each exterior wall in a room. Two or more are often needed in large rooms or along extended walls beneath long picture windows. What's needed is a "curtain" of warm air or cold air thrown up (or down) from the registers around the perimeter of the house, between the occupants and the outside temperatures. The heat sources work best when they're located along the bottom of a wall, beneath the windows. This position counteracts the most likely places where cold might seep into the room. In warm-climate areas the registers can instead be located in the ceiling, directly over windows on exterior walls, because cool air falls.

Here are some other considerations:

1. Make sure that no registers contribute to unwanted noise transmission between rooms. Two-sided or double-opening registers that serve two rooms at

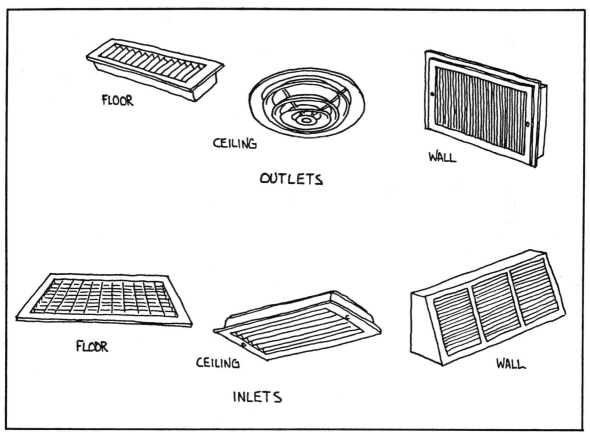

FLOOR

CEILING

WALL

OUTLETS

FLOOR

CEILING

WALL

INLETS

Fig. 19-7. Registers.

once are fine in certain instances, but be on guard against any doubling up that violates someone's privacy. These double registers can be a big source of inter-room noise transmission.

2. Floor-mounted supply registers are more efficient than the baseboard type. It remains a good practice, however, to use the baseboard registers in rooms such as kitchens, bathrooms, and laundries to prevent water, wax, or other materials from entering the duct system.

3. Return registers are best mounted on the walls in the interior of a home. They should be positioned near the floor in cold-climate areas and along the ceiling in warm-climate locations.

4. After the completion of the heating/cooling rough-in, the heating contractor should install temporary covers over all outlets and registers, both supply and return, to prevent debris, dust, and other materials from getting into the ductwork during the remaining construction.

5. The more areas that can be tapped for return air, the faster and more efficient the system will be. The best systems include many more than the minimum number of inches required for return air.

THERMOSTATS

Thermostats shouldn't be placed where they'll be affected by drafts, or in the hottest or coldest parts of a home, or in places where they'll interfere with furnishings and decor. Place thermostats on inside, not outside walls. If you go with zoned hot-water heat or have adjustable fans on your forced-air system (dual-control fans), consider having more than one thermostat strategically located in the house.

NOISE

A noisy heating or cooling system can be annoying. There's no justification for such noise, except that it costs the manufacturer a trifle less, and saves the builder a trifle more.

Ask to be taken to a home that already employs the heating/cooling system that you're considering. Listen to it start, operate, and stop. Be sure you do all three because sometimes a burner fires with a bang, or pops when the flame goes out. Sometimes the motor that runs a circulating pump or blower is noisy. Fan noises can be transmitted through the ductwork (like through an old-fashioned speaking tube), and quite often the noises created in boilers are carried throughout the house in the metal piping.

How much attention you pay to these noises depends on your sensitivity to disturbing sounds.

SIZE AND LOCATION

You shouldn't go with an oversized furnace or air conditioner unless it's in anticipation of a room or space you plan on adding at a later date. An oversized unit will run inefficiently in a smaller space than it was engineered for.

Try to locate the furnace/air conditioner in as central a location as possible to get an even distribution of heat and coolness. The air conditioner's outside compressor, as mentioned earlier, should be installed in an out-of-the-way place in the shade where it won't annoy anyone.

AIR CLEANERS

Forced-air heating and air conditioning systems are ideal for the application of electrostatic air cleaners. They not only clean the air, but tend to make it smell fresher too, which is a feature that's becoming more important with today's comprehensive sealing of houses. The accompanying reduction of air infiltration has actually reduced the amount of fresh air that used to move into homes through various uninsulated surfaces and cracks in the outer shell.

The other types of heating systems aren't so accommodating to the air cleaners, and separate units must be installed in individual rooms.

HUMIDIFIERS

Heat tends to dry out the air inside a house and make it uncomfortable. This is particularly true of forced-air systems. If mechanical means for supplying moisture are not available, the humidity inside the home can drop to the level of dryness found in a desert. This condition causes shrinking of wood and other materials, which might open gaps in the house structure and shell and permit air infiltration.

As with air cleaners, humidifiers can be installed directly onto a duct system. They can be hooked up to a constant water supply so they'll continually and automatically put needed moisture into the air. For houses heated with systems other than forced-air, humidifiers must be placed in appropriate areas throughout the home to accomplish the same thing.

Dehumidifiers can also be needed in certain locations and times of the year, usually in basements during hot and wet weather conditions. Individual units run by electricity will usually do the job.

Chapter 20

Insulation

There aren't many people who live in a climate where they don't have to worry about protecting themselves from temperatures that are periodically too hot or too cold. Since the oil embargoes and the realization that many of the energy sources on which we depend are not unlimited, insulation in houses has taken a position of high priority no matter what the location.

In its most universal application, insulation belongs inside or against any barrier located between a heated space and an unheated space, or between a cooled space and an uncooled space. Applied to the structure of a house, this means that insulation should be within all exterior walls, in attics or under roofs, beneath floors exposed to the outside as well as floors covering unheated crawl spaces or slabs, and in extremely cold locations, on walls in a heated basement.

In short, insulation should envelope all living areas of your home, leaving no openings except doors, windows, and necessary vents. As discussed in the heating and plumbing chapters, the heating or cooling ducts and hot water heater and pipes must be insulated to provide energy efficiencies, especially when such ducts and pipes pass through unheated or uncooled spaces.

Fortunately, sealing a home against heat conduction and air infiltration is a relatively simple process if

done during construction of the house, and it's fairly inexpensive, considering the energy saved in the long run. If insulation is not properly planned, and must be added at a later date, after the house has been completed—now *that* can run into big problems and big expenses.

Most insulation is installed after the framing is complete and the electrical, plumbing, and heating systems have been roughed in and inspected.

R-VALUES

Insulation quality is often expressed in "R-values." An R-value is merely a quantitative expression of the ability of any material to resist the passage of heat. For example, a fiberglass batt 6 inches thick has an R-value of approximately 19. The same material in batts 12 inches thick possess an R-value of about 38. The greater R-value, the greater the material's resistance to the passage of heat (or coolness), and the better its insulation value.

Table 20-1 gives an insulation materials comparison, Table 20-2 lists some of the same kinds of insulations, and shows what their most common thicknesses equal in insulating values.

Table 20-1. Comparison of Insulation Materials.

Form	Type	Approximate R-Value Per Inch of Thickness	Relative Cost 1 = least, 5 = most
Blankets &	Fiberglass	3.1	1
Batts	Rock Wool	3.7	1
Boards	Fiberglass	4.5	5
	Polystyrene	3.5 to 5.4	5
	Urethane	4.5	5
Foam	Urea-Formaldehyde	4.2	5
	Urethane	4.5	5
Loose Fill	Cellulose (blown)	3.6	2
	Rock Wool (blown)	2.9	1
	Fiberglass (blown)	2.2	1
	Vermiculite (poured)	2.1	4
	Perlite (poured)	2.7	4

Because heat rises, the potential heat loss in a house is greatest through the roof, and the least through the floors. Therefore, different recommendations exist for insulating those areas as well as the walls—each with a different, appropriate R-factor.

Here are some conservative recommended R-values for your house: walls—R-20; floors—R-27 (above grade), which includes overhangs, cantilevers and below projecting windows; and ceilings—R-40.

U-VALUES

On occasion you might find the thermal qualities of an insulation material expressed in terms of "U". The U-value is the reciprocal of the R-value, and can be determined by dividing the R-value into the numeral value. For example, a fiberglass batt having an R-value of 19 has a U-value of 1/19, or .053. Basically, the lower the U-value, the greater the thermal resistance of the material, and the better its insulation quality.

TYPES OF INSULATION

While it's true that each material used to make up your house possesses some insulating value, the effectiveness of individual types of materials vary greatly. For example, a 1-inch-thick blanket of fiberglass insulation has the same insulation value as approximately a 3 1/2-inch-thick layer of pine wood planking, a 22-inch-thick wall of common brick, a 40-inch-thick layer of solid concrete, or a 54-inch-thick (4 1/2 feet!) layer of stone.

The most popular types of insulation used in modern homes are blankets and batts of fiberglass and rock wool, rigid boards of polystyrene and urethane, asphalt-

Table 20-2. Thicknesses of Various Insulations and Insulating Values.

R-Values:	R-11	R-19	R-22	R-30	R-38
Fiberglass blankets/batts	3 1/2"-4"	6"-6 1/2"	7"-7 1/2"	9 1/2"-10"	12"-13"
Rock wool blankets/batts	3"-3 1/2"	5"-6"	6"-7"	8"-9 1/2"	10 1/2"-12"
Fiberglass loose/blown	5"	8 1/2"	10"	13 1/2"	17"
Rock wool loose/blown	4"	6 1/2"	7 1/2"	10"	13"
Cellulose fiber loose/blown	3"	5 1/2"	6"	8 1/2"	10 1/2"

Fig. 20-1. Blanket insulation.

impregnated fiberboard, fiberglass sheathing, rigid foam boards, foams of urea-formaldehyde and urethane, and various loose forms of cellulose, fiberglass, rock wool, vermiculite, and perlite.

Fiberglass and Rock Wool Blankets and Batts

These two products make up about 90 percent of all homeowner insulation (Figs. 20-1, and 20-2). They're made of compressed fibers that come in a continuous

Fig. 20-2. Batt insulation.

roll form (blanket) or in rolls having perforations along every few feet or yards so you can pull off regular rectangular pieces (batts). Both blankets and batts are available in various thicknesses and widths, and with or without facing material or vapor barrier material on one side.

Fiberglass and rock wool also come in shredded forms for hand pouring or machine blowing applications. Be aware that both of these materials can be irritating if they come into direct contact with your skin. Rock wool is the less bothersome of the two.

Fiberglass and rock wool insulation can be used throughout the entire house, for ceilings, walls, floors, basements, around windows and doors, or anywhere else energy might otherwise be lost to conductivity and air infiltration. They both have good insulation values and are among the most economical types to purchase.

Rigid Boards

These are primarily used as sheathing beneath exterior siding, as underlayments for roofs, and around foundations (Fig. 20-3). Rigid boards are manufactured in a variety of materials:

Polystyrene Rigid Boards. Polystyrene is a plastic that, in rigid board form, dents easily and is highly combustible, but is also very weather and moisture resistant. It's excellent for below-grade or exterior wall applications. Used indoors, it's a fire hazard unless covered by 1/2-inch-thick sheets of gypsum wallboard.

Urethane Rigid Boards. Urethane is a compound that, in rigid board form, has good insulative value, but gives off poisonous gases if ignited.

Asphalt-Impregnated Fiberboard. This rigid insulation is one of the oldest types available. It has a relatively low insulation value when compared to other insulations. Old-timers might advise you to use it, but you'll be better off with one of the more modern varieties.

Beadboard. This rigid insulation consists of polystyrene beads fused together. It's easily dented and must be handled carefully. In this form, it has a average insulating value when compared to the other rigid boards.

Fiberglass Sheathing. Fiberglass sheathing is made of compressed fiberglass wool sandwiched between tough facing materials that together form a semi-rigid board. Its insulation value is average among the rest of the rigid board insulation family.

Rigid Foam Boards. These boards have the highest insulating qualities of all rigid board materials. They form their own vapor barriers, too. Foam board dents easily and must be handled gently. Like many other rigid board insulations, because of its flammability, it should be covered by 1/2-inch-thick gypsum board on interior walls and ceilings.

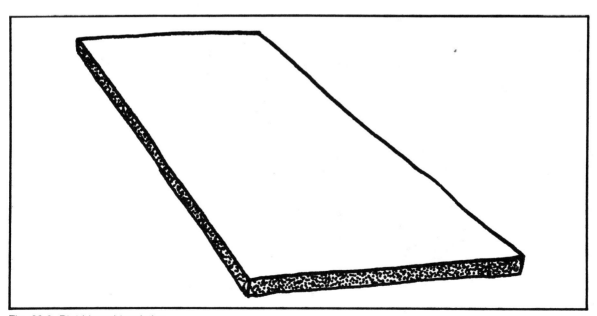

Fig. 20-3. Rigid board insulation.

Foams

There are two primary types of foam used for insulation: urea-formaldehyde and urethane.

Urea-Formaldehyde Foam. This clinical-sounding material is both expensive and effective, with some reservations. It's excellent for insulating walls because not only does it have high R-values (the measurement of the effectiveness of insulation), but it fills wall cavities totally and, once inside, creates its own vapor barrier. It's very fire resistant, too. On the minus side, urea-formaldehyde foam carries a slight embalming-type odor that can linger for a considerable time. Some engineers say that its insulating value tends to slowly decrease with time as the compound deteriorates.

Urethane Foam. This type of insulation also completely fills the spaces wherever it's installed, and also creates its own vapor barrier. Although urethane foam has one of the highest R-values of all insulation, because of the toxic gases it emits when burned, it's not recommended for use in walls.

Loose Insulation

There are three main types of loose insulation that can be hand poured or blown into walls and other voids: fiberglass and rock wools, cellulose, and vermiculite and perlite (Fig. 20-4).

Fiberglass and Rock Wools. By far the most popular loose insulation. It's efficient and inexpensive.

Cellulose. Cellulose is made from recycled paper products. It's a shredded insulation that's usually machine blown throughout the voids targeted for filling. Because cellulose fibers have natural air cells or pockets, cellulose has relatively high R-values. On the negative side, in walls where gravity can play a continuing role, this loose insulation tends to settle over the years and might leave the upper reaches of a wall uninsulated.

Vermiculite and Perlite. These materials are not commonly available, nor are they as commonly used as other kinds of insulations. Vermiculite is made from expanded mica, while perlite is a type of volcanic rock.

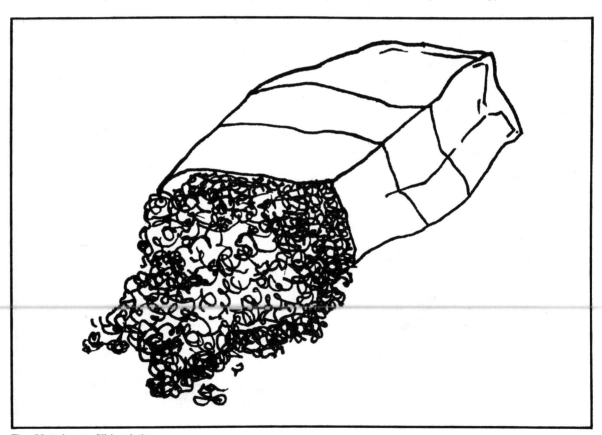

Fig. 20-4. Loose fill insulation.

They come in loose-fill granules (the smallest available) to be poured between and into small, hard-to-reach places. Although ideal materials for filling difficult voids, their R-values are relatively low. At the same time, vermiculite tends to absorb moisture and becomes mushy—thus decreasing its already low insulation value.

PLACES TO INSULATE

Even if you live in an area having a mild climate, your home will be more comfortable if it's well insulated. It's not possible to have too much insulation in the walls and top floor ceiling, or in floors over carports, garages, porches, and other areas exposed to the weather, provided the insulation is installed correctly.

There are six primary areas that should not be overlooked: basements, floors, walls, attics, ducts and plumbing, and cracks and joints that require caulking.

Basements

Basement or masonry walls may be insulated by first putting up furring strips and then inserting insulation blankets or batts in the usual way. Another method is to install rigid board insulation that's attached to wood nailing strips, which are bolted to the walls. After basement insulation is put up, it should be covered with a gypsum wallboard or another approved wall covering.

In general, rigid sheets of foamed cellular plastic or cemented-together particles or fibers are the most efficient per unit of thickness. They're good to add to any masonry or concrete surface.

Floors

Prudent insulation of floors can save 5 to 15 percent of your heating costs. It's easy to insulate under floors during construction, and not very expensive. Here's what to consider:

1. Insulate between the top of the foundation wall and the still plate under the first floor decking (Fig. 20-5). This will stop air infiltration more than it "insulates," which is important when it comes to overall energy efficiency and savings.

2. The subfloor junction with the sole plate should be sealed. The framing crew can place a double bead of caulk on the subfloor before the sole plate is put in place.

3. Underfloor insulation is important unless the first floor sits over a heated basement. A floor over an

Fig. 20-5. Sill plate sealer.

Fig. 20-6. Insulation cutaway.

unheated basement or crawl space can be easily insulated by laying blanket insulation with a vapor barrier that faces up (Fig. 20-6). To insulate over a concrete slab, lay a hard, rigid board such as plywood over the top of the slab.

4. Seal where pipes, electric wires, or telephone and television cables penetrate the sole and top plates.

5. Seal around plumbing drains, and beneath bath tubs by blocking large openings with pieces of construction sheathing and sealing the remaining cracks with expanding foam.

6. For soundproofing, insulate the floors between a first and second story. You can use blanket, batt, loose fill, or foam insulation for sound control.

Walls

1. Insulation should be stuffed into the cracks and small spaces between rough framing and the jambs, heads, and sills of windows and doors. Also fill spaces behind conduit, electrical outlets (notorious energy wasters), switch boxes, and other obstructions and built-ins.

2. The openings formed near corner studs, cavities and T-junctions created during the framing construction should all be filled with insulation before the entire framing is completed. Otherwise it will be very difficult, if not impossible, to get insulation into some of those places.

3. The spaces between wall studs should be lined or filled with blankets or batts during framing, or with loose fill or foam once the dry wall or plasterboard is up. The insulation should be placed in the stud cavities with a vapor barrier facing the heated or cooled area.

4. For additional R-value use insulated sheathing or apply rigid board insulation to a regular sheathing that might have a low insulation value. An insulated sheathing, though, is by far more economical.

5. Interior walls should be insulated if you want soundproofing. Almost any insulation will work for controlling sound between rooms.

Fig. 20-7. Ceiling insulation.

6. If the exterior walls of a house are wrapped with perforated aluminum material or building paper prior to the application of the siding, air infiltration will be greatly reduced.

Attics

Particularly in a single-story house, most heat loss occurs through the attic. By installing sufficient attic insulation you can cut up to 30 percent of your fuel bill.

1. The most common method of insulating an at-tic is to staple blanket insulation between the ceiling joists with the vapor barrier on the side closest to the home's living areas (Fig. 20-7). Another popular way to insulate an attic is to put down loose fill or foam insulation once the ceiling dry wall or plasterboard has been put up.

2. If you have cathedral ceilings, the proper way to insulate is to install batt material between the roof rafters against the underside of the roof sheathing or deck before the installation of dry wall or plasterboard (Fig. 20-8). There should be an air space between the

Fig. 20-8. Cathedral ceiling insulation.

top of the insulation and the underside of the roof sheathing or deck. This space is needed to ensure proper ventilation.

3. Any attic overhead pull-down doors should be hinged on one side of the door frame and equipped with springs so that when closed, the door can be snugly secured. The insulation that fits over the attic side of such a door should be able to be pulled over the opening before the door is closed, unless it's permanently attached to the top of the attic door itself. There should also be a rubber seal around the door frame to stop any possible air infiltration.

Ducts and Plumbing

By insulating heating and cooling ducts, pipes, and water heaters, you can reduce your annual heating costs by as much as 10 percent.

1. Wherever heating ducts run through unheated parts of a house, such as through attics, basements, or garages, they can waste as much heat or more than they deliver. That means the occupants pay for a lot more heat than they receive. One- or 2-inch-thick blanket insulation is manufactured in sizes designed for wrapping around ductwork and pipes (Fig. 20-9).

2. Water pipes need insulation, too, especially cold-water pipes exposed to freezing temperatures in winter and hot-water pipes installed between a water heater and fixtures. The pipe insulation will also muffle gurgling noises and will prevent the irritating problem of condensation during warm temperatures when cold-water pipes tend to "sweat" and drip condensate onto basement floors, carpeting, and furnishings (Fig. 20-10).

3. A 2-inch-thick blanket insulation, available in a kit form, can easily be applied to any water heater to reduce heat loss in a cold basement.

Caulking

A house is made of many different materials. With age, temperature changes, vibration and general wear, cracks will develop where the different construction materials meet. Even when doors and windows are weatherstripped, air can infiltrate through these other cracks and joints in the floors, walls, and roof. Individually, these small cracks might seem insignificant, but together they can cause chilly drafts and raise heating or cooling costs. They also invite insects, leaks, and rot-causing moisture. Although you can't expect to seal up your house completely, by caulking it thoroughly you can curb most of the drafts and heat loss. It's a simple and inexpensive process, yet it can cut up to 10 percent off your heating bills.

The caulking itself is an ideal solution for sealing narrow gaps and cracks. It's a soft, rubberlike material that will conform to any opening, will stay supple, and won't crack or deteriorate for at least several years, or—depending on the type of caulk used and the application—for many years.

The following list presents key areas that should be caulked in the typical house:

■ Around window and door frames.
■ Where different siding materials meet.
■ Anyplace where a crack occurs between brick or pieces of siding.
■ Between the foundation and the house framing walls.
■ Around the outside of an air-conditioning unit that protrudes from a window.
■ Around heating and cooling ducts and pipes running from a separate outside central air-conditioning unit.

DUCT   DUCT TAPE   VAPOR BARRIER

Fig. 20-9. Duct insulation.

Fig. 20-10. Pipe insulation.

- Around roof flashing, vents, and pipes.
- In gutter and downspout joints.
- Around vents and fans.

VAPOR BARRIERS

In addition to insulation, a home's living areas should also be sealed with appropriate material (called a vapor barrier) applied along the inside of the studs, the ceiling joists, and the floor joists—mainly to prevent the movement of moisture from the living areas into the insulation. Insulation loses some of its efficiency when it becomes damp or wet, and if moisture is not retained inside the living spaces of a home, the occupants won't be as comfortable and will need more heat to achieve a satisfying inside temperature. In humid locations, however, where air conditioning is often running to remove moisture from the inside air, the vapor barrier will prevent outside moisture from entering the living areas.

Vapor Barriers For Walls And Ceilings

There are three methods for applying a vapor barrier to exterior walls or ceilings:

1. Install insulating blankets or batts having faces of vapor barrier backing such as treated kraft paper or aluminum foil placed on the living area side of the wall or ceiling. If loose poured or blown insulation is used in the ceiling (usually less costly than blankets and batts), then a vapor barrier should be arranged using one of the two remaining methods.

2. When unfaced blanket or batt insulation is used, a separate vapor barrier must be provided.

Aluminum-backed plaster or drywall is a good choice. It provides an effective vapor barrier and is simple to put up. In rooms where gypsum paneling is not employed as a wall finish, use either this or the next method.

3. Staple or nail polyethylene sheet material to the interior of the studs and ceiling joists. This is probably the most effective vapor barrier, but there is some question among experts whether or not the increased efficiency of this system compared to the aluminum-backed drywall is worth the extra installation step and accompanying costs. In addition, if this system is used, then drywall, wood, or other paneling cannot be applied with adhesive.

Vapor Barriers For Floors

The usual procedure is to place 15-pound building paper between the subfloor and the finished flooring. In addition, if the floor is located above an unheated basement or crawl space, blanket or batt insulation can be applied between the floor joists with a vapor barrier positioned against the subfloor.

A Safety Note

Peel back flammable vapor barriers at least 3 inches away from the heat-producing components or features in a house. Chimneys, flues, stoves, electric fans, and heating sources can all be potential fire starters.

SUMMER AND WINTER WEATHER PROTECTION

Where possible, provide shade for sunny east, south,

and west walls and windows. Use roof overhangs, shade trees, sun screens, shades, or interior draperies. Try to be flexible; shade makes for easier cooling in summer, yet sunshine streaming in through a window in winter can be a welcome sight.

Consider placing evergreen trees on the north side of the house to help block the cold northern winter winds.

Chapter 21

Wall Covering and Trim

The materials you select to cover your inside walls, plus the trim that's installed along corners and joints, are features that should weigh heavily in your decorating scheme. They're part of a home's inside cosmetics—what you and your visitors will see—and part of the home's structure.

WALL COVERINGS

There are a number of possibilities to select from when it comes to covering or finishing off the inside walls. The most popular materials are drywall, plaster, and paneling. If drywall or plaster is selected, then specify if you want the finished surfaces painted or wallpapered.

Drywall

Drywall consists of large (usually 4- x -8-foot) panels of prefabricated gypsum plaster sheathed on both sides in paper (Fig. 21-1). The panels are measured to fit, cut to fit, and nailed to the rough walls. The nails are driven to slightly countersink into the dry wall. A smooth continuity between panels is obtained by filling in and troweling over the joints with a plasterlike joint compound after a thin water-saturated paper tape has been

applied to each crack or joint line. The nail depressions are then filled in with the same joint compound and troweled smooth. Light but sturdy metal angles are installed wherever two pieces of drywall form right angles, to give the corners strength. The "plastered" edges are then troweled smooth with a thin layer of joint compound. Once the joint compound dries, it's smoothed out with sandpaper.

Final finishing is accomplished with more sanding and by last-minute "point-up" work—taking care of any minor irregularities and making edges sharper. When everything is dry and sanded smooth, the entire surface is painted, usually white. Once painted, there should be no evidence of panel joints or nail indentations.

At an additional cost, special finishes may be applied to dry-wall (and ceilings) in "skim" coatings of topping compounds similar to plaster. A skim coat can be left smooth, swished, dappled, ridged, or otherwise be finished to obtain a variety of unique textures (Fig. 21-2). Some contractors will apply a coating of material, and while it's still wet, take a large sponge and press it against the surface, drawing out the plasterlike material in many tiny points that look like miniature stalactites. It's a way to have your own walls and ceilings

Fig. 21-1. Drywall—vertical application.

finished in a manner that you'll find nowhere else.

Advantages:

1. Drywall is relatively inexpensive to purchase and install.

2. It's less inclined to crack than plaster is.

3. It's easier to repair than plaster is.

4. Drywall can be installed quickly and doesn't take as long to set up as plaster does. The rest of the finishing work can proceed without much delay.

Disadvantages:

1. Drywall must be painted because its finished appearance is not uniform. All joints, nail indentations, and repairs are whiter—due to the joint compound—than the rest of the drywall surfacing, which is a grey or buff color (whatever the shade of the drywall's outer paper layer).

2. The modular construction and the taping of drywall sections can sometimes be discerned, especially if it's a less-than-perfect installation. This can be worse than an occasional crack in a plaster wall.

PLASTER

Before drywall was invented, plaster was the number

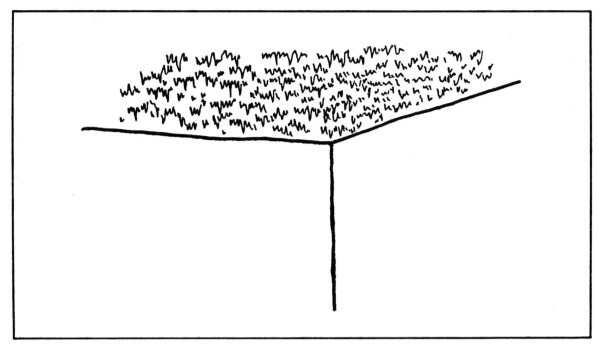

Fig. 21-2. A textured ceiling—dappled finish.

one way to finish off a home's interior walls. Plaster it-self is a mudlike material that's put over plasterboard lathing (which gives the plaster a secure base to ad-here to). It's applied in either a single 1/2-inch-thick layer, or two 3/8-inch-thick coats one on top of the other. Plaster can be finished in smooth or textured surfaces to complement any decor.

Advantages:

1. Together with its lathing, a plaster wall is thicker and provides better insulation values than drywall.

2. A plaster wall is more fire resistant than a sur-face finished with drywall.

3. It's more rigid than drywall, and less likely to bend or buckle.

4. It provides better soundproofing than drywall.

5. With plaster, a wider variety of creative artis-tic effects are possible. The topping layers on drywall cannot create such startling designs and surface modifi-cations.

Disadvantages:

1. Plaster costs more to install.

2. Plaster cracks easier than drywall does, and is more difficult to repair.

3. Plaster takes longer to install. It takes time for

it to dry or set up that could otherwise be used for com-pleting more of the wall finishing work.

PANELING

Prefabricated panels of wood and other materials are used to cover walls because of the panels' beauty, va-riety, low-maintenance qualities, and simple installation methods (Fig. 21-3). In fact, the finishing job on drywall or plaster walls that will be covered by paneling can be less than perfect, thus saving time.

You can find panels constructed to look and feel like a wide selection of building materials, including mar-ble, stone, brick, and stucco. There are even panels that simulate a wallpapered surface. These handy prefabricated sheets are becoming increasingly popu-lar in kitchens, bathrooms, foyers, and even in main liv-ing areas planned in today's modern dwellings. They are still not, however, as widely accepted as wood and wood-simulated paneling designs.

Plywood makes very good paneling for walls. It's durable and will add a rich look to a family room, recre-ation room, study, den, or dining area. Wood species such as elm, pecan, birch, and certain kinds of walnut lend themselves nicely for use on the plywood veneers.

Paneling should be installed against walls that have

already been roughly finished in drywall or plaster, so the paneling will have sufficient backing to prevent waviness or buckling. Care must be taken to ensure that all sections of paneling are plumb and fitted tightly together, with adequate nailing along their edges.

Advantages:

1. Paneling will not show cracks in the walls.
2. It is easy to install.
3. It adds to the insulation and soundproof values of the walls.
4. It can create many different moods to a room or area.

Disadvantages:

1. Making repairs to damaged paneling can be very difficult.
2. Paneling adds to the overall expenses of wall finishing beyond the costs of drywall or plaster.
3. It's another step in the construction of a house that takes additional time.

TRIM

Look in any home and you'll see wood or other trim installed wherever different construction finishing materials intersect: along joints for instance, and corners, door frames, windows, and other built-in house features. Trim makes an otherwise unattractive meeting place of two planes or surfaces (such as plastered walls and wood flooring, drywall and carpeting, wood paneling and vinyl flooring) into an attractive border accent.

Trim can also have functional applications: it can tightly hold down edges of paneling, carpeting, and linoleum, and, when placed in dining areas at chair-back height from the floor, will prevent the chairs from being pushed back from the table and scratching the walls.

The cost of a house's interior trimwork will vary, depending on what you specify. Here are some general points to consider before planning the trim for your house:

1. Even though hardwood trim is more expen-

Fig. 21-3. Paneling.

sive, it's the best and most common choice because it's more durable than softwood trim.

2. Decide whether you want the trim to be stained, sealed, varnished, or painted. Some contractors will charge extra, depending on how the trim will be finished.

3. Specify where you want trim to be used. Some places in a house are not necessarily trimmed unless you make specific arrangements for them, such as around open doorways or around closet exteriors.

4. Be careful not to mix or have too many different molding shapes. Match up the kind of trim material you select with the basic style of your home.

TYPES OF TRIM AND MOLDINGS

These types of trim are a good representation of what's available on the market and what each is most commonly used for (Figs. 21-4 through 21-11):

Casing Trim. Casing trim is the molding applied around doors and windows to cover the junctions of the door or window frames and finished wall. It's available in several styles and sizes. Doorway casing runs all the way to the floor, past and adjacent to baseboard trim that stops snugly against it.

Base or Baseboard Trim. Base trim or molding covers the gaps between the floor and the finished walls. Its style is usually the same as the door and window casing, but larger in size. Plasterers and drywall finishers will assume that the baseboard is at least 2 1/2 inches high and will leave unfinished that part of the wall that is less than that distance from the floor. If you select base that's less than 2 1/2 inches, check with your contractor to ensure that the drywall or plaster walls will be finished near enough to the floor so no unfinished wall will show above the baseboard.

Base Shoe Trim. Base shoe trim is molding applied between the finished floor and the baseboard. It's very flexible and can fit tightly along both the floor and baseboard trim, despite any irregularities of the structure that can be present even in the best of wood construction.

Base Cap Trim. These narrow sections of trim may be used to handsomely "cap" the tops of plain flat baseboard moldings. If you choose to use them, they'll also close any gaps that might exist between the wall and the baseboard.

Crown-and-Bed Trim. Crown-and-bed trim is a decorative molding used to soften sharp lines where two planes meet. Usual applications are at corners where walls and ceilings intersect. Such moldings are ideal for decorative trimwork, around a fireplace mantel for instance, or to make picture frames. The backs on most crown trim sections are hollow.

Stop Trim. Stop trim is molding nailed to a door jamb to stop a closing trim. It's also used on windows with sliding sashes.

Stool Trim. Stool trim is molding used at the bot-

Fig. 21-4. Trim/molding.

Fig. 21-5. Trim/molding.

tom of windows to provide a snug joint with the lowered sash.

Picture Trim. Picture trim molding was so named because it was originally designed as a perimeter trim for which pictures could be hung. It can still be used that way, but a more modern application is to use it as a substitute for crown trim.

Shelf Edge or Screen Trim. This molding is designed to cover the raw edges of screening on doors or windows, to decorate the edges of wood members such as shelves, or to conceal exposed plywood edges.

Corner Guard Trim. Corner guard trim is used to protect and finish outside corners.

Drip Cap Trim. Drip cap trim can be installed for use at top edges on the exterior side of doors and windows to prevent moisture from getting inside the walls.

Chair Rail Trim. Chair rail trim is a decorative and functional molding installed about 32 inches above the finished floor or at whatever height is best for a particular application that would protect the walls in a room from dining room chair or other furniture damage.

Wainscoting. Wainscoting can be wood planks

CROWN · AND · BED TRIM

Fig. 21-6. Trim/molding.

FRAMING STUDS

JAMB

DOOR STOP

CASING

Fig. 21-7. Trim/molding.

CROWN   COVE   BED

Fig. 21-8. Trim/molding.

Fig. 21-9. Trim/molding.

Fig. 21-10. Trim/molding.

Fig. 21-11. Trim/molding.

with tongue and groove arrangements, paneling, or other similar materials installed on the lower portions of interior walls, often between the baseboard and a chair rail. The wainscoting is applied first, then the chair and baseboard trims are fastened over the wainscoting's bottom and top edges.

Ply Caps. Ply caps are moldings used at the top of wainscoting to provide a smooth finish. They're also effective for edging plywood and for framing any panel, especially if the panel will be used as a slab for, say,

a table top.

Rounds. Rounds can be purchased as quarter-rounds, half-rounds, and full rounds. Typical applications for full rounds are as closet poles, curtain rods, and bannisters; for half-rounds, as decorative surface trim or seam covers; for quarter-rounds, as decorative trim for inside corners and as shelf cleats.

Handrails. Handrails are installed along one or both sides of a staircase (Figs. 21-12 and 21-13).

Fig. 21-12. Rails.

287

Fig. 21-13. A stairway with a handrail.

Chapter 22

Bathrooms

Bathrooms rank right up there next to kitchens when it comes to the most important rooms in a home. As such, the quality and number of bathrooms in a house can greatly affect the dwelling's resale value. Naturally, your family's needs will also help determine how many bathrooms you'll want, and how they'll be appointed.

In a nutshell, bathrooms usually consist of a hand sink or lavatory, a toilet, and a tub/shower unit (Fig. 22-1). Bathroom lavatories or sinks are available in numerous types, styles, sizes, and colors.

TYPES OF BATHROOM SINKS

The main types of bathroom sinks or lavs are four: vitreous china, porcelain/enamel over cast iron or steel, marbleized bowls or counter units, and Corian.

Vitreous China Sinks

These sinks are inherently acid resistant, and although they can be stained by rust, copper corrosion, and certain other persistent staining substances, vitreous china is considered an excellent material, with good resistance to chipping (Fig. 22-2). Vitreous china can also be formed into very fancy bowl and even into original works of art that are substantially more expensive than most other bathroom sink units.

Porcelain/Enamel Over Cast Iron or Steel

These are practically the least expensive sinks you can purchase. They're less durable, and more susceptible to chipping and normal wear than the others mentioned here.

Marbleized Bowl/Counter Units

There are two kinds of marbleized bowl and counter units: acrylic plastic and cast methacrylate, both of which can contain veining that simulates, in a highly polished surface, that of marble (Fig. 22-3). They're even referred to as synthetic marble. Both come in various colors and finishes, including high gloss "onyx" models.

These sinks are frequently integrated into a vanity countertop, with one-piece construction. They have a popular appearance, are easy to clean, and can be molded into beautiful patterned bowls (in the form of a sea shell, for example).

The main drawback to marbleized surfaces is that the finished layers are very thin. A heavy scratch can

Fig. 22-1. A full bath.

penetrate the coating and expose the white material underneath, which is difficult to cover up or repair.

Corian

Sinks and counters made of Corian are made with the color, often a bone white, all the way through the material. If scratched there's no color change and the mark can be easily removed with steel wool or fine sandpaper. Corian can also be sawn, drilled, and filed with standard carpentry tools for custom applications. It's one of the better materials for bathroom (and kitchen) counters and sinks, but it's also one of the most expensive.

STYLES OF BATHROOM SINKS

More and more people are turning to designer and continental-style lavs and lav sets. The following styles of sinks can be purchased in a wide range of materials and prices in round, oval, or sculptured bowls.

■ Wall-hung lavs requiring no floor supports or pedestal.

■ Lavs supported by cabinets or pedestals.

■ Lavs supported by metal legs.

■ Lavs built into ceramic-tile surfaced counters.

■ Lavs recessed into countertops of plastic laminate China sinks are popular with plastic laminate countertops.

■ Single-piece sink and counter units, either modeled or sculptured.

BATHROOM SINK INSTALLATION

A bathroom sink should be large enough for comfortable use, especially if it must suffice to wash your hair or bathe an infant. Try not to select anything smaller than 20 × 24 inches. In master bedrooms consider a double-bowl arrangement.

Every sink should have at least a narrow backsplash along the back of the wall it's mounted against, to protect the wall from spilled or splashing water. A drip edge should be included at the front and side ends of the countertop or around the tops of basin lavatories to prevent water from overrunning the top surfaces onto the floor. Sinks should also have an overflow prevention catch hole that drains water after it reaches a certain upper level in the bowl.

TOILETS

Toilets or "water closets" are made in various grades, from low-cost models to luxurious units with gold-leaf seats. The best buys are usually in the middle of any manufacturer's range. Toilets can also come in water-saving designs that operate with less water than standard models—an especially helpful feature if you're in an area having metered water or water shortages.

Most toilets are made of vitreous china, although some manufacturers carry injection-molded ABS plastic tanks as well.

The important features to look for in a water closet are:

- Sanitary self-cleaning action.
- Size of the free-flow water passage.
- Quietness of the flushing action.

Fig. 22-2. A vitreous China sink.

Fig. 22-3. A marbleized sinktop.

Types of Water Closets

There are only three basic types of toilets that you should be aware of: siphon-jet, reverse-trap, and washdown. The first two are the only ones you should consider for inclusion in your home.

Siphon-Jet Bowls. The free water surface in a siphon-jet toilet bowl covers practically the entire visible bowl area (Fig. 22-4). Water and wastes exit the siphon-jet from the rear bowl.

1. A siphon-jet bowl is usually manufactured in an elongated oval form that's attractive and comfortable.

2. It's the most sanitary of the three types mentioned.

3. The free-flow water passage is the largest of the three, with less chance of becoming obstructed.

4. It's the quietest type of the three.

5. It's available in a one-piece wall-mounted unit that makes cleaning the bowl itself and the bathroom surfaces around it simple.

6. When made in a single piece, however, the cost is quite high because any chip or flaw destroys the entire unit at the factory and in your home. When the bowl and tank are separate units the cost for the combination drops.

Reverse-Trap Bowls. A reverse-trap bowl is a shorter type of toilet than the siphon-jet (Fig. 22-5). Like the siphon-jet, water and wastes exit the reverse-trap from the rear of the bowl. The reverse-trap toilet has a free water surface covering roughly two-thirds of the bowl area and has a smaller water outflow passage.

1. It's a little less sanitary than the siphon-jet.
2. It's moderately quiet in flushing action, and the

Fig. 22-4. A siphon-jet toilet.

Fig. 22-5. A reverse-trap toilet.

most popular type with builders of middle-priced houses because it costs quite a bit less than siphon-jet models in similar combinations of tank and bowls.

Washdown Bowls. Washdown bowls are old-fashioned type toilets that can be distinguished by their vertical front profiles. They have very small free water areas, half that of the siphon-jet and the smallest of the three types mentioned. Water and wastes exit the washdown models from the front of their bowls.

1. It has a noisy flushing action.
2. It's not effective in self-cleaning.
3. It's the least sanitary of the three types mentioned.

BATHTUB AND SHOWER UNITS

A typical home features one bathroom having a combination bathtub and shower, and another bathroom with only a shower (Fig. 22-6).

Bathtub and Shower Sizes

Choose bathtub and shower units you'll feel comfortable in, and plan them in advance so the bathroom framing can be erected to size.

Many sizes of bathtubs are available. The usual widths are 2 1/2 to 4 feet, and lengths, 4 to 6 feet. Typical tub depth is 12 to 15 inches above its floor surface. Numerous special models such as sunken units and old-fashioned tubs with legs are on the market. Remember that the deeper bathtubs will result in less water splashing.

A small shower stall is only about 30 inches square. It's preferable to go with one that's between 36 and 48 inches in both width and depth.

Types of Bathtubs and Showers

Fiberglass Tubs and Showers. Fiberglass tubs and showers are extremely popular units that come either as a single molded piece complete with walls, or with separate walls that are assembled around the tub (Fig. 22-7). Most new construction uses the single unit tubs/showers having molded walls.

Advantages:

1. They're easier to clean than other types.
2. They tend to resist mildew better than ceramic tile surfaces.

3. They solve the age-old problem of getting a watertight joint between the tub's edges and the adjoining wall surfaces.
4. These units can be purchased with their own ceilings attached.
5. Fiberglass units are warmer to the touch than the harder models, especially during cold weather.
6. They don't soil as quickly as conventional porcelain.
7. They can be purchased with molded-in soap holders and seats.
8. They're very attractive and come in numerous colors.

Disadvantages:

1. They flex under the weight of even a medium body. Sufficient support must be placed beneath the units during installation to prevent the fiberglass floor from "giving."
2. When the shower water strikes the bottom of a fiberglass unit, it tends to be somewhat noisy unless insulation is installed between the bottom of the shower and the top of the floor, to absorb sound.

Enameled Cast-Iron or Steel Bathtubs. Cast-iron or steel tubs finished in regular enamel rarely come in any other color than white. The same tubs finished with acid-resisting enamel, however, come in white and colors. Because you can't tell by observation which grade of enamel is used on white tubs, the manufacturer's warranty must be requested. Elect the models with acid-resistant enamel if a cast-iron or steel bathtub best suits your decor.

The walls above a cast-iron or steel bathtub can be finished with a number of coverings including ceramic tile and glazelike hardboards. Tiles come in a huge variety of colors, sizes, and patterns. The glazed hardboards work well when installed with adhesive over a suitable backing material such as gypsum board. Their joints, corners, and edges are trimmed in color-matching metal mouldings in various shapes to fit the junctures. A few manufacturer's lines of glazed hardboards have unusual patterns and themes such as of ferns, lace, multiple lines, metallics, antiqued and textured woodgrains, and other modern surface representations.

Advantages:

1. They're exceptionally sturdy.
2. They're a lot quieter than fiberglass.

Fig. 22-6. A bathtub/shower unit.

Fig. 22-7. A fiberglass tub and shower.

3. The adjoining walls can be finished off in practically any way that would best suit your decor.

Disadvantages:

1. The enameled or porcelain surfaces chip very easily.

2. Sometimes these tubs will distort after installation. If a ceramic tile wall is connected to the edge of a tub that distorts, the tile might become loose.

3. These tubs tend to show water stains after a period of time, and must be caulked wherever their edges meet a wall.

Ceramic Tile Bathtubs and Showers. Ceramic tile bathtubs and showers are just that—tubs and showers built in place with walls and floors of ceramic tile. Often people going with this type of bathing facilities elect to have the tub sunken beneath the floor level. Ceramic tile applied with cement is the most permanent installation. Tile applied with mastic glues is second best because its performance is highly dependent on well-

maintained caulking. If water works its way through the joints, the mastic base could deteriorate.

Advantages:

1. They're unique and colorful.

2. They offer unlimited patterns, designs, sizes, and shapes. They can provide the ultimate customizing possibilities.

Disadvantages:

1. They're expensive.

2. The tile installation must be of superior workmanship or the tiles will eventually loosen and the tub will leak.

3. Tiles are more difficult to clean than the other types of tub and shower surfaces (especially the small mosaic tiles).

Bathtub and Shower General Considerations

A number of points should be considered while you

plan your home's bathing facilities.

1. The plumbers should set any tub or shower unit in place before the framing is completed (Fig. 22-8). Some of the larger molded fiberglass units won't fit into the house once the framing work is finished.

2. If a tub/shower unit comes from the manufac-turer covered with protective plastic or paper, fine. The covers will protect the unit's finish while construction work goes on around it. If not, it's advisable to tape plas-tic around the inside of the tub, with some thick paper on the bottom for padding, so somebody can step in-side and work around the tub without damaging the finish.

Fig. 22-8. Wall framing around tub/shower unit.

3. A tub/shower unit should be securely fastened to the walls. This is especially important with tubs that use ceramic tile for wall protection. Any movement later on will cause the tub to crack away from the tile.

4. The floor of the tub/shower should have a non-slip surface.

5. The tub/shower unit should have handles that you can live with.

6. Remember not to place the bathtub beneath a window. This design fault will mean providing some form of shower curtain to cover the window, especially if the tub is equipped with a shower head. In cold weather, a window over the tub can create unpleasant drafts, and whenever the window must be opened, you'll either have to reach across the tub or step into the tub to get enough opening leverage.

FAUCETS

The selection of faucets is a matter of personal taste. Many, many styles are available, from high-tech to old-fashioned, constructed with many different materials. One thing for sure, the quality faucets will last longer and retain their appearance over the long haul.

There are three basic grades of faucets: good, marginal, and cheap. Good faucets are constructed with solid brass innards having tough coatings of chrome, nickel, brushed or polished brass, or other durable materials. Marginal faucets are usually made of lightweight zinc or aluminum castings that will tarnish quickly, drip, and look dreadful within a short time. Cheap faucets can be identified by their crosslike handles, with four horizontal spokes coming out from their center. Faucets really get a good workout from the average family. Cheap faucets just won't do the trick.

In general, faucets having luxury features cost more than the straight models. Push-pull or dual-control handles, or single-lever-control faucets are certainly convenient, though, and are made in models of high quality.

You might realize a better price if you select all your plumbing fixtures, sink, tub, and shower, from the line of single manufacturer.

Once the faucets are installed, check to see if the cold water is on the right hand side of each fixture. Believe it or not, people have been scalded while discovering that their faucets handles had been reversed by mistake.

SHOWER NOZZLES

The shower nozzle you choose should have a flexible ball joint for directional control, plus a control to adjust the spray. Self-cleaning shower heads make the most sense to buy. Cheap nozzles offer little or no control of spray direction or quality.

An automatic diverter control should come with a combination tub/shower. It automatically diverts the water back to the tub faucets after someone has taken a shower. Such a setup prevents the next person who might want to take a bath from being pelted with hot or cold water. Omission of the diverter, an inexpensive item, can also cause accidental scalding of children.

BATHROOM CABINETS

There are two types of cabinets that you should consider for installation in your bathrooms: medicine cabinets and lavatory cabinets.

Medicine Cabinets

Medicine cabinets are used for storing medicines, razors, toothpaste, and other items of personal hygiene. These cabinets come in a variety of styles. Four of the most popular:

■ A rectangular door and mirror with diffusion lighting fixtures above the mirror (Fig. 22-9).

■ A sliding door operation, again with a light fixture positioned above the mirrored doors (Fig. 22-10).

■ Hinged mirror door cabinets with lighting fixtures positioned on both sides (Fig. 22-11).

■ And more economical cabinets consisting of single mirror doors on recessed cabinets without lights (Fig. 22-12).

Lavatory Cabinets

Lavatory cabinets, also referred to as vanity bases, frequently support the sink and sink counter in addition to providing storage space below (Fig. 22-13).

Lav cabinets, when compared to the smaller medicine cabinets, offer considerably more storage space. They also eliminate unsightly plumbing connections below the sink, help develop the bathroom decor, and can even be used to accommodate a built-in clothes hamper.

LIGHTING FOR THE BATHROOM

All bathroom lighting, especially lighting in bathrooms without windows, has to be well thought out. You need

Fig. 22-9. A medicine cabinet.

Fig. 22-10. A medicine cabinet.

Fig. 22-11. A medicine cabinet.

general plus very good task lighting for shaving and applying make-up. And don't forget lighting to read by in the bathtub.

Some of the most successful lighting for bathrooms is created by indirect rather than direct light sources. Light can be bounced off a white ceiling, concealed behind battens, and doubled off of mirrors. Where ceilings are low, choose flush recessed fittings so you won't hit them while toweling off.

Lighting can also create interesting focalpoints that are particularly important for internal bathrooms without windows. In fact, lighting can even be employed to create a fake window effect.

BATHROOM CHECKLIST

The following is a checklist of items that shouldn't be overlooked when you're planning bathrooms:

—A waterproof shower light.
—Electrical outlets well out of reach of the shower, tub, and lavatory water.
—Soap holders at the lavatory, tub, and shower.
—Toothbrush holders.
—Shutoff valves at each fixture so water can be

turned off for repairs without turning off the main water supply to the house.

—Consider having a built-in linen closet.
—Consider that the plumbing plan include the installation of a lead or PVC pan underneath the shower as a protection against leaks.

—A toilet paper holder located so it's not subject to splashing water from sinks, tubs, or showers, or to dripping from washcloths hanging on a towel bar above.

—A grab-bar or safety bar should be installed to serve two purposes: to help a person using the tub to get up easily, and to help the bather step out of the tub safely.

—One 36-inch towel bar (the longest standard bar available) for each person using the bathroom. This gives ample space for a folded bath towel, hand towel, and washcloth. Use towel rings only for supplemental towel or washcloth storage, or for drying off wet towels.

—Install the shower soap dish as far from the shower head and as high up on the back wall as practical. Centering the dish between the ends of the tub

Fig. 22-12. A medicine cabinet.

Fig. 22-13. A lavatory cabinet.

or shower stall results in the rapid melting of the bar soap under prolonged exposure to the water spray.

—A combination ceiling unit that includes an exhaust fan, an electric radiant heater or sunlamp, and a light. A good ventilation fan or exhaust fan is important for removing odors and excess humidity in the air. Such humidity can be harmful to the walls, paint, paper, and insulation.

—Consider installing several tasteful garment hooks.

—Specify whether you want glass sliding doors or a shower curtain for your shower. Doors cost considerably more, but to some people, doors are marks of luxury. To others they represent more surfaces to clean.

—The best way to select fixtures is to compare brands and see what line of fixtures most appeals to you. Stick to one brand for each line of fixtures. In selecting colored fixtures such as non-white tubs and toilets, remember that there is a good color match between fixtures of the same brand, but noticeable variations in the same color when fixtures of different manufacturers are compared.

301

Chapter 23

Kitchens

If there's one overall most important room in a typical house, it's the kitchen. Just look through the shelves of any library, bookstore, or magazine rack. Dozens and dozens of books have been written entirely about kitchens, and countless magazines devote their pages to the same subject. Kitchens have spawned enormous manufacturing invention, variety, and capacity with seemingly limitless amounts of materials, appliances, and other items designed and marketed specifically for home kitchen use.

Ask any real estate agent which room in a home will best up the dwelling's market value, and he or she will invariably answer with "the kitchen." As discussed in Chapter 1, a kitchen functions as the hub in any household. In kitchens we prepare our meals, eat our meals, socialized, relax, and work.

One way to look at kitchens is to think of their four major components: dining facilities, countertops, appliances, and cabinets.

DINING FACILITIES

Even if you plan to have a formal dining room, you'll still find dining accommodations in the kitchen a convenient necessity. Consider one or both of the following two options:

1. A table placed in the kitchen, out of the traffic flow (Fig. 23-1). Such an arrangement, also referred to as a breakfast nook, can be openly situated between the kitchen and another room (a family room, for example). In some homes the kitchen table is simply put in a corner out of the way. Wherever it's located, the kitchen table and chairs should always be as close as possible to food preparation and cleanup areas.

2. A snack bar or counter is frequently used instead of a table with chairs (Fig. 23-2). The counter itself can be employed to separate the kitchen from an adjacent family room or eating area, whether it be a nook or formal dining room. A counter is especially nice because it encourages excellent communications and socializing because it allows whoever is working in the kitchen to carry conversations with family members or friends while those guests comfortably enjoy food and drink at the counter.

COUNTERTOPS

Of all parts in a kitchen, countertops get the most use. And they're not only used regularly, but are also subjected to the greatest abuse and the most frequent cleaning. To be truly serviceable, a countertop surface must be able to resist moisture, heat, color fading, sharp

Fig. 23-1. A kitchen with dining space.

knives and blows, scratching, and staining from all sorts of nasty foods and substances such as grape juice, beets, and Easter-egg dyes.

The most popular countertop materials are plastic laminates, wood, ceramic tile, corian, and stainless steel.

Plastic Laminates

Plastic laminate countertops are by far more popular—due to a combination of price, availability, and utility—than all the rest. They come in many different styles, designs, and colors. The best ones are factory-made in a single laminated piece consisting of a rigid base covered by a hard, heat-resistant plastic topping about 1/16 inch thick. This topping is formed or molded under heat and pressure with a rolled front edge and a vertical "back-splash" that abuts whatever wall the countertop is mounted against.

Plastic laminate countertops come in two types of construction: custom and postformed. The custom-made top has an advantage of greater flexibility in shape—it's made to fit. But it also has a joint between the top and back-splash section that is difficult to keep clean, plus a seam in front that's exposed to wear and, like all laminate seams, might come apart if the cement fails. A well-made top, though, even with seams, will last indefinitely with proper care.

The postformed top is molded under high pressure by machinery designed specifically for producing countertops. The result is a smooth countertop with a seamless rounded back joint that's easy to clean, and a seamless smooth lip or drip edge on the front that prevents spilled liquids from running off the countertop onto the floor.

Although it's not as nice a setup as factory-integrated laminated models, the same kind of hard, heat-resistant types of plastic can be adhered to flat plywood countertops, with square edges finished off in the same plastic through the use of narrow strips that are cemented onto the edge surfaces. The exposed edge corners are dark, showing the thickness of the plastic laminate itself. The edges can also be covered with stainless steel or aluminum trim. In the best countertop installations there are no joints in the worktop surface.

Less expensive polyester and vinyl plastics can be used instead of the durable plastic mentioned above.

Fig. 23-2. A kitchen with a snack bar.

But both polyester and vinyl are more susceptible to heat damage from cigarettes, hot skillets, or other hazards.

Wood

Wood countertops are constructed of hardwood blocks or boards of maple, beech, birch, or oak glued together in a thickness of about 1 1/8 inches or more. It makes an excellent, attractive surface for cutting on and for handling hot dishes, but requires more care in clean-

ing. Plywood doesn't make an acceptable countertop by itself, but is commonly used as a base for other finishing materials. Wood in the form of compressed hardboards treated with a protective coating of tung or other oil is a low-cost alternative suitable for inexpensive kitchen designs.

Ceramic Tile

Ceramic tile makes a beautiful, easy-care countertop if well laid. It's colorful, distinctive, and heat- and fade-proof. It's also hard and noisy, and can cause dish and glass breakage. Neither does it provide a good cutting surface (a wood or plastic chopping/cutting board can take care of that problem easy enough). Tiles can be large or small, glazed or unglazed with non-porous vitreous bodies, and laid plain or in patterns. Stone can also be used in place of ceramic tile.

Corian

Corian is a tough, nonporous material that resembles alabaster. It's exceptionally durable and easy to work with. Because it's a solid substance with a homogeneous color, accidental burns or scratches can be rubbed out of the countertop with common household abrasive cleaners. Corian countertops also come with integrated sinks. This material is more expensive than laminated plastic, and the choice of colors is limited.

Stainless Steel

Stainless steel is another option. It's the first choice of commercial food service institutions such as restaurants and hotels. Stainless steel is hard, durable, sanitary, heatproof, attractive, and expensive. It's usually satin-finished to avoid any evidence of scratches, but constant cleaning will eventually polish the work area anyway. On the negative side, stainless steel is noisy and tough on dishes and glassware. And when dirty with gritty substances, it gives some cooks the shivers when they touch it.

Countertop Placement

Give careful thought to where you locate your counterspace.

1. You'll want some near the entrance you use from the garage, to set groceries on.
2. Some should be placed adjacent to the refrigerator where you must put away groceries and take out food for meals.
3. If possible, try to arrange for continuous counterspace from cooking range to sink to refrigerator. The refrigerator, however, should never be placed between the sink and the range because it would divide the counter and impede the kitchen work flow.
4. A countertop next to the range is convenient to temporarily put hot foods on before serving them.
5. There should be counterspace on one side of the sink or dishwasher for rough cleaning and stacking of dishes and glassware and for placing food

brought back from the dining area. Another clear countertop should be planned on the other side of the sink or dishwasher (usually to the left) for placing the washed and drying dishes and utensils.

APPLIANCES

The most basic kitchen planning concepts revolve around a very simple arrangement of three common-to-all-kitchen areas: the food preservation and storage area (refrigerator), the food preparation and cooking areas (range, oven, and microwave), and the food mixing and cleanup areas (sink, disposal, and dishwasher) (Fig. 23-3).

Refrigerators

A refrigerator should be large enough to accommodate your expected family size and still be able to take care of all the entertaining you plan. Consider which modern options, if any, you'd like, such as automatic ice maker, fast freezer, or ice water spout. Make sure the model you want will fit into the spot allocated on the plans. And whenever possible, the refrigerator door should be reversible, so with a simple changeover it can open from the right or the left, in case you want to remodel some day, or relocate.

Ovens, Ranges, and Microwaves

The oven can be a part of the cooking range, situated under the burners, or it can be a separate unit built into the kitchen cabinets. Consider a double oven if you do a lot of baking or entertaining. Make sure the ovens aren't placed next to the refrigerator where cooking heat could affect the operation of the refrigerator.

The range or top burners can be a separate unit built into a countertop, with at least 4 individual burners. In some makes the burner tops can be temporarily removed so other cooking devices such as grills and rotisseries can be substituted.

Don't overlook the inclusion of a microwave oven. Inexpensive models are available for placing on a countertop, or mounting in a wall cabinet or beneath a hanging cabinet. They're particularly useful in conjunction with freezers, because they eliminate the chore of defrosting food beforehand. Combination microwave/convection ovens are available that offer the best in both cooking methods.

Sinks, Disposals, and Dishwashers

The best kitchen sinks are made of either stainless steel

Fig. 23-3. Kitchen appliances.

or enameled cast iron. Both are easy to keep clean and will retain their good looks over the years. There are also porcelain-on-steel sinks that look like cast iron but are hard to keep clean, chip easily, and lose their gloss quickly. The more expensive line of stainless steel sinks contain higher percentages of chrome alloys, which provide better appearances and a reduced tendency toward "water spotting."

The style of sink you choose should have at least two wash bays or bowls and a rinse spray gun.

If desired, an optional garbage disposal makes cleaning dirty dishes and cookware faster and easier, but can be a headache if not properly used and maintained.

It's not absolutely necessary, but it sure is handy to include an automatic dishwasher. If you decide on one, place it next to the sink and be sure to allow at least 18 inches between any side wall or counter running at right angles to the machine's front. If you cram a dishwasher all the way into a corner, then two people won't be able to load or unload it at the same time.

CABINETS

Kitchen cabinets have always been the focalpoint of any kitchen (Fig. 23-4). More than anything, they set the flavor of the room and make or break its appearance. It's no coincidence that they're usually the first to be

Fig. 23-4. Kitchen cabinets.

changed when kitchens are remodeled, and countless home improvement contractors earn their living by replacing or giving facelifts to kitchen cabinets. Nowhere else in the typical house are so many different items kept. If it weren't for kitchen cabinets and drawers, chaos would prevent numerous cooks from staying organized and efficient in the kitchen.

Most kitchen and other cabinets (bathroom cabinets, for instance) are manufactured in one of three ways: custom-made, factory-made, and custom factory-made.

Custom-made cabinets are constructed in a local cabinet shop according to your exact plans. Such a method offers practically unlimited selection as to material, design, and finish, but it's also the most expensive. If your needs are unusual, though, custom-made cabinets could be your best option. Your builder should get bids from several cabinet makers before you make a final decision.

Factory-made cabinets are built to the specifications of the different lines the factory carries. Your choices are thus limited in the selection of material, design, and finishes. These cabinets range in price from inexpensive to very expensive, depending on the individual manufacturers and lines.

Custom factory-made cabinets are assembled at a centralized plant in response to orders taken by local kitchen cabinet outlets. You can expect a much greater choice in material, design, and finish than with factory-made cabinets. Prices are usually less than those of custom-made cabinets and more than those of the medium and economy grades of factory-made models.

For your kitchen cabinets give careful thought not only to cost, but to what cabinet style, type, sizes, quantities, and accessories will best suit your needs.

Types of Cabinets

There are four basic types of cabinets you can select from: unfinished wood cabinets, stained wood cabinets, painted wood cabinets, and wood and plastic laminate cabinets.

Unfinished Wood Cabinets. Like any unfinished product, unfinished wood cabinets can be purchased at a substantial savings if you decide to finish them yourself or have the house stainers do the job for you, before or after the cabinets are installed. If this method is selected, the cabinets can be finished off to exactly match other woodwork and trim throughout the house.

Stained Wood Cabinets. Stained wood cabinets

are finished at the shop or factory in a carefully controlled environment that ensures an excellent application and curing of the many finishing coats that are available.

Wood and Plastic Laminate Cabinets. Wood and plastic laminate cabinets are finished either at the factory or at a local shop. They're constructed of wood or a wood product such as particleboard, and then covered with a plastic laminate. The plastic laminate offers a wide variety of bright and wood-grain colors and patterns, and is easy to clean.

Special Cabinets

Apart from the standard cabinets that are set squarely against a wall, other units are available to fill special storage space needs.

■ Two-way cabinets have doors on both their fronts and backs. When suspended from a soffit over a counter peninsula or island, they permit you to put dishes away from the sink side and to take the same dishes out from the dining-room side when setting the table. Matching two-way base cabinets can also be purchased.

■ Quarter-round cabinets are used at open ends of cabinet rows to give a rounded-off look and to permit items to be stored on the front or side of the cabinet from several angles. Similar half-round cabinets and shelves can also be purchased when needed.

■ Mixer cabinets are used for storing large electric mixers. A mixer cabinet is a base unit with a shelf that pulls out and up.

■ A bottle cabinet is a base cabinet with pull-out trays that are egg-crated to hold bottles.

■ When wall and floor cabinets are located on both sides of an interior corner, a revolving round "lazy Susan" cabinet tray arrangement can make good use of otherwise hard-to-access space.

Cabinet Storage

Kitchen storage space is obviously most efficient when it's conveniently located. This is particularly true in the cooking area, where most kitchen work is done. It follows that, when planning a kitchen layout you must arrange for enough storage space of the right kinds to be made available near each of the major appliances for utensils, foods, seasonings, cleaning supplies, and other items that are used at the various work stations.

Here are some sample checklists of items you

should plan storage space for near the kitchen cooking range, refrigerator, and sink.

Near the range:

—Seasonings, instant coffee, tea and cocoa, flour, cornstarch, cooking oils, and shortening
—Saucepans, skillets, griddles, roasters, frying pans
—Knives, large forks and spoons, ladles, tongs, shears, carving tools, measuring cups and spoons
—Platters, serving dishes and trays
—Hand mixers and blenders
—Cooling racks

Near the refrigerator (this area usually serves as the food preparation area):

—Baking sheets, pie and muffin pans, rolling pins, sifters
—Casserole dishes, measuring cups and spoons, mixing bowls
—Paper towels, sandwich bags, aluminum foil, waxed paper, plastic wrap
—Seasonings
—Bottle and can openers
—Spatulas, ice cream scoops, refrigerator dishes
—Ice bucket
—Sandwich grill, waffle iron
—Bowl covers
—Bread and cake boxes, cookie jars

Near the sink (and dishwasher if you have one):

—Soaps, cleansers, detergents
—Paper and garbage bags
—Paper towels and napkins
—Scouring pads
—Silver polishing supplies
—Window cleaning supplies
—Appliance waxes
—Scrub brushes
—Coffee pots
—Colanders and strainers
—Cutting boards
—Dishcloths, towels, mop
—Everyday china and glassware
—Draining rack
—Double boiler
—Juicers
—Funnels

—Saucepans
—Tea kettle, pitchers
—Garbage can and trash can

Cabinet Slide-Out Shelves

Consider having some of your lower shelves as pullout trays that will give greater visibility and ease for reaching their contents. Heavy pots and items can be brought into full view and easy reach by simply pulling a tray forward. Vertical drawers are particularly satisfactory below the sink, where often-used dishpans, strainers, and colanders are kept.

Special drawers are also available with dividers for canned goods if there is no pantry. Bread box and slide-out vegetable bins are good for storing onions and potatoes.

VENTILATION

A poorly ventilated kitchen will suffer from chronic condensation and the odors of your last meal. In addition to strategically placed windows and sliding doors with screens, an overhead lighted hood with a fan that's vented directly to the outdoors is one of the best overall solutions. But a fan built into a counter range top and vented to the outdoors is also very popular and effective.

LIGHTING

The kitchen should have enough illumination, day or night, for cooking, work, or socializing. If possible, locate a window over the sink.

Consider fluorescent lights underneath cabinets that hang from a soffit over countertops to illuminate counter work surfaces. Plan recessed lights in plain soffits above work counters and ranges (if a light isn't already in a range hood).

There should be separate lighting over the main kitchen area. A light over a table with chairs or over an eating nook should be dimmer-controlled.

KITCHEN WALL FINISHES

The walls of a kitchen, especially the wall surfaces above the work counters, should be finished with an easy-to-clean material. Three popular, functional wall finishes are:

■ Fabric and vinyl wallpapers—these have excellent surface washability and durability characteristics. They come in a wide variety of colors and patterns.

■ Predecorated wallboards—these colored and patterned vinyls come in factory manufactured sheets that are easily cut to fit and applied to the walls.

■ Ceramic tile—ceramic tile is durable and attractive, but it requires a little more maintenance and effort to keep clean.

MISCELLANEOUS

1. Make sure there are enough electrical outlets along the countertops.

2. Consider locating broom and pantry closets in or near the kitchen.

3. Specify on your prints the make, model, color, and style of the sink, refrigerator/freezer, dishwasher, range, oven(s), microwave, garbage disposal, and exhaust system.

4. Consider having a small desk in the kitchen where you can sit down to thumb through cookbooks, make out shopping lists, check bills, and make telephone calls.

5. A telephone in the kitchen is a must. It should have an extra-long cord so if the phone rings while you're cooking supper, you can pull it over to the range top and keep stirring that sauce or frying those potatoes while talking.

6. Every surface in the kitchen should be easy to clean with a damp sponge: countertops, walls, floors, appliances, and cabinets.

Chapter 24

Floor Coverings

The selection of a particular floor covering depends on where it will be used, its appearance (available styles and colors), durability, ease of maintenance, price, and the buyer's personal taste. There are seven basic floor coverings to consider when planning your home: carpeting, linoleum and vinyl flooring, wood, tile, slate, stone, and brick.

CARPETING

An almost limitless array of carpeting is manufactured for homeowners to choose from. Practically any quality, basic material, texture, color, weave, or price range can be had for the asking. It's an all-around excellent floor covering that can literally be used anywhere in a home.

Certainly, carpeting is an interior decorator's dream because it can create as many moods as colors can influence. It can stretch small spaces into large, and shrink large spaces small. It can supply a rich, modern look to a room or create a statement of cool neutrality. It feels wonderful beneath bare feet and is equally enjoyed by small children who can tumble and fall on it without injuring themselves. It helps control sound from echoing through a home by absorbing noises, and is

relatively simple to replace when worn out.

On the other hand, it's tough to remove a cupful of spilled grape juice from a lily-white shag. Carpeting shows various stains and soils, and can pose big cleaning problems if chewing gum, grease, ink, or various foods and drinks become embedded in its yarn tufts. Pet odors can also find refuge in carpeting, and during times of low humidity, static electricity can collect and shock people who touch metal lamps or even one another. Also, when carpeting is installed in high-traffic areas of a house, no matter what its quality, it will eventually show wear paths, can be marred by dropped cigarettes, and cannot be refinished in any way. When soiled, carpeting needs either time-consuming shampooing or must be professionally cleaned.

There are carpets custom-tailored to solve a wide variety of problem situations: fibers designed to control the static electricity that draws dust or gives shock on cold, dry days; carpets having soil resistance built right into the fibers themselves; industrial grade carpeting designed to take unbelievable punishment, indoors and out; textured carpets with a mixture of sheared yarns and loops engineered to camouflage stains and look good while doing so; and lustrous sheared velours and velvets that look so plush that people are afraid to step on them.

You can tell quality carpet by its material and nap or yarn density. Wool makes excellent carpet, but it's expensive. Nylon is the most popular, accounting for over 75 percent of all carpeting manufactured today. Acrylics, polyesters, rayon, and polypropylene or olefin are no longer as popular as they once were, but are all still used for specific applications, the latter, for instance, in kitchens and wherever indoor-outdoor carpeting is needed.

Carpets used to be woven, but are rarely woven anymore. Instead they're tufted—a process by which yarns are looped through a woven backing then locked into that backing with an adhesive and then another thin layer of backing. Then the yarn loops are cut or sheared to various lengths. The higher quality carpet contains a higher density of yarn, or more yarn per square inch. The higher the density of yarn, the greater the carpet wearability.

Because so many varieties of carpeting exist, here are some guidelines to help you plan the carpeting for your house:

1. The first decision to make is one of style. By area, do you want a plush, a shag, or some other style better suited for each application. After you narrow down the style, begin comparing fibers available in each of the style groups you've selected. A general rule is to seek deeper and denser piles (the configurations of the fibers) than you think you'll need. Remember that shag carpeting is intentionally manufactured with low-density yarn to achieve the "shag" look.

Two of the most popular styles of surface texture for carpeting are the sculptured and the plush. The sculptured style is composed of designs created by alternate areas with and without heavy pile. The plush type has a constant pile thickness and is more likely to show footprint indentations.

Another texture is the level loop pile, which wears well and hides footprints nicely.

A fourth style is the frieze or twist type, which also thwarts footprints and hides dirt and dust fairly well. Like the level loop pile, the frieze or twist carpet stands up to rough use. And if it has a level surface, it's easy to keep clean.

2. When checking for carpet density, watch out for crimp. Crimp is exactly what it sounds like: mechanically induced zigs and zags in the individual carpet fibers that add bulk and fullness. A given amount of crimped fiber will fill more space than can otherwise be filled with the same quantity of regular, straight fibers. The actual crimping of the fibers is on such a tiny scale, it can't be seen at a casual glance. When crimped fiber is spun into yarn, air is captured between the zigzags; the yarn then looks straight and solid, but at the same time has a fuller, fluffier look than the non-crimped versions.

This crimping technique is used to make "high bulk" yarns that are fluffed or tufted into some very stylish, elegant carpeting. It's one way to obtain a plusher-looking carpet without encountering a higher expense in the process. Crimped-fiber carpeting might look good at first. It might feel and even sound good, but all a consumer really gets is the same amount of fiber that's available in more "honest" versions with thinner piles—plus a lot of air.

3. When choosing carpet styles and types, consider the wear that they're likely to receive. You'll probably want the most beautiful carpet in your living room, but if the room will see heavy family traffic instead of being just a visitor's parlor, then you might be further ahead to avoid light-toned solid colors and favor an anti-soiling plush or textured carpet instead. In general, when it comes to the busier areas in a home—the living room, family room, halls, and stairways, for instance, you'll do better with carpeting that will stand up to the traffic it receives. Bedroom carpets receive relatively light duty and thus can be good places to either economize or go luxury.

4. Most carpeting should have a padding of some type laid beneath the entire surface covered. Padding helps length a carpet's useful life by absorbing much of the footfall pressure that would otherwise grind the carpet backing against the hard and potentially raspy surface of the floor decking. It also makes a carpet seem plusher than the carpet really is by softening the impact of a person's footfalls so that walking on such a surface is more comfortable.

Various paddings are available, manufactured from hair, felt, rubber-coated jute, cellular rubber, high-density sponge, latex foam rubber, and urethane foam. All are acceptable when purchased in 1/4- to 1/2-inch-thick layers.

5. Carpeting should be installed on only smooth surfaces that are relatively uniform and ridgeless. Whenever a separate cushion or padding will be used, such undersupport should have all necessary seams covered with tape.

6. Carpeting should be planned so the least number of sections and seams will be needed. When possible, the seams should be positioned in low-traffic areas. Carpeting should always be laid in the same direction; that is, a butt end of a fresh roll shouldn't be seamed to a side of a previous roll.

Types of Carpeting

Here are brief introductions to the various types of carpeting materials available.

Nylon Carpeting. Nylon can go by various nicknames such as Antorn, Anso, Ultron, and Enkalure II. By any name, nylon is the most popular carpet available. It's especially good for entrance halls and stairways where traffic is heaviest. It can be purchased in many bright colors, has excellent resistance to abrasion and overall strength, and is relatively inexpensive.

Advantages:

1. Fairly inexpensive to produce and purchase.
2. Available in a huge quantity of styles, textures, and colors.
3. Good to excellent texture retention.
4. Good wet cleanability.
5. Excellent durability.
6. Good to excellent appearance retention.
7. Easy to maintain.
8. Excellent resistance to abrasion.
9. Good to excellent resistance to alkalis and acids.
10. Excellent resistance to insects and mildew.
11. Good resistance to compression and crushing forces.
12. Good resistance to staining.
13. Long life expectancy.

Disadvantages:

1. Nylon carpeting in particular needs a good backing material to attain dimensional stability.
2. Tends to pull and fuzz when abused.
3. Can retain oil and soil and "look dirty" easier than several other man-made fibers can.
4. Some discoloration might result from prolonged exposure to sunlight.
5. Nylon has more sheen than wool and some of the other synthetics.
6. Can develop static build-up.
7. Offers little protection against cigarette burns.

Acrylic Carpeting. These synthetic fibers closely resemble wool in texture, appearance, and abrasion resistance. Some acrylic carpeting, though, is very flammable. Due to their resistance to staining and soiling, and their low-maintenance demands, acrylic carpeting works well in kitchens, bathrooms, basements, porches, patios, and poolside areas—wherever dampness could be a problem, or spills are likely to occur.

Advantages:

1. Good colorfastness, with good resistance to sunlight.
2. Resembles wool in appearance and texture.
3. Resists aging well.
4. Good texture retention.
5. Good to excellent wet cleanability.
6. Good resistance to static build-up.
7. Good to excellent resistance to staining and soiling.
8. Excellent resistance to insects and mildew.
9. Good resistance to alkalis and acids.
10. Good resistance to abrasion.
11. Good resistance to compression and crushing forces.

Disadvantages:

1. Offers little protection against cigarette burns. Some acrylics burn very easily.
2. Tends to pull and fuzz when abused.
3. Some loss in textile strength upon prolonged exposure to sun.

Modacrylic Carpeting. Modacrylic carpet fibers are acrylic fibers, chemically modified to reduce their flammability. They wear and look like acrylic carpeting.

Polypropylene Olefin Carpeting. Polypropylene Olefin is synthetic material that has one of the lowest moisture absorption rates of all carpet fibers. It's easy to clean and very resistant to stains and soils. It can be used instead of acrylics in kitchens, bathrooms, basements, porches, patios, and poolside areas. Be aware, though, that they aren't as resistant to compression or crushing as the acrylics, and they don't retain their texture as well.

Advantages:

1. Good appearance retention.

2. Easy to maintain, with excellent wet cleanability.

3. Good durability and resistance to aging.

4. Excellent resistance to staining and soiling.

5. Excellent resistance to abrasion.

6. Good to excellent resistance to alkalis and acids.

7. Excellent resistance to insects and mildew.

8. Can be treated by the factory to give good resistance to direct sunlight.

Disadvantages:

1. Does not afford much protection against cigarette and other burns.

2. Has only a fair resistance to compression and crushing forces.

3. Has only a fair retention of texture.

Polyester Carpeting. Polyester makes an attractive carpet with many fine features, but it's less favored than nylon, acrylics, and polypropylene because of its resiliency deficiencies.

Advantages:

1. A soft luxurious appearance.

2. Good colorfastness.

3. Excellent resistance to insects and mildew.

4. Less static prone than wool.

Disadvantages:

1. Resiliency of polyester is somewhat less than that of nylon and some other types so that dense, deeper pile construction is needed for the same performance.

2. Stains easily with oily materials.

3. Prolonged exposure to sunlight will result in some loss of strength.

Rayon Carpeting. Rayon is not generally recommended over any of the others. It soils easily and gives poor resistance to abrasive wear.

Advantages:

1. Unaffected by most acids and solvents.

2. Can be an attractive flooring in areas of light usage.

Disadvantages:

1. Poor resistance to abrasion.

2. Soils rapidly.

3. Fuzzy types are very flammable.

Wool Carpeting. When it comes to beautiful, durable, top-of-the-line carpeting, there's no doubt that natural wool fiber leads the pack. Wool is a wonderfully durable fiber that has yet to be surpassed for its excellent appearance retention and resilience. Carpet wools are as varied as the different types of sheep across the globe, with fibers that range from fine and lustrous to coarse and springy.

Advantages:

1. Excellent appearance and texture retention.

2. Excellent durability.

3. Good to excellent ease of maintenance.

4. Warm and comfortable to the touch.

5. Dyes well.

6. Good resistance to abrasion.

7. Good protection against cigarette and other burns.

8. Excellent resistance to compression and crushing forces.

9. Good resistance to staining and soiling.

Disadvantages:

1. Expensive.

2. Offers only fair resistance to alkalis and acids.

3. Must be treated to resist insects and mildew.

4. Possesses only fair to good wet cleanability (but many other cleaning methods have been developed over the years).

5. Can promote static build-up.

Carpet Warranties

Any carpet you purchase should be covered by a warranty against manufacturing defects. The warranty comes directly from the manufacturer and is an assurance that the carpeting wasn't made in a slipshod manner (or if it *was*, by accident, the company will replace the damaged carpet or reimburse the owner). It's the consumer's protection against defects such as if the tufts pull out of the carpet, or if the face of the carpet comes apart from the backing, or if the dye bleaches out when shampooed.

Carpet warranties are very important, so important that you should never even consider buying a carpet that's not guaranteed against manufacturing defects. As important as such warranties are, though, remember that they don't guarantee performance over the long haul.

LINOLEUM OR VINYL FLOORING

Linoleum, or better known today as vinyl flooring, makes a fine floor covering for bathrooms, kitchens, recreation rooms, sunrooms, laundry rooms, or any other rooms for that matter. Vinyl flooring offers wide selections of patterns, colors, quality, and even textures. It can be laid out with very few seams, and is easy to maintain. Its resistance to grease and other staining materials is excellent, and overall, this flooring is fairly durable. Some vinyl flooring never needs waxing, while other vinyls might need waxing from time to time.

WOOD

Wood floor covering has been used for centuries. It comes in many types and styles and can be used in almost every part of a house.

Wood flooring is available in hardwood and softwood varieties. The hardwoods have excellent wearing qualities. The most popular species are white and red oaks. Standard hardwood flooring boards come in widths of 2 1/4 inches and random lengths. To achieve a more custom look, oak flooring can also be supplied in random widths with or without a distinct V-groove between the lengthwise side edges of each plank.

Softwood floors really mean pine wood planks. They come in standard and random widths and lengths. It's a lot less expensive than oak, but pine will show wear more quickly than oak. On the plus side, pine has a beautiful grain and takes staining very well.

Parquet flooring is another variety. It consists of small pieces of wood pressed and bonded together in a square tile. A parquet floor can be a thing of beauty, but it's more troublesome to take care of than plain plank flooring.

Wood flooring will not wear appreciably if its surface pores are filled in and the entire top is sanded, sealed, and periodically resealed with a protective finish of either layers of wax, varnish, shellac, lacquer, or hard polyurethane. If desired, wood stain can be applied before the polyurethane. Polyurethane is available in a matte (soft gloss) or a hard shiny finish. The choice is one of personal taste. In addition to giving wood flooring a pleasant appearance, polyurethane requires no waxing. Eventually, though, even polyurethane will have to be reapplied.

Wood floors have good grease and other stain resistance, and are fairly simple to maintain.

While linoleum and carpeting must be torn out and replaced when worn, wood flooring can be resurfaced. It's the only floor that can actually improve with age and successive refinishings.

On the negative side, the periodic refinishings, with their laborious sanding, can be expensive and time consuming. Hardwood floors can also be very noisy. You can soften up their effects by placing large area rugs in the living spaces, and by hanging substantial draperies in the same room to absorb sound. And lastly, wood flooring not securely nailed will squeak when walked upon.

TILE

Tile flooring—individual square or other multi-sided tiles—come in a full range of colors, finishes, and sizes in two popular types: ceramic and quarry.

Ceramic tiles are generally glazed in bright, shiny colors and are used to create custom floor designs or mosaics that set or follow a room's decorating scheme. Many ceramic tiles are made in Italy and Portugal.

The term "quarry tile" is misleading. It isn't actually tile that comes from a quarry. Rather, it's made from a mixture of clay, shale, and grit that's baked at high temperatures. Quarry tiles feature muted, earthen tones that are not glazed or shiny. Their dull surfaces help create a natural look inside homes that are planned with unsurfaced or natural building materials and decor. The term "quarry" really hails from the old-fashioned word "quarrel," which means a four-sided stone.

Tile makes a floor covering that possesses a lot of character, and is often used in hard-service areas such as entrance foyers and patios, and is equally at home in bathrooms, kitchens, and sunrooms. Tile is ideal to use on floors that will be on the receiving end of passive solar heating designs. During the day the tiles will soak up heat that will, in turn, be gradually released during the evening.

Although tile is very durable and will withstand wear and tear well, it has many seams and joints that are susceptible to cracking and staining. These seams and joints should be periodically recoated with an impervious surface sealer. Aside from the grout or other joint cement, the tile itself requires little maintenance. For best results, with the exception of bathrooms, tile should be laid in cement mortar over a dropped or sunken subfloor.

STONE

If you elect to specify a natural stone floor in a foyer

315

or room, there's no doubt that such a floor will be one of a kind. There are dozens of colors and sizes of natural stones, from purple, red, and black slate, to multi-toned granites, colorful quartzites, and the softer polished marbles.

Practically speaking, most stone used in today's homes is either slate or bluestone. Slate is a smooth gray, red, blue, or black sedimentary stone that splits easily and evenly into convenient slabs. Bluestone is a sandstone that can be gray, green, blue, or buff colored. Their most popular applications are foyers, hallways, kitchens, and sunrooms.

Although marble is the most luxurious stone that's widely available, it's a soft stone that tends (especially in the darker colors) to scratch easily and to absorb stains readily unless covered with a few coats of a good stone sealer. It's also slippery when wet.

What all well-cared-for stone offers is natural beauty and durability, plus ease of maintenance. The seams and joints, however, will need the same treatment as they receive with tile flooring—an impervious surface sealer.

Because natural stone is very heavy, depending on the thickness of stones selected, floor joists should be installed 12-inch on center for additional strength.

BRICK

If you're considering tile and stone, you might as well look at brick, too. It also comes in many colors, textures, and types. Brick is installed and cared for in a similar manner to tile and stone. All three can add a lot of character to select areas in your home.

Part 3

Where To Build It?

Where you decide to live is largely a combination of personal and practical choice. City, suburbia, and country all have their good points and bad. Plus within each of those areas you'll find potential building sites that should each be evaluated according to the lay of the land in relation to the sun's orientation, the prevailing winds, and the natural or man-made features of the surrounding territory.

Chapter 25

City,

Suburbia, or Country?

With some people it's like asking them if they prefer blondes, brunettes, or redheads. A person attached to a particular blonde might imagine that a brunette would be better. Or a redhead. Others are satisfied with the partner they've got, and never even think of approaching another. It's the same idea with city, suburbia, and country building sites.

Some city dwellers, it seems, have always longed to move to the country, or at least to a wide-open suburb where they can stretch out from the small, crowded lot they've grown up on. Other city residents shudder at the very thought of such isolation and opt to remain where they are. The determination boils down to a combination of work and social activities, practical and financial aspects, and personal preference. A country dweller can long to shed his drafty old farmhouse and 40 acres for a tiny ranch full of modern conveniences on an easy-to-maintain 60- x -120-foot lot. Or he might not.

THE CITY

Before nations came into existence there were cities. Living in cities is nothing new. At first they were places where primitive men and women congregated for convenience's sake—places they could farm at, or gather

from. During medieval times the cities had walls around them to provide protection for the inhabitants. Then cities developed into much more sophisticated settlements, and before long, the city existed as a central unifying influence on large groups of people.

Cities and towns are still the most important gathering places for people. The very term for city in Latin is "civitas," which helps derive the English words "citizen" and "civilization." In cities we manufacture, receive, and provide goods and services. We do research, teach, and learn. We encounter slums and crime and corruption. We wrestle with inadequate public services and enjoy the efficiently run ones.

A city is really a community in which everything there is drawn together in hopes of creating a desirable way of life for its inhabitants. In a city you have—or should have—rapid and inexpensive transit systems, effective law enforcement, proper disposal of sewage and garbage, fire protection, provisions for good housing, jobs for the people, education for children and adults, plenty of libraries and museums, theaters, and places for concerts, plays, and public events. There should be recreational sport facilities, adequate care for the sick and poor, and a government that's not overly corrupt.

We avoid isolation in cities, we take in a full range of enriching cultural activities, and we enter a vibrant night life. To some extent, it's a "fast" existence, where the action is and where the greatest number of opportunities in the business world are. It's where big-time radio and television stations are based and where major newspapers and other publications keep track of society's pulse.

All of this highlights the main advantage of city life—convenience. Everything is close, including schools, churches, medical services, supermarkets, restaurants, and corner bars. Plus, within a city it's even possible to find a location that *seems* like a piece of country—an isolated lot bordered by rows of mature pines, for instance.

Why then do people move out of cities, or elect to build elsewhere? One reason is because city land, with its high real estate taxes, is scarce and expensive, especially in major metropolitan areas. Another reason is that cities tend to have physical atmospheres more polluted by vehicle exhaust and industrial emissions than other locations do. Some say cities are too fast-paced, and too noisy. Other people shy away from city life because of high crime rates. Some make a deliberate choice to live somewhere else to be near people who are closer to their own social status or income level.

Cities have changed substantially since pre-World War II days when the rich, the poor, and the developing middle class all lived relatively close together in the same or adjacent neighborhoods. Back then there was a sense of social strength in cities that has been largely lost since the masses of middle and upper classes migrated to suburbia.

SUBURBIA

Suburbia is a relatively recent phenomenon which has been studied from every angle by planners, sociologists, and psychologists. In fact, entire books have been written about the suburbanite and his/her environment.

It's true that the very existence of suburbs was created by a wide variety of technological advances: automobiles, delivery trucks, rapid transit systems, miles and miles of concrete and asphalt paving, plus sundry inventions such as septic tanks, sewer mains, telephone lines and miscellaneous energy delivery systems have all contributed toward the establishment of satellite neighborhoods located "so many minutes or hours away" from the nearest city.

And what's more, suburbanites have developed certain characteristics that tend to bind them together, and particular habits that distinguish them from city and country dwellers. The main difference is their mobility: suburbanites are primarily commuters who tend to own their homes in areas that are on or near open spaces, away from crowded urban locations.

When you think of suburbia, you're likely to think of large yards, modern houses, swimming pools, two-car garages, incoming and outgoing transferred executives, couples jogging along the streets in color-coordinated jogging outfits, and long drives to work.

On the surface, suburbs sound like fine, clean places to live, and many of them are. On the other hand, quite a few suburbs consist of subdivisions having residents of roughly the same age group, social strata, and even income level. At one time neighborhoods were differentiated from each other by race or national origin, and within those neighborhoods you had a healthy cross section of the classes—from rich to poor. Suburbia changed all that. Suburbanites have demanded zoning controls that now completely rule the modern economics of suburban land development and make it impossible for the less affluent to join them. Whenever such middle-to-upper class members of the same age group and income level band together, the healthy diversity found in a neighborhood of large and small houses, young and old residents, and rich and poor families cannot flourish.

Contemporary subdivisions sort of lack the kind of all-aroundness that's characteristic of older communities, and in doing so perhaps sacrifice the good of the overall community in order to guarantee that the subdivision itself consists of essentially the same type of occupants, all with similar ages, incomes, and interests.

In subdivisions in the suburbs you're likely to have people more interested in maintaining their own property values, at times through political and zoning maneuvers, than being overly concerned about problems of the entire community or the success of community-wide programs developed for charitable causes, cultural enrichment, or the public good. While the older communities could be described as possessing a beneficial diversity within an overall unity, modern subdivisions are much more homogenous and self-centered. Ironically, all of this tends to reflect on the bottom line of suburban housing values in a positive way. Well-planned subdivisions have been a financial boon for the homeowners who live in them. They have been and continue to be places where housing appreciates considerably, unlike the many changing neighborhoods that feature nothing but housing units that are steadily depreciating, and losing their owners' life savings at the same time.

Suburbia has also been called "instant living environments," so named after the people who simultaneously move into a newly developed area. When this happens, the area has no tradition to rely upon, no familiar patterns of living for the residents to follow.

Now that the drawbacks of many suburban subdivisions have been touched upon, it's still safe to say that many *other* suburban subdivisions are exceptionally well-rounded places to live, and don't fit into the sociologist's definition of suburbia. It's just another case of many exceptions to a rule.

THE COUNTRY

Country living means many things to many people. To some it means wide-open acreage. It means fields and forests, ponds, meandering streams, corn and wheat and alfalfa, odors of sweet-smelling hay and manure, red barns sporting advertisements for chewing tobacco, snakelike tar and chip roads, dug or drilled water wells, spectacular thunderstorms and full moons, tractors, farm machinery and animals, plenty of fair-weather weekend visitors (especially when vegetable and fruit crops mature), and long drives to practically anywhere—to shopping, to school, to church, to work, to visit relatives. It means children do not have many playmates nearby. Although there's peace and quiet most of the time, there are also hunters in the fall and winter, and maybe in parts of spring and summer as well.

To others, the country offers little but boredom. Some people feel comfortable cohabiting with nature, while others need the fast pace of city life. To novice country dwellers, weeding a garden or feeding a flock of chickens can quickly become tiresome, and routine chores inevitably lose the novelty they at first possessed.

In the country, people generally have time to think, to contemplate. They're likely to have more leisure hours than they would elsewhere, because for some reason, country people tend to get up earlier, despite going to bed with a healthy kind of tiredness.

There's freedom in the country for pets to run loose, and there are plenty of birds and other wildlife for naturalists to feed, study, and even manage. There's room to blast a stereo up without aggravating neighbors, and there are places to ride horses and to gather wild mushrooms and blackberries.

AREA CHECKLISTS

City Living:

—Benefits of community life
—Feelings of security from living close to neighbors
—Fast-moving environment
—A center of media attention
—Night life and entertainment
—Less feelings of isolation
—It's where the jobs are
—Provisions for all types of housing
—Heterogeneous neighborhoods
—Good fire protection
—Good police protection
—Hospitals nearby
—Close to schools and churches
—Close to shopping
—Convenient utilities and public services
—Rapid and inexpensive transit systems
—Libraries and museums
—Noisy environments
—Slums and crime
—Polluted air
—Pockets of declining property values
—Occupants range from very poor to very rich, all ages and income levels
—Children will have friends to play with

Suburban living:

—Security of neighbors with less crowding
—A quieter place than a city
—Stable and appreciating property values
—Less crime and slums
—Cleaner air
—More privacy
—Larger lots
—Modern houses similar in size and types
—Open playgrounds and parks
—Modern utility systems
—Mostly middle class and upper class occupants of similar age and income level
—High property taxes
—A preoccupation with houses and household items
—Not much night life nearby, or cultural activities
—Further away from jobs, shopping, medical services, fire and police protection
—Children have friends to play with

Country living:

—Lots of isolation and quiet
—Good for pets and communing with nature
—Good for farming and raising gardens
—Good for thinking and contemplating
—Low real estate taxes
—Clean air
—Outdoors activities: hiking, horseback riding, hunting, fishing, gathering berries
—Not much crime
—Opportunities to study and manage wildlife and land

—Freedom from prying eyes of neighbors
—More time required for grounds maintenance chores
—More equipment needed for grounds maintenance
—Long drives to practically anything: school, church, libraries, hospitals, shopping, recreation, and work
—Private water and septic systems
—Lack of stringent zoning regulations
—Not many children nearby for youngsters to play with
—Can receive unwanted lengthy visits (often unannounced) from friends and relatives

Chapter 26

Selecting a Building Site

If you know the general kind of area you want to live in: country, suburbia, or city, then you've already made several of the most important decisions along the way toward selecting a suitable building site. Now it's time to narrow things down further. A number of considerations should weigh heavily in your search for the best individual site available for your house.

ZONING

Zoning is one of the most elementary considerations to be made when viewing a potential homesite. Local zoning codes are ordinances that divide all the property in a city, town, or county into a number of land-use classifications such as single-family residential, multi-family residential, agricultural, business, commercial, and industrial. In other words, zoning codes state what can and can't be done with all the property within the code's boundaries.

How would you like to invest your life savings into a dream house only to have neighbors move a trailer in across the street, start raising chickens and minks out back, erect big "Eggs for sale" signs along the road, plus have their four sons run a small engine repair shop in summer and hold snowmobile races during winter?

Zoning helps prevent unhealthy mixes of properties and owners having conflicting interests.

Of course, you can take a chance that even though the property across the street is zoned to permit a trailer park, Old Man Peters has owned the land for years and he'll probably never move. He *could* have a heart attack, though. And then his daughter *could* sell out to someone who would leap at the chance of putting in a 200-unit mobile home park.

One way to guarantee that nearby land won't be used for something that bothers you is to purchase enough of it behind, in front, and to the side of your location. That can be costly, however, and there probably won't be that much property available anyway. Even if you do protect your location, you can't be sure that a few hundred yards down the road—if it's a wide-open zoned area—a trailer court won't still be started up. Those are the chances you take when settling in a loosely regulated location.

When considering such a site, you can, of course, use your best judgement and the opinions of real estate professionals. For instance, if nice houses—the kind you're planning—already dominate such a loosely zoned area, chances are they they'll have a positive, inflationary effect on the remaining parcels, and consequently only nice houses will likely be added to the area because that's the highest and most practical use for the land.

Pay particular attention when considering property that's on the borderline of a residential area, or is zoned "transitional." Transitional areas are often dominated by residential dwellings, but they permit some business or commercial uses as well. It's a simple task to check any property by consulting the latest version of the zoning code. Just call the applicable city, town, or county zoning officer. By reviewing a zoning map the zoning officer can tell what classification the parcel falls under, and the classifications of the surrounding properties.

THE NEIGHBORHOOD

Ask any realtor what the three most important exponents of real estate value are and he or she will likely counter with "Location, Location, Location." In a very real sense, the location of a house can set the dollar range of its market value, both the minimum and maximum. This means that you should seek a neighborhood that lends itself to the kind of house you are planning if you want to ensure the highest possible market value for your home. To attain this maximum value and resale potential, your house should not be the largest and most expensive dwelling on the block. If it is, you will be doing your neighbors a good deed, and yourself a disservice. Your house will tend to buoy up the resale value of the less-expensive dwellings that surround it, while those less-expensive, smaller houses will pull the market value of your house down closer to their levels.

On the other hand, if you're willing to totally ignore the ideas of market value and resale potential—say you're planning to live in the house forever, and your children and theirs will one day take it over—and if you genuinely like the neighborhood and know it's not in a declining area, it could be a different story. In that case, by "over building," by matching a larger-than-average, more expensive house to a lot in a less expensive area, you could probably obtain a relatively low cost building site compared with other lots where larger, more-expensive houses are being built. Also, if you'll consider lots with odd locations, near busy streets, perhaps even next to a ramshackle Tobacco-Road house that sticks out in the neighborhood like a sore thumb and looks ready to fall over in the slightest breeze, you're likely to find bargain building sites.

Regarding the Tobacco-Road house (and there are many of them out there), you *hope* that it falls to the ground because as soon as it does, the value of your house and lot would immediately skyrocket, especially if something nice is erected in the old shack's place.

The location of the entire neighborhood in relation to modern necessities and conveniences is also important. You'll have to carefully consider the pros and cons. An advantage to a newly married couple might be to be very close to grade schools. To a couple not planning to have children, or having children long grown up and gone, that same point could well be a disadvantage.

Is the neighborhood reached by a hilly road that will be difficult to negotiate in winter? Is the potential site far west of town, while your workplace is to the east? If so, remember that you'll not only have to drive twice through the entire city everyday, but you'll have to head into the blinding sun each morning—and then into it again at the end of each day on the return trip. Will you be close to a favorite golf course, tennis club, library, or other high-frequency activity location? Will you be uncomfortably far away from good friends and relatives you like to visit frequently?

Is the location in a high-crime area? Be careful when new to an entire city or area. A brief phone call to a local law enforcement agency will be to your advantage. Realtors can be helpful, too, for revealing neighborhood information and for estimating resale potential.

LOT CONFIGURATION

While shopping for a building site, keep in mind the type of house you're planning. As discussed in Chapter 2, building sites with certain configurations naturally lend themselves to specific house types and are impractical for others. Of course, in some situations you can alter the lot contours to suit your house, but usually at substantial cost. In many cases, due to the lay of the land, it's not possible to successfully change the site's topography.

In a nutshell, here again are the matches and mismatches between lots and house types:

■ Single-story ranches are ideally suited to flat building sites, or lots that slope gently to the sides or rear, particularly if the plans call for a walkout from a basement or lower living area.

■ One and one-half stories are best matched with relatively flat lots or sites sloping slightly to the rear.

■ Two-stories can be efficiently situated on small, flat lots, and sites having slight grades.

■ Split foyers are good on sites with front-to-back or back-to-front medium to steep slopes. They're totally incompatible with flat lots.

■ Multi-levels are ideally suited to side-sloping lots on hilly terrain where the bottom level of the house faces and opens toward the downhill side and the upstairs level opens toward the uphill side. A flat lot won't work for a multi-level.

LOT SIZE

You begin to establish your building site's order of magnitude when selecting either country property (large building site), suburbia (medium building site), or city (relatively small lot), and then you further limit the possibilities by choosing one or several neighborhoods or areas to make your decision from. Finally, it's a combination of what's available, what's suitable, what's affordable, and your personal preference.

Do you like to garden, or grow fruit trees? Will you be putting in elaborate swing sets and play areas for children? What about an in- or above-ground swimming pool? Will you be doing a lot of outdoor entertaining, with parties and picnics? Do you prefer to maintain your own landscaping in a big way? Or you could hate the idea of cutting grass and tending bushes, shrubs, and trees. You might want the privacy afforded by a large lot, or you might be nervous not having a neighbor within calling distance. Maybe you don't want to be bothered with a long driveway that has to be shoveled or plowed during winter. Or you might want to be set back from the road to distance yourself from traffic noise.

Some people like to spread out on a roomy parcel, or acreage. Others who spend most of their time away from the house on work or recreational activities might find a large lot too superfluous and a bother to maintain. A larger lot can also mean higher real estate taxes.

HOUSE ORIENTATION

The orientation or positioning of your house on any particular lot depends on how the lot is situated along the street, what the topography of the lot and surrounding land is, which direction it faces (North, South, East or West), local building property line setbacks and regulations, and the house's floorplan.

When a choice of lots exists, orientation potential can become a major factor in the decision of which site to purchase. Window selection and positioning are closely related to orientation theories.

Remember too that a well-planned house can be constructed with any orientation, as long as the dwelling fits the lot. It might mean you will want to use more or less glass than you would have otherwise, or you'll beef up the insulation, or plan extra-wide overhangs. You must weigh the pros and cons of each particular site to come up with the best orientation possible.

Your plans for such a house positioning should address which rooms you want to receive maximum exposure to the sun, and protection from the cold winter winds and driving rains. In some cases, it might be better to use the mirror copy of your original plans. By flip-flopping the floorplan in certain situations you can place the garage on the side of the house that shields the living areas from winter storms and sub-zero chill factors.

SPECIAL LOCATIONS

Some people dream of living on a scenic ocean or lake bluff, or out on their own island in a picturesque bay. Trouble is, once they try such locations they might not like them. The weather can be brutal, house maintenance frequent and even the views—intense as they are—can become tiresome. Be careful of locations frequented by sightseers, sportsmen, photography buffs, or any special interest groups. Houses close to steep cliffs or ridges can be hazardous to children. Be aware of long-distance travel time required to reach many special out-of-the-way locations.

BUYING THE LOT

If the zoning is correct, the lot configuration fits your house plan, the orientation is satisfactory, the area meets all your personal qualifications, plus you like the site more than anything else you've seen, then go ahead and buy it. But when you do, take care of the following points:

1. Get a clear title. If you're purchasing the house with a mortgage, the lending institution will require you to have the property's deed searched. An attorney (or someone representing an attorney or title company) will go to the county clerk's office and look through the records, checking for any possible claims, liens, judgements, or clouds on the deed that would prevent you from receiving a free and clear title. You should have the records searched even if you pay cash for the property, and there's no mortgage involved, to protect your own interests. Attorneys, or title guarantee companies

that perform the same services, are insured in the unlikely event that they make a mistake.

2. If necessary, obtain a septic permit. The ability to secure a septic permit if there is no public or private sanitary system is a must. The site will be worthless if you can't have a septic system on it. If there's any doubt at all, you should specify in the property's purchase agreement that the transaction be contingent upon the ability to obtain a permit for the type of septic system that you want—perhaps a below-ground system instead of a more expensive and unwieldy above-ground "mound" arrangement.

A percolation test will be made when you apply for a permit. It determines how fast sewage liquids can seep into the ground. If the soil is primarily clay, a septic system might not be approved for below ground because the sewage liquids will not readily pass through the impermeable clay for proper disposal. Even when building on a large site, the size of which (10 acres, for example) exempts you from the necessity of securing a septic permit, you should still test the soil permeability to determine which type of septic system will be the most efficient and safest for you.

3. If needed, procure a water well. If no public or private water supply systems are available, you'll need to provide your own well. If there's any doubt about the prospect of digging or drilling an adequate well, you can include such a contingency in the property's purchase agreement. The well should provide good, drinkable water, in quantities (expressed in gallons per minute) that satisfy local and Federal Housing Administration standards. The well must be dug or drilled a safe distance away from any septic system leaching field.

4. Find out about any subdivision restrictions. If you select a building site in a new or established subdivision, make sure you find out if any building restrictions are in effect. Building restrictions are rules of a subdivision governing various construction and maintenance aspects to make certain that the houses in the subdivision will all be of relatively equal size, condition, and value. They regulate such things as how soon construction must begin and be completed after the lot's deed is transferred to the buyer, how the dwellings must be positioned on each site, what can be stored on the properties and what can't, how the yards must be maintained, and what will happen if those rules and others are disregarded.

In essence, building restrictions are designed to protect the integrity and value of each individual dwelling in a subdivision by prohibiting any house's improper or unusual construction or lack of maintenance that could have derogatory effects on the rest. The theory is, with strict building restrictions you won't end up with Tobacco-Road type houses ruining the neighborhood.

What follows are sample subdivision restrictions that are representative (written in the same kind of jargon) of what you'll find in thousand of similar sets of restrictions throughout the country. They address most of the major concerns you'll encounter anywhere. Keep in mind, though, that each set of subdivision restrictions is slightly different, custom-tailored by the people who began and later live in the subdivision in question.

Declaration of Restrictions for Oakland Hills Subdivision

(A plat of which is recorded in Smith County Map Book 8, page 21)

The following restrictions on property in subdivision to be known as Oakland Hills Subdivision, which is a part of Index No. 202-490-23, located in Crosscreek Township, Smith County, Pennsylvania, and part of Index 7344-700 and 7346-700, located in the City of Dalemont, Pennsylvania, shall govern all lots in said subdivision and they are covenants running with the land and binding upon all owners and their grantees, heirs, legal representatives, successors and assigns:

1. The property shall be used for private residence purposes only and shall be used only for single one-family residences, together with customary garage with space for not less than two or more than three cars, for sole use of owner or occupant of lot upon which said building is located.
2. No buildings shall be erected nearer to the front line nor nearer to the side street line than the building set-back lines of 30 feet shown on the recorded plot, except that where topographical features make it desirable an attached garage on Lots 22 through 37 may, with the approval of the designated Architectural Consultants designated by Realto Corporation, be located no closer than 20 feet from the front property line.
3. Homes built on lots fronting on Ash Street shall have at least the following areas: 1-story—shall have at least 2000 square feet; 1 1/2 story—shall have at least 2000 square feet at ground level and at least 700 square feet upstairs; 2-story—shall have at least 2000 square feet at ground level and at least 1200 square feet upstairs; split-level—shall have at least 2000 square feet at ground level and at least 600 square feet total on other levels. Homes

built on lots fronting on Douglas Street shall have at least the following areas: 1-story—shall have at least 1800 square feet; 1 1/2-story—shall have at least 1800 square feet at ground level and at least 600 square feet upstairs; 2-story—shall have at least 1800 square feet at ground level and at least 1000 square feet upstairs; split-level—shall have at least 1800 square feet at ground level and at least 500 square feet total on other levels.

4. No building, wall, fence, hedge, or other structure shall be erected or maintained unless plans including all floor plans, elevation, plot plan showing proposed grading, location of buildings, fences, hedges, lamp posts, outdoor fireplaces or major planting, such plans and elevations to show clearly the design, height, materials, color scheme, shall be submitted to, and approved by the Consulting Architectural Firm designated, at the time of lot purchase, by Realto Corporation. The fee for this consulting service, established at the time of purchase, shall be paid by the Purchaser.

5. No weeds, underbrush or other unsightly growth shall be permitted to grow or remain anywhere upon this property nor upon adjoining right of way between property and street. Until property is actually occupied through construction of a building, permission is given to Realto Corporation to mow the grass and remove the weeds or unsightly growth.

6. Property is subject to an annual charge or assessment of 60¢ per front foot adjusted annually to reflect cost-of-living change, to be paid by the owner of the property into a fund administered by Realto Corporation, its successors or assigns, such charges or assessments to be applied toward payment of the cost of maintaining right-of-ways, planting, or beautifying and caring for parks, plots, and other open spaces owned by the Corporation and maintained for the general use of owners of the property, caring for vacant and improved lots, removing grass and weeds therefrom and other things necessary or desirable to keep the property neat and in good order.

7. No billboards or advertising signs of any character whatever shall be erected, placed, permitted or maintained on such property. This shall not be construed, however, to prevent Realto Corporation from maintaining upon the property, at such locations as it may choose, billboards, signs, or sales office on an unsold lot during the initial sale of lots, nor from maintaining attractive signs of moderate size at the entrances to the subdivision identifying it.

8. Any building started on any lot in the subdivision shall be completed within six months.

9. The keeping of any animals or poultry other than ordinary domestic pet animals is prohibited.

10. A house and garage shall be commenced on a lot within one year after the delivery of the Deed to the lot. If the purchaser shall fail to construct the house and garage within this time, Realto Corporation shall have the option of extending the period for construction for six months *or* repurchasing the lot at the same price it was sold to the purchaser, and the purchaser shall give a good and merchantable deed free and clear of all encumbrances to Realto Corporation or its nominee. Resale of the lot by the purchaser shall not change the requirements for construction expressed herein in any way, and such re-sale must be made subject to the initial or optionally extended date.

11. Areas designated as Private Parks in the subdivision plan are the property of Realto Corporation, and use of such park areas by the residents of Oakland Hills or its extension is a privilege extended by the corporation, subject to its reasonable and proper regulations to maintain the character of the subdivision, and subject to withdrawal of the right of access and use by any individual(s), at any time for violation of said regulations.

12. No noxious or offensive activity shall be carried on upon any lot, nor shall anything be done, placed or stored thereon which may be or become an annoyance or nuisance to the neighborhood, or occasion any noise or order which will or might disturb the peace, comfort or serenity of the occupants of neighboring properties. Nothing shall be permitted or maintained on the premises or adjoining street or streets unless specifically approved by Realto Corporation, or its successors or assigns. During construction trucks shall use Old State Road exclusively for access in Crosscreek Township.

13. All provisions, conditions, restrictions, and covenants herein shall be binding on all lots and parcels or real estate and the owners thereof, regardless of the source of title of such owners, and any breach thereof, if continued for a period of 30 days from and after the date that the owner or other property owner shall have notified in writing the owner or lessee in possession of the lot upon which such breach has been committed to refrain from a continuance of such action and to correct such breach, shall warrant the undersigned or other lot owner to apply to any court of law or equity having jurisdiction thereof for an injunction or other proper relief, and if such relief be granted, the plaintiff in such action shall be entitled to receive his reasonable expenses in prosecuting such suit, including attorney's fees, as part of

the Decree, Order of Court or Judgement.

Provided, that any violation of the foregoing provisions, conditions, restrictions or covenants shall not defeat or render invalid the lien of any mortgage or deed of trust made in good faith for value as to any portion of said property, but such provisions, conditions, restrictions and covenants shall be enforceable, against any portion of said property acquired by any person through foreclosure or by deed in lieu of foreclosure, for any violation of the provisions, conditions, restrictions and covenants herein contained occurring after the acquisition of said property through foreclosure or deed of lieu of foreclosure.

14. In the event that any one or more of the provisions, conditions, restrictions and covenants herein set forth shall be held by any court of competent jurisdiction to be null and void, all remaining provisions, conditions, restrictions and covenants herein set forth shall continue unimpaired and in full force and effect.

15. Any and all of the rights and powers herein of Oakland Corporation may be assigned to any Corporation, Authority or Association, and such Corporation, Authority or Association shall to the extent of such assignment have the same rights and powers assumed and retained by the owner herein.

16. The aforesaid provisions, conditions, restrictions and covenants and each and all thereof, shall run with the land and continue and remain in full force and effect at all times and against all persons until the following January 1, at which time they shall be automatically extended for a period of ten years and thereafter for successive 10-year periods unless on or before the end of one of such extension periods the owners of 70% of the lots in said subdivision shall by written instrument, duly recorded, declare a termination or modification of the same. The aforesaid provisions, conditions, restrictions and covenants and each and all thereof may at any time be amended or modified by the owners of 70% of the lots in said subdivision by written instrument duly recorded.

17. No delay or omission on the part of the owner or the owners of any lot or lots in said property in exercising any right, power, or remedy herein provided for in the event of any breach of any of the provisions, conditions, restrictions and covenants herein contained shall be construed as a waiver thereof or acquiescence therein; and no right of action shall accrue nor shall any action be brought or maintained by anyone whomsoever against the undersigned owner or an account of the failure or neglect of the undersigned owner to exercise any right, power or remedy herein provided for in the event of any such breach of any said provisions, conditions, restrictions or covenants which may be unenforceable.

18. The "Owner" herein referred to shall include the present owner of the land in Oakland Hills Subdivision and any extensions thereof, their successors in interest and authorized agents. "Oakland Hills Subdivision" as referred to shall include its successors and assigns.

A BUILDING SITE CHECKLIST

—Is the site properly zoned?
—Is it close enough to:

Work
Schools
Shopping
Neighbors
Friends
Family
Church
Entertainment Centers
Medical Services

—What's the reputation of the school system?
—Are there any parks or playgrounds for children?
—How do the real estate taxes compare with those in other areas?
—Are there any home association dues?
—Are there other houses similar to yours in size and type and value in the neighborhood?
—Have you inquired about the crime rate compared with crime rates in other locations?
—Is the site in an area which is declining, remaining stable, or improving?
—Are there any potentially irritating factories, dragstrips, or seasonal attractions nearby?
—Is the location private enough for you?
—How are the garbage collection services?
—What kind of fire protection is there?
—What kind of police protection?
—What kind of emergency transport systems?
—Does the lot suit the type of house you are planning?
—Does any portion of the site consist of fill dirt?
—Is there or will there be good drainage on the site?
—Is the lot large enough for your house and your activities?

—Is the lot small enough for your house and your activities?

—Did you see a plot plan of the site?

—Has a recent property survey been completed? Are the lot's boundaries staked out so you can see them?

—Do you know at what level or height the sewer line will connect the house plumbing?

—Will the site provide your house with a good orientation?

—Are the views available from the location good or bad?

—Can you get a free and clear title to the land?

—Will you need a septic permit? Can you get one?

—Do you need a water well? Can you obtain one?

—How are the water wells nearby?

—Do any subdivision or other restrictions apply? And if so do you understand them?

—Have you inquired into the probable resale value of your house and property in case you must unexpectedly move in the near future?

Chapter 27

House Orientation, Positioning, and Landscaping

The way you orient and position your home on a building site and how you landscape its surroundings will greatly affect the benefits or drawbacks your home will receive from local weather conditions, will affect the home's market value and saleability, and will even affect the liveability of the dwelling in many subtle ways.

ORIENTATION

Four factors play important roles in determining the best orientation for your house:

◾ The location of the building site in relation to surrounding typographical features, other houses, and the street.
◾ The sun.
◾ The wind.
◾ The available views.

Location

If you purchase a building site that's the last one available in a subdivision, or is positioned between two existing dwellings, or anywhere within a subdivision having strict building restrictions, there's nothing dramatic you can do with the orientation. you already know where

the house has to sit, which way it has to face, and even where it must be positioned at the front and sides on any one particular lot.

But, if you have a wide choice of building sites with several different orientations possible, or if you're planning to build in the country, where few restrictions are in force, it's another story. If so, consider how the following three factors might influence your selection.

The Sun

There's nothing in life quite as regular and dependable as the rising and setting of the sun (clouds are the capricious variables of weather). You can take advantage of our sun's free heat and light by planning and facing the side of your house with the most glass toward the south. That way low-angled sunrays will penetrate into the rooms during winter, bringing warmth and illumination. Then in summer your roof overhangs will block out high-angled rays that will otherwise make your air conditioning work overtime.

In general, south-facing areas will be warm, north-facing areas will be cold, east-facing areas will get the pleasant early morning sun, and west-facing rooms will bear the brunt of hot afternoon rays.

The Wind

Depending on the circumstances, wind can be a help or hinderance. It can rob you of heat during the winter, with its icy chill factors. Or it can just as easily get rid of unwanted heat during stifling hot summer days.

In North America the prevailing winds blow from west to east, although seasonal and regional variations frequently find warm, moist breezes lofting up from the south, and cold, dry winds howling out of the north.

Because of the predictable nature of large weather patterns, whenever possible you should minimize the number of windows you place in the north and northwest sections of your house, since they'll be the hardest hit by the most unfavorable winds. Try to locate your garage there instead, to absorb most of the wind's punch before it reaches the rest of your home.

The View

To some people, available views will take precedence over anything else. If your lot is on the south side of a lake and you want to face the water, you can't help but face some of your living quarters toward the north. If so, compensate for such a handicapped orientation by paying more attention to selective landscaping, privacy walls, thermal glass, and similar features.

Be aware of the land or water that surrounds you, and how a nice view can be changed overnight by neighbors who cut down or plant trees, pile up garbage cans, or park a junk car in their back yard.

POSITIONING

Again, the house that's matched to a regular subdivision lot has few options as to its positioning—the only choice might be to construct the house either the actual way the plans are drawn versus the way a mirror image of those plans would appear. For example, a two-story house might have the garage on the left-hand side of the original set of plans. The same house could be "mirrored" so the garage would be positioned to the right.

Keep in mind that any mistake or poor choice on your part could create a house that's obviously mismatched with its lot. Ideally your home will blend into the site as if it "grew" in place. Several factors determining the best positioning potential follows.

Vehicle Access

You want a practical exit/entrance to your garage from the street. This is especially important in cold-climate locations where snow and ice make driving and walking a chore. Be careful of how steep the driveway's fall will be because it could add to the difficulty of getting in and out. At the same time, watch out for driveways that slope toward their garage. Rain water could run into the garage and ice coatings will make such a sloped driveway a definite safety hazard.

The amount of space to allow on the two sides of the house should be considered. Most homes are simply situated at the center of regular city and subdivision lots—sometimes to satisfy subdivision restrictions or to provide minimum clearances from the property lines as dictated by township or other applicable municipalities. If there is an option, it might be best for you to put the house toward one or the other side of the lot. What if someday a large dump truck or backhoe needs to drive into your back yard? What if you decide that your back yard is sloped too much, and you want to level it off with clean fill dirt?

Remember to leave ample space on the side that is most passable for trucks—preferably the garage side of the house. That way heavy equipment will also be able to reach your back yard, in case you someday decide to install an inground swimming pool, or put up an addition there.

Pedestrian Use

A second point to remember is one that if ignored, can become detrimental to the house's value and saleability. Depending on how your house is oriented, try to minimize the difficulty of having to climb numerous steps when entering or exiting the home.

Utility Connections

The third point to consider when setting up the positioning of your house is utility connections. A dwelling's corner or side where the utilities need to be connected should be accommodating. A lengthy concrete patio in this area would add difficulty to the hookups and would mean extra length and costs for making those connections.

Outdoors Functions

Some exclusive subdivisions feature houses that have practically no back yards. Instead the homes and their front yards are situated to give visitors a showcase effect—beautiful front lawns, stunning flowers and well cared for trees, all surrounding homes that look architecturally designed for each individual building site.

Other houses and lots are planned to permit a variety of practical outdoor activities. Review your interests and those of your family and friends. Do you like to:

■ Grow your own food? Then what about small "kitchen" gardens or large family gardens? What about fruit trees and berry bushes? Do you want to grow your own Christmas trees?

■ Keep pets or farm animals? Is there enough room somewhere for a dog to run and a cat to prowl? If permitted, do you plan to provide spaces and buildings for farm animals?

■ Participate in recreational activities? Badminton, croquet, tennis, horseshoes, swimming, basketball, swing sets for children, canoe storage shed, rope swings and hammocks, sand boxes, or raising roses and tulips?

■ Plan large outdoor parties and picnics?

■ Work at your main business or a part-time business? What about home workshops behind the garage, or those white boxes for beekeeping?

Setback From the Street

If there are already houses on both sides of your lot, you don't have much room to maneuver here. The same applies if your building site is part of a strictly regulated subdivision.

On the other hand, if you're building where there's no one else around, consider keeping the house a reasonable distance away from the road to reduce the amount of dust and noise generated from passing traffic. If you drive through the country you'll notice that some people build very close to the road, while others build quite far from it. In cold-climate areas, if you elect to go way back from the road, be prepared to handle the snow removal in some fashion. The larger and longer the driveway, the more snow you'll have to contend with.

When a choice exists, weight the advantages and disadvantages of privacy versus convenience, and arrive at a happy medium.

LANDSCAPING

Any realtor will tell you: Take two identical houses, put one on a nicely landscaped lot and the other on a landscape that has weeds erupting everywhere, dying trees, overgrown shrubs, and ankle-length grass. Can you guess which property would be easier to sell? And

which one would bring a higher market price? And also, which one would be more pleasant to live in?

Landscaping is important, no doubt. It can help you to reduce energy costs. It can make your home a more comfortable place to live. It can eliminate the need for manufactured fences or privacy screens. It can help reduce your grocery bills. It can attract interesting wild birds and animals within sight. It can define a play area for children or a picnic area for adults. It can block off an unsightly view or frame a pleasant pastoral scene. Landscaping is certainly one of the finishing touches required to make a new house an attractive property.

Clearing the Lot

If you're lucky enough to have a wooded or partly wooded building site, remove only those trees necessary to permit construction—unless you already have definite landscaping plans for the entire parcel. It takes only a short time to remove a full-grown tree, but years and years to replace it. Before even a single specimen is removed you should determine the boundaries needed for construction of the house, driveway, and septic system if one must be installed. Then you can selectively remove and save trees in a calculated manner.

Soil

The quality of the site's soil should be checked before the lot is excavated and bulldozed. If there's a layer of good topsoil, have it pushed into a pile on the side to be saved until after the house is up and ready for landscaping. To be sure your grass will be healthy, request a minimum of 3 inches of topsoil. Have at least that much in flower, shrubbery, and garden beds as well.

Plantings

Your lot's final grade should be established and stabilized as soon as possible through the planting of grass or other ground covers to protect the soil from water and wind erosion. A thin layer of straw spread over the seeded areas will minimize harmful water and wind action until the ground cover has grown enough to do the job.

Try to arrange for overall, well-balanced plantings that will alternately bloom or turn throughout the growing seasons, and not burst into magnificent "peacock" displays for two weeks and then disappear.

Remember that landscaping, and especially plantings, can be a gradual process. There's no need to rush things. Don't let professional landscapers pressure you into an all-or-nothing strategy. You can first put in a lawn and a few shrubs. Next year you can add a hardwood tree and some ground cover. And the next year a row of flowering hedges . . .

Evergreen trees retain their leaves during winter. They can block wind and provide shade year around. Try to place them between the house and the prevailing winds (Fig. 27-1).

Deciduous trees drop their leaves for the winter. They'll let sunlight filter through in the coldest months, when it's sorely needed. They're most advantageously placed to the south and west, where they'll also—because their leaves are full during summer and fall—block out the sun during the warmest part of the year.

Also arrange for outside air conditioning units to receive shade during the summer cooling season.

In general, trees offer beauty and provide homes for songbirds. Large trees lend the property that "established community" appearance.

Be aware that some of the landscape suppliers might not guarantee their products if you plant the items yourself. On the other hand, many of them will, and it's not very difficult to do the plantings yourself. Just follow the nursery's instructions to the letter.

If there's not much topsoil to start out with, you might want to consider grass sod. The advantage with sod is that you'll have an instant lawn that comes with some of its own topsoil. Naturally, sod costs more than a seeded lawn.

Fig. 27-1. Landscaping with evergreens.

Fig. 27-2. A decorative landscaping accent.

Drainage

The landscaping should slope away from the home's foundation and drain all the water from the gutters, downspouts, and sump pump out to the street or to some other harmless place. Make sure the first 10 feet surrounding the house perimeter tapers away from the foundation with at least a 6-inch drop all the way around.

Overall there should be a minimum of 1 to 2 feet of fall from the house to the street, and this includes the driveway.

Some lots need a culvert pipe beneath the end of the driveway for proper drainage along the street. Depending on where your lot is, check with the appropriate government office for the right size and method of installation. In some communities the highway department will install the culvert pipe at no cost to the owner. If not, see that its installation is included in the site work contract.

Erosion

To control minor problems with surface erosion on a bank, plant earth-holding shrubbery or ground cover. If an embankment is going to be steep due to the lay of the land, consider building a retaining wall out of railroad ties, rocks, or solid masonry. This will control erosion and simplify your lawn maintenance. If you employ solid masonry retaining walls, make sure they have weep holes to drain groundwater and release hydrostatic pressure that builds up on the earthen side of the wall.

Decorative and Finishing Touches

Browse through any of the numerous mail-order catalogs for an idea of how many unusual, exotic, practi-

Fig. 27-3. Landscaping decorations.

cal, and frivolous plants and trees are available. See if any decorative items such as fountains, bird baths, trellaces, pots, garden fences, ornaments, and automatic sprinklers will catch your fancy (Figs. 27-2 and 27-3). The selection of landscaping options practically defies description.

Final Landscaping Payment

If you happen to move into your newly constructed house before the grading is done and the lawn is planted (perhaps due to inclement weather, or the season in which the house was completed), make sure that some of the builder's funds are held back in escrow to be released to the builder after he completes what was promised in the contract. The contract should clearly state whose responsibility it is to provide a final grading with topsoil and a lawn.

Chapter 28

Driveways, Sidewalks, and Patios

These are three outdoors features of a home that can either add to or detract from a dwelling's appearance and livability, depending on how the driveway, sidewalks, and patios are planned and constructed.

DRIVEWAYS

Unless you know a legal way of locating your garage only a few inches away from a street, you're going to need a driveway.

Gravel Driveways. Gravel driveways having small pea-size stones up to material about 1 inch in diameter can do the job out in the country and in certain less-expensive areas of the city or suburbia, but not without numerous drawbacks. First of all, the stone will tend to disappear little by little. Some of it will catch on vehicle tires and get trucked on down the road. Some of it will get pushed into the driveway's sub surface. Some of it will get ground into power. Young children will throw it at one another and into the street or yard. You yourself will likely shovel it into your lawn along with winter snows in cold-climate areas (gravel is tough on snowblowers, too), and you'll also inevitably dull your lawn mower blades on it during summer. Ruts and muddy spots will develop, and weeds can sprout through the gravel's edges. From time to time the stone will need refreshing with another full or partial truckload.

Had enough? Gravel driveways should be your last resort.

Asphalt Driveways. Asphalt driveways are much, much better than driveways made of simply gravel. Asphalt driveways are fairly durable. They can look attractive and they're generally less expensive than concrete versions. Drawbacks are that they need periodic resealing or they'll deteriorate substantially faster than concrete will. Asphalt also softens up during hot temperatures, and can be pressed out of shape or indented by heavy vehicles or objects such as the legs of a camper/trailer. Although the black color of asphalt can look very attractive with certain homes, it generally clashes with concrete sidewalks when sidewalks are specified at the front of the property by local building codes.

Concrete Driveways. A concrete drive is the best, most permanent and trouble-free driveway you can have. It's durable and strong and can be poured in any configuration desired. It matches concrete sidewalks, steps, and patios nicely.

Driveway Configuration and Location

To a certain extent, these characteristics are determined by the house type, orientation, and position on the building site. Hopefully you're reading this material *before* you've completed your house and garage, so you can still make any adjustments that are appropriate. Consider the following points when planning your driveway:

1. Make sure the driveway will be wide enough. Measure driveways at other houses to get an idea of what you want. A single-car drive should be a minimum of 11 feet wide, with its width at the street (the curb entrance) not less than 17 feet to permit cars to easily turn into and back out of it (Fig. 28-1). If you elect to go with a narrower drive, it might be sufficient to drive a car on, but when visitors have to park on the driveway they'll have to step out of their vehicles into grass or a planting area. A double driveway should be at least 22 feet wide with an entrance at the street of 28 feet. Where any curvature is involved, a few extra feet of width is desirable to prevent drivers, especially visitors, from driving onto the lawn.

Fig. 28-1. A single-car driveway.

337

Fig. 28-2. A driveway with a turnaround.

2. Your driveway's length will be determined by your house's setback from the street and by the shape of your driveway's approach: straight, curved, or half-circle. In cold-climate areas consider that long driveways will have to be plowed out by someone: by you or by a hired plow jockey. Whoever plows will need somewhere to put the snow, too.

3. If your lot is large enough, consider having an automobile turnaround (Fig. 28-2). It takes extra surfacing but offers real convenience. It's difficult to realize how often you come and go by car until you have to back out of a driveway onto a busy street every time. When properly laid out, turnarounds can also serve as handy parking areas for visitors, and provide a good

spot for car washing and maintenance. With multiple-car families, the turnaround can ease the inevitable and irritating shuffle of cars to get the right one out of the rotation.

4. A step further than the turnaround driveway is the semi-circular drive having two accesses to the same street (Fig. 28-3). If you have enough space, and you have a front-entrance garage, this in-one-way and out-the-other pattern can also work to avoid people hav-ing to back out of the driveway.

5. Garages that open toward a side of a lot are best served by a wide driveway that extends beyond the garage entrance to furnish turnaround and parking space to the rear.

6. Don't accept a driveway that forces you to back out onto a blind curve or hill. It's far too dangerous for yourself, your family, and unsuspecting visitors.

7. Should you plan a straight or curved approach

Fig. 28-3. A semi-circular driveway.

to your garage? If you have the option, a curved approach often results in a better appearance. It adds flavor and character to a property.

8. Make sure the driveway isn't too steep to negotiate safely. This is important in all climates. It's dangerous to constantly keep cars parked on a steep grade—especially when children are in the neighborhood. Plus it's inconvenient. People can slip on wet or icy sloped pavement. At the same time, don't plan a driveway that slopes down into a garage. It'll provide a runway for rain water and snow melt, and will be tough to drive up during the winter. If need be, adjust the garage up or down a few blocks.

Driveway Construction Specifications

1. Driveways shouldn't be dug, graveled, then poured or paved overnight. Rather, it's a two-stage process. First comes the excavation of the ground to permit placement of a base layer of 6 to 10 inches of crushed stone. Next comes the pouring of concrete or laying of blacktop (asphalt) surface layers *after* the base stage has had 8 to 12 months to settle and compact. This waiting period while cars and trucks are compacting the gravel base will result in a more stable concrete or asphalt surface that will show considerably less mi-

nor cracking and almost no pavement break-off areas where base settling would otherwise be the cause. The waiting period also provides enough time for trouble spots to appear in the base so they can be repaired before the paving occurs.

2. Have the crushed stone or fill material placed about a foot wider on each side of the driveway than the pavement's eventual width. This prevents undercutting of the slab and subsequent breakage by surface water draining off the drive.

3. To ensure proper drainage the driveway must slope away from the garage toward the street about 2 or 3 inches downhill for every 10 linear feet of driveway (Fig. 28-4). Of course, this applies to only a typical driveway; the long winding drive applies only where it nears the house. If you can't arrange for that much slope on a regular driveway, before the driveway is surfaced have some fine stone added to the base so the gravel layer is crowned in the center. An 11-foot-wide drive should have a crowned or raised center 2 to 2 1/2 inches high, and a 22-foot wide drive should have a crown reaching 4 to 5 inches in the center, so water will drain off to the sides.

4. Check out any unusual street problems such as a fireplug, utility pole, or tree that might be in the

Fig. 28-4. Driveway slope.

Fig. 28-5. Driveway drainage.

way of your driveway. Consider that there could be a city ordinance covering the number, size, and placement of driveway approaches.

5. Where water causes problems around driveways, a length of plastic pipe might help improve drainage at low spots or water collection points. For instance, if rainwater tends to collect along one side of a drive-

way, it might cause trouble in the form of surface ice and frost heaval during winter. A length of pipe installed underground, running from a small rock basin at the lowest collection point on one side of the driveway to a screen-covered outlet on the other side, will alleviate the situation (Fig. 28-5).

6. A driveway should be at least 5 inches thick.

The builder typically will use 2-x-6s or 1-x-6s to obtain the 5-inch depth, keeping the bottom edges slightly below grade (Fig. 28-6). Specify that the concrete mix should be a minimum of 5 1/2 bags of cement per cubic yard of concrete.

7. The concrete should have either steel rod or wire mesh running through it for strength. Approximately every 11 feet a groove should be cut into the concrete to allow the pads room to react to temperature changes and other stressful conditions. Spacer felt—thin strips of felt—should be placed between the individual concrete pads or blocks to allow for contraction and expansion due to temperature changes.

8. The finish on a concrete drive can be either a swirl or a broomed effect. It's a matter of personal preference.

9. Freshly poured concrete driveways should be barricaded until they cure and have a chance to be coated with a good concrete sealer. As mentioned in the chapter on garages, concrete sealer will prevent the penetration of any oil or grease into the concrete and will make the surface easier to clean.

10. Asphalt should ideally be applied in two layers over the stone base. Roughly two-thirds (or 3 1/2 inches) of the total thickness should be laid down as a binder course containing larger stone or aggregate, then the

Fig. 28-6. Concrete driveway construction.

other third (1 1/2 inches) can be laid on top. The wearing course contains finer material so it can be tamped and rolled into a smooth, watertight surface.

SIDEWALKS

Sidewalks are another necessity. You have to get from the driveway to the front, side, and rear entrances somehow. Sidewalks are the accepted, civilized way so you and your visitors won't have to walk through wet grass or dusty ground.

Sidewalk Planning

Here are some considerations to help you plan your sidewalks:

1. Make the lead sidewalk from your driveway to the front entrance steps at least 4 feet wide. Traditional 36-inch-wide sidewalks are fine for side or rear doors, where only one person at a time must be accommodated. Front entrances should be able to be approached by two visitors walking side-by-side instead of single file.

2. Following the rationale used with driveways, a curved sidewalk adds a certain character and flavor to the appearance of a home (Fig. 28-7).

3. Have adequate lighting along sidewalks, especially where steps are located.

4. Many people are tempted to opt for sidewalks, entrance steps, and patios made of brick, flagstone, masonry, or patio blocks. These materials might sound like a good idea, but they're usually more trouble than they're worth. If you go with anything other than concrete, be prepared to do a considerable amount of maintenance work every 2 or 3 years to correct surface deterioration that will occur. It's possible to construct sidewalks out of asphalt, but it's not advisable unless the walk is extremely long and winding, such as a bike path through a wooded area.

Sidewalk Construction

The first step in sidewalk construction (Fig. 28-8) is to excavate the walkway and spread 4 or 5 inches of crushed stone in the excavation, which is to be used as a base layer. The stone base should then be either compacted with a hand-held tamping machine or left alone for 6 to 8 months for the ground and gravel to

Fig. 28-7. A curved sidewalk.

Fig. 28-8. Sidewalk construction.

settle naturally. If the ground and stone base are not allowed enough time or compacting effort to settle, the base beneath the sidewalk will sink and move, causing voids that will no longer support the sidewalk in those areas, and cracks or heaved concrete will result. Follow these guidelines:

1. At least 2- x -4-foot lumber should be used to form the sidewalks to provide a 3-inch-thick walk.

2. The sidewalk forms should give walks a slight side slope or length slope that falls away from the house for proper rainwater and snow melt surface drainage (Fig. 28-9).

3. The finish on concrete sidewalks can be either the swirl or the broomed type.

4. As with the garage floor and driveway, make sure that a clear protective sealer is applied over the freshly-dried concrete sidewalks.

5. Keep sidewalks above ground level. Don't let weeds and grass sod encroach over sidewalk edges. The surrounding lawn roots should be several inches below the sidewalk surface for easy lawn mowing and a neat appearance.

PATIOS

There are few strict rules when it comes to the size, type, and location of patios. Some people will find a small concrete patio or wood deck to their liking, positioned off the dining room. Others plan huge, sprawling wood decks that completely encircle their dwelling. Some people like to have roofed-over patios, or prefer to close the sides in too. Certain individuals count on using covered patios year round and will heat them in cold-climate locations.

Patio Planning

Here are some points to consider when planning patios:

1. The patio is similar to the garage in that it's one of the least expensive parts of a house per square foot of usable area. If you're going to have one, it might as well be a nice, roomy one.

2. It's best to construct patios with below-ground footers and foundations, so if you decide to cover it some day you're sure to have sufficient support and you won't have to worry about it heaving in response to freezing temperatures.

3. Whatever your outdoor patio construction materials are, provide some height to the patio by keeping the patio floor surface at least 4 to 6 inches above the surrounding ground. This will make cleaning much simpler.

4. Consider building in some above-the-floor,

Fig. 28-9. Sidewalk slope.

raised planter areas, walls, dividers, or other permanent features that will provide horizontal or flat surfaces wide enough for people to comfortably sit on (Fig. 28-10).

5. If a patio is more than a few steps above the ground level it's advisable that step handrails be installed for safety (Fig. 28-11).

6. Make sure your patio is properly positioned for privacy and easy access to the living areas of the house, especially the kitchen.

7. Arrange adequate lighting for patio evening use and enjoyment.

8. Unless you plan a year-round enclosed patio, consider what storage facilities you'll be able to use for storing patio furniture, planters, and other seasonal items. If you have the space, one alternative is to construct an outdoor weatherproof closet or storage enclosure adjacent to the patio or near the garage or rear of the house.

Patio Construction

There are two primary types of patio foundation constructions: concrete and wood. The concrete patio employs the idea of a concrete footer below the frost line with a concrete block foundation built up from there. The inside is backfilled and brought up to the top of the foundation with a 12-inch-thick layer of bank gravel. Then a concrete pad reinforced with wire mesh or rod is poured over the top. Once cured, the concrete patio floor should be coated with a good concrete sealer.

Fig. 28-10. An outdoor patio deck.

Wood patios or decks have certain attractions to people who like the warm, soft feel and rustic look of wood. Wood can also be stained or painted to match existing color schemes. All wood decks should be constructed with lumber that can withstand the rigors of constant exposure to the weather. Redwood, cypress, and cedar are all excellent choices, but they're expensive. Salt-treated southern yellow pine is an alternative. It's a durable type of wood at a bargain price. Order grades of salt-treated lumber labeled .25CCA for lumber that will be used above ground level, and .40CCA if it will be used on grade or below.

If your patio will be enclosed with treated wood,

follow the suggestions given in the previous paragraph. If it's going to be enclosed as part of the house and covered the same way, then follow the specifications outlined in earlier chapters on interior framing and exterior coverings.

If you plan to go with a treated lumber deck or enclosure, check into the hardware and fastening devices on the market that are manufactured with rust-resistant metals. They include both galvanized steel and aluminum nails.

The probability of warping will be reduced if spans of railing are kept at or less than 8 feet in length. Built-in benches will also add to the stability of the structure.

Fig. 28-11. A patio deck with steps and handrails.

347

Part 4
Who Should Build It?

Once you know which house you want, and you have a fair idea of how it should be put together, and you think you know where you want it built, it's time to find a contractor.

Finding the *right* contractor is a task to be taken seriously. To make the entire housebuilding process a pleasant and rewarding experience, it's paramount to select only an established, skilled, reliable builder who knows what he's doing and who's satisfied many a previous client.

Once found, there are ways to work along with a builder that will virtually ensure that a helpful and trusting give-and-take relationship develops.

Chapter 29

Selecting a Contractor

By now you should relax and pat yourself on the back. Surely, you are miles ahead of most people who will, regretfully, *begin* by selecting a contractor as their initial step toward acquiring a new house. Consider that these trusting souls choose a contractor without really knowing which house would be best, without understanding building jargon, without realizing the degree to which they can and should participate in the planning of their new home. Too, they have little knowledge of which questions to ask the builder, and often spend time and energy worrying about practically meaningless—at least to the overall worth and enjoyment of the house—details. Rather, they are likely to be led by the nose into purchasing a house that is merely convenient and cost-efficient for the contractor to build, instead of a customized compatible house that will provide the owners with more living enjoyment and value for their money.

But you, by now, understand what houses are all about. You have the knowledge to be able to make the correct choices between the various types of homes that will best suit your needs. You know their advantages and disadvantages, and how each will affect your own situation. Such information is clearly required in order to arrive at an intelligent decision. It's a personal

thing, the planning and erecting of a house, and to do it right, you needed to consider the pros and cons of traffic patterns, floorplans, and even of certain building materials. You needed to understand, at least in a fun, general way, how and even why a house is constructed the way it is. You needed to know the importance of the drawings and prints, and about low-grade or minimum-spec materials and construction methods that are being pushed at the unwary by a certain percentage of contractors. And you needed to understand why quality materials and methods should be insisted upon at contract time.

So, in a sense, you've already completed your most critical homework. You know what you want to build. Now you have to determine *who* you want to build it. Select a minimum of four potential contractors from whom you will solicit competitive bids, and then narrow those builders down to one. Of course, at this point you still might not have located a suitable building site, but that can come later.

Be aware that you are on the verge of entering into a unique relationship that should be mutually rewarding with a man, woman, or any combination of, including a large company or corporation in business to build houses.

The contractor will thus be a pivotal figure around which the success or failure of a housebuilding project can occur. Anyone who thinks about the possible consequences—good or bad—will agree that it's a relationship that must be carefully entered into, not formed on the basis of skimpy information or a casual referral. Ask any real estate agent, housing inspector, or mortgage lending officer what are the overriding reasons for bad houses, and you'll likely as not receive an answer blaming inept or dishonest builders.

Naturally, builders vary in quality, just like individuals in any profession. You have good doctors and bad, capable attorneys as well as incompetent ones; and you have skilled, sincere builders . . . and shysters. Be aware that, although the proportion of good, skillful, reliable builders is high—just as in the rest of the professions—there seems to be a disproportionally large number of shoddy houses or marginal quality homes being built by a relatively few inept or/and dishonest builders who lend their calling a bad name and draw a lot of harmful publicity to their field.

BUILDER TYPES

Because it's so important to make a correct choice when selecting a builder—you certainly don't want to change builders in midstream—here are the major types of builders you're likely to encounter while searching through the marketplace.

Established, Skilled, Reputable Builders

First and foremost is the established, skilled, reputable builder. This can mean a large company, a middle-size company, a small company, or a partnership. It can also mean a one-man operation. The larger established, reputable builders are "old faithful" contractors who are often associated with prestigious, well-planned subdivisions and might also have built rental complexes along the way. These contractors have been building houses for years and they tend to specialize in certain types, styles, and price ranges leaning toward mid- to most-expensive houses in an area. These builders are so well respected that real estate agents even use the contractors' names in advertising, as a "Bebell Ranch" or a "Luciano Two-Story."

Other highly respected builders specialize in low- to mid-priced houses. They've learned how to build as much quality as possible into their dwellings for the lowest possible cost. They purchase large quantities of similar materials at a discount and use the same items in all of their homes. They build on less-expensive lots. Their floorplans tend to be smaller and simpler. These builders are exceptionally good at the range of homes they're working in, and the need for their product is great. Builders of inexpensive houses frequently work with customers who cannot afford much, but who can take advantage of government assistance and financing programs; consequently, these contractors become very good at creative financing.

All of these builders are proven pros. They have the building skills, the business acumen, and the people skills needed to survive in their cut-throat, vicious, tumultuous trade. The established builder has seen it all. He's aware of mortgage financing options, of decorating styles, of landscaping solutions, of energy saving ideas. And if he doesn't build a good, solid house and back it up with professional service, the word will get around. It spreads through the business community like wildfire. Real estate agents will "talk up" for resale the houses built by reputable contractors, boasting so much that they will even raise the market value of such houses in the same way that name brand recognition will ring more for a supermarket product.

This established, skilled, reputable builder category also includes those "semi-retired" individuals who build only a few quality houses per year, methodically walking around with burnished cherry pipes, slowly and surely, with few wasted motions, constructing most of the house themselves, stick by stick, often with the homebuyer helping as a laborer to reduce costs. This type of builder can take a long time to complete a house, but it's worth it if the homebuyer can wait. The most specialized parts such as the foundation, roof, plumbing, and electrical systems are often subcontracted out. The rest of the house is frequently charged in a cost plus arrangement, with the builder receiving an hourly wage and the cost of all materials used. It's a good alternative for people who like a lot of individual attention and contact with their builder, and who most definitely aren't in a big hurry.

New Skilled Builders

Next come the relatively new skilled builders with good intentions who have recently begun building houses on their own. These are individuals trying to make a name for themselves who have probably "paid their dues" while working for the builders previously described—the ones who already have become established and reputable. In fact, these newer skilled contractors are where the established and reputable builders come from.

At first, the new skilled builders are happy to build a handful of dwellings per year, and might well build them as soundly as their previous employer. But what many of the new skilled builders lack, in addition to a proven, solid reputation, are the financial reserves to weather bad times and costly errors, plus the ability to run the business end of construction as the artistic science it is. They can be likened to flamboyant rookies coming up through the farm system. Some of them who might be initially productive will fail to weather the distance. These builders often construct their own houses and those of relatives and close friends to get a start and to have finished houses to show potential clients.

Marginal New Builders

The third major type of contractor is the marginal new builder. Note that you won't find any marginal *established* builders around, because the very definition defies longevity. These are builders who begin by remodeling their own kitchens or helping someone else do theirs. They might have worked a summer or two on a framing crew or putting up roofs for a good builder before one day deciding to paint their name on a truck, get an answering machine, and run an ad in the classifieds. They could enlist the support of a relative who happens to work for a real estate agency.

These are the "contractors" who aren't all that concerned with erecting a quality house. They keep a disheveled building site and hire the kind of workers who drink beer on the job and casually toss empty cans and bottles into the foundation excavation. They'd never dream of parting with the dues to join a local builder's organization or chapter of the National Home Building Association. They can afford to keep their prices competitive though, because they tend to build on inexpensive lots, purchase marginal quality materials, employ construction method shortcuts, and build simple, no-frill floorplans. These are builders who are more interested in making a profit than ensuring customer satisfaction. They might not be dishonest (or they don't mean to be) but they don't have the necessary skills, business sense, and reputation as the first or even the second type of builder. A new company formed specifically to develop a large piece of raw land into a subdivision could fall into this category. These individuals, although they might believe they are skilled builders, and can talk a good game, should not be relied upon to come through with a quality house.

Inept, Dishonest Builders

The fourth and last major builder category is that of the inept, dishonest builders. A.M. Watkins, author of a number of fine volumes on house buying, house planning, and home maintenance, refers to this worst kind of contractors as "vanishing builders." "Vanishing" is an appropriate monicker because even shortly after these builders erect a house, no matter if small or large problems arise, the vanishing builders simply disappear from the scene.

The vanishing builder is the one you *must* guard against. He might seem completely trustworthy when first approached, and even when a house is completed, everything might look in order—but just try to get him back to take care of a downspout that comes loose from a gutter, or heaven forbid, a major problem like a leaky roof or basement. The vanishing builder lives up to his name and allows his customers no recourse, no means of exacting repair or retribution.

The reasons for this builder's disappearances are legion, and his excuses many. He might have simply pulled up stakes and departed for a more favorable climate. He might have disbanded and gone back to work for someone else because he couldn't take the headaches that come with self-employment. His phone number is likely to be disconnected or changed and unlisted. Or he might declare bankruptcy.

In any event, these builders have neither the skills nor the intentions needed to become established, reputable contractors. They chase the fast dollar. Even the best scenarios will find the vanishing builders starting out as subcontractor helpers who understand the basics of how houses are constructed, observe that successful builders earn a good living, and long for a piece of the action. Several form a partnership and paint their names on their trucks and boom-build a house or two for unsuspecting friends (who don't remain friends for long) and later for gullible relatives. At first the vanishing builders take care in their initial few houses so they have something they think is nice to show potential customers, and then, when competitively quoting, give low bids, get several jobs, and begin building the houses. They select subcontractors they have been friends with, and aren't firm enough to make them do a good job so the subcontractors make mistakes that don't show up until later—such as a cellar wall that collapses inwards because it isn't properly supported. Meanwhile, the vanishing contractors, after running into irate customers, zoning problems, unreliable subcontractors, slow building periods due to the economy, weather delays, and expert competition from the skilled established and new contractors, cannot last.

In fact, few vanishing contractors even know how much money they will make on a particular house. They might actually bid—without fully realizing their expenses—less than it costs them to construct the dwelling. No company, no matter how skilled at building houses, can last or exist unless it generates a fair profit.

Sometimes vanishing builders find backers who put up front money for land they hope to develop. Together the vanishing builders and their representatives can be flashy individuals with persuasive sales skills and wildly inaccurate promises. These people don't care about the satisfaction and security gained by developing a good reputation in the community. Chances are they won't be around long. This can be exceptionally irritating to the homebuyers who invest in the first few houses in a promised subdivision that never quite gets off the ground and is saddled with unfinished roads, inadequate utilities and surroundings scarred by bulldozers—all from poor overall planning.

Now that you know what kinds of contractors are out there, you should be able to decide on which ones you'll consider. If possible, discount the latter two: the new marginal builder and the inept dishonest builder. That leaves the established, skilled, reputable builder and the skilled new builder. The first one is a sure thing. Go with one of them and you'll be guaranteed (or just about) of a satisfying, well-constructed dwelling that the builder will stand behind. Go with the newer builder and you're running a risk, however calculated, that something could happen to put him out of business in the near future.

Certain factors could influence you to go with a newer builder despite the increased risk. Perhaps you know one personally. Or you prefer to work with someone closer to your own age. Maybe he has bid substantially lower than the established builders, on the exact same specifications. All good builders were new contractors when they started, and if you believe one such builder is a rising star on the local scene, fine.

On the other hand, consider that relatively few builders starting out will have the staying power to make a long-range success in the business. Competition is tough, and the home marketplace is not very forgiving. It certainly takes a while to develop the depth of resources, the tricks of the trade, the knowledge of how to handle customers, the required contacts with sub-

contractors and housing inspectors, and the convenient credit lines at building material supply houses.

On yet another hand, if you happen to hear that any builder you are considering is being sued for something, it doesn't necessarily mean that he's dishonest or incompetent. At one point or another even the best, most trustworthy, most reliable builders are likely to become involved with some kind of legal actions brought on by unreasonable customers or conditions beyond their control.

SELECTING YOUR BIDDING CONTRACTORS

When approaching contractors for bids on your house, four is a good number of builders to insist upon. If you would choose only two, they *could* be two of the highest bidders available. With four, at least one of them is likely to be significantly lower. And builders, no matter what they say, are influenced by particulars such as the amount of work they already have waiting for them, if they're familiar with and prefer to build the floorplan in question, the time of year, and a host of idiosyncratic reasons too varied to list. In any event, let at least four contractors submit bids: the four you feel are the cream of the crop. That doesn't mean you can't go to more than four. Go ahead if you have the inclination, time, and energy. But on the whole, four will give you a good representation.

While deciding which contractors to select as your bidders, find out what you can about the likely candidates by investigating them in the following manner:

1. Ask real estate brokers and associates. However, be careful to ask only established real estate people. They know who builds the best houses and who gives the best service. They'll give you excellent leads that you can follow up later on. Mention beforehand what type of house you're looking to build, with a price range and if possible, a setting (rural or city).

2. Ask bankers and lending officers. They know which contractors to avoid, and can also refer you to reputable builders. Financial people tend to be slightly more objective than real estate agents, but they can also have slightly less knowledge about current construction particulars in the community.

3. Ask building material suppliers and subcontractors, especially plumbers, electricians, roofers, and siding contractors. If they hedge, tell them you'll keep the information confidential. These individuals know which builders are erecting a lot of homes and which builders are floundering (not paying or unable to pay

their bills). They know which builders go with quality materials and which keep coming back for marginal- or poor-quality supplies.

4. Inquire at local chapters of the National Home Builder's Association, the Chamber of Commerce, and the Better Business Bureau. The Home Builder's Association is a good place to start. It's a sharp organization that encourages builders to keep up with the latest designs and technology. More often than not, if you describe the type of house you're planning and the approximate cost, the representative (probably a local builder himself) will shuffle his feet and in a low voice suggest a few possibilities—but not officially, of course. He'll probably give you a membership listing and speak "off the record" in order not to ignore and slight other members by making a publicized referral. For a builder, being a member of an organization like the Home Builder's Association is a plus, but it needn't be mandatory for your purposes.

5. Follow up your leads and meet the builders. Select the first candidate, call him at his place of business and identify yourself as a potential customer who would like to meet him for a few minutes (it's important to stress a *few* minutes, because as you'll quickly find out, builders are extremely busy people). Preferably, meet him at his office so you can see firsthand how he takes care of the business end of his work. Find out how long he's been a contractor, how he prefers to work with his customers, and what warranties and guarantees does he make. What responsibility does he assume for subcontractors he will hire? Ask for a list of previous houses he built from five years ago to the present. Naturally, he's not going to steer you to any trouble spots, but any example will be helpful to gauge the quality of his work.

At this point, don't supply a lot of detailed information about the house you want to build. Tell him the type, the size, and perhaps a general idea of the floorplan.

It's a bad sign if a builder has jumped in and out of several construction companies during the past year or two, or if he's changed his company's name several times, or if he went from a one-man operation to a partnership to a corporation, and then back to a one-man show again. A frequent indicator of pride in workmanship (but not 100 percent accurate) is that most good, reputable builders use their names in their company's title.

Ask him to tell you what makes his houses unique. How are they well constructed? What are their selling features? Is it attention to detail? The quality of workmanship? Is it energy efficiency using the latest technological advances? Let him prove to you why you should consider his organization. If he's close-mouthed, irritated, and short of patience with you it could mean either of two things: that's just his basic personality or he's not thrilled about receiving the work. If it's the former, remember that if your personality doesn't mesh with his, the relationship will probably be an uncomfortable one. If the latter, maybe he's got all the work he can handle already, and if so, his bid will likely be high anyway. You're usually better off considering only those contractors who you can develop an open, friendly rapport with, and who also seem to really want the work.

6. Now it's time to start looking at some of the houses the builders have recently erected. Attend any of their open houses you can find, and tour some of the examples you've been given as references. It's also a good idea to visit a few open houses constructed by builders you've been told to steer clear of, in order to get a first-hand look at marginal or poor construction. Using the information you've learned so far, you can rate the houses you visit and see how they stack up against one another.

7. Contact some of each contractor's past customers. Building a house is a big deal. You want to select someone who will listen to any problems that crop up later, *after* the house is finished and the builder has been paid. Ask the owners of several of his houses how the builder followed up on problems that might have occurred. Did he act promptly and courteously on anything that needed replaced or repaired? Were the owners satisfied with his construction results and after-sale services? Even total strangers like to talk about their houses when approached in a low-key kind of way. From them you can receive lengthy outpourings of information.

Don't discuss the houses with the owners while a contractor is present or you'll get clouded information. Owners won't speak freely for fear of insulting or making the contractor angry. If you have to, go back at another time, alone.

8. Have the contractor take you through one of his brand-new or under-construction houses. This is only after scoring him high on the referrals, the personal interview, and the quality of his construction. If possible, have him take you through a house similar to the one you are planning. Let him do most of the talking. Let him demonstrate why you would be better off going with his services instead of those from somebody else.

Visting the contractor's job site will also tell you

much about his construction habits. If the place is messy and disorganized, it's reasonable to wonder whether he should be trusted to handle your house.

9. Find out if any of the potential contractors possess traits that would give them advantages over other builders. Do any of them own lumber supply stores, or any other building supply companies? If that's the case, you might be able to obtain lumber, windows, insulation, fixtures, siding, and roofing at or very close to cost. Do any of the builders own land that you'd consider as ideal for your building site? Or will any of them provide their own financing, at below-open-market rates and terms?

SOLICITING THE BIDS

Now it's time to take the four or more contractors you select and go to them with your drawings, sketches, specifications and ideas, and ask them to figure out prices for you. Don't go to your bid meetings unprepared. Instead, make up a copy of what specifications you'd like, and have one for yourself and one for each builder. Give a copy to each of the four or more builders who are in the bidding process for your house, so they're all bidding on the same job.

What follows is a fairly complete listing of individual specifications for a 2400-square-foot 2-story house with an attached 2-car garage. It contains references to most of the specifications you should be concerned with. You can make up a similar set of your own before you approach your builders for bids.

Now all you do is wait for the bids to come back for evaluation. Chances are that at least two of them will be significantly lower than the other, and will warrant your closest inspection. Are they talking apples to apples? Review their quotes with the contractors in person to clarify any points that are confusing or too general.

When your evaluations are through, just pick one, the contractor you think you should go with based on his bid, his reputation, his construction examples, and any other factors you're considering. If you've done your homework already, any of the builders you asked to bid would be a safe choice.

Choosing a good contractor will save you money, make you money, and ensure piece of mind and a comfortable, well-built home. It will make the entire house-building experience fun. Selecting a wrong contractor will definitely cost you more money, and could even destroy your sanity and faith in Mankind.

Before we're off the subject, here are a few additional don'ts to remember when dealing with the selection of your contractor:

1. Don't hire someone who is based out of town.

2. Don't hire a contractor based on his price alone.

3. Be suspicious of extremely low bids. A bid that is miles away from the rest could mean that the builder doesn't know his business costs or that a fraudulent builder is counting on breaking the contract midway through construction by enticing you to change some small detail—consequently rendering the agreement null and void. Once that happens the builder will renegotiate, charging you additional costs and blackmailing you into accepting either more house or more costs than you originally bargained for.

4. Don't accept oral agreements. Get the specifications in writing.

5. Don't believe that you're going to get excellent construction at discount prices. The best you should hope for is sound value at a fair price.

6. Avoid the marginal and vanishing builders like the plague.

AFTER YOU'VE SELECTED YOUR CONTRACTOR

After you've selected your contractor it's smart to call the unsuccessful bidders and tell them they didn't get the job. Not many people have the courage or consideration to do so, and the contractors will appreciate your honesty and straightforwardness.

Foundation:
    a. Footers: 8″ × 20″ footer
                                             concrete mix of 1-3-5
                                             strength of 2400 psi

    b. Foundation Wall: 10″ concrete blocks
                                               11 course high
                                           reinforced with rods and cement
                                           filled holes every 6 feet

    c. Basement Floor: concrete slab
                                             4″ of gravel under slab
                                           4″ thick concrete slab
                                           concrete mix of 1-2-4

    d. Footer Drains: 3″ plastic pipe in gravel
    e. Supports: 8″ steel I-beam
                                             4″ steel columns

    f. Windows: steel, (2)
    g. Sills: 2″ × 10″
                                           1/4″ foam under sill plate
                                           1/2″ bolts for anchors

Flooring:
    a. Floor Framing: SPF 2″ × 10″
                                             bridging 1″ × 3″

    b. Subflooring: plywood 4′ × 8′ × 1/2″ CDX
                                           first and second floor
                                         right angles
                                       glued to floor joists

    c. Finish Flooring:
       First Floor: 5/8″ particle board
                                       all rooms
                                       15-lb. felt between layers

       Second Floor: 1/2″ particleboard
                                       all rooms
                                     15-lb. felt between layers

    d. Ceiling Framing: SPF 2″ × 10″ second floor
                                       bridging 1″ × 3″

    e. Roof Framing: yellow pine
                                       24″ o.c.
                                     truss construction

Exterior Walls:
    a. Wood Frame: No. 2 BTR SPF 2″ × 6″ (16″ o.c.)
                                   waferboard (7/16 thick)
                                   foil covering
                                   vinyl coated aluminum siding (light gray), grade A, bevel type, double 4 (8″), aluminum nails, woodgrain finish
                                   6″ fiberglass insulation
                                   outside cellwood shutters, front only, color: smoke
                                   12″ overhangs

Inside Walls:

SPF 2″ × 4″ (16″ o.c.)
1/2″ drywall walls

oak trim throughout (stain, sealer, varnish)

Roofing:

solid plywood 4′ × 8′ × 3/4″ CDX
asphalt shingles, 250-lb.
15-lb. felt under shingles
tin flashing

Gutters and Downspouts:
  a. Gutters:

aluminum
.025 gauge
4″ size
shape 06

  b. Downspouts:

aluminum
.025 gauge
3″ size
shape—square
4 drops total

Insulation:

ceilings 10″ fiberglass
walls 6″ fiberglass

Miscellaneous:
  a. Closets: 2 rods/2 shelves per closet
  b. Other on-site improvements: rough grading only
  c. Landscaping and finish grading: by owner
  d. Walks and driveway: quote separate price for this
    6″ thick with mesh and rod reinforcement

  e. Hardware (doors):   brass

Interior Doors and Trim:
  a. Doors: flush
    oak
    1 3/8″ thick

  b. Door Trim: modern
    oak

  c. Base: modern
    oak
    3″ size

d. Finish: doors and trim
stain-sealer-varnish

e. Dining Room: chair rail

Windows: double-hung and casement

vinyl
sash thickness 1 3/8″
insulated grade
head flashing (vinyl)
weatherstripping (vinyl)
triple-track storm windows on the outside
oak trim (stain, sealer, varnish)

Entrances and Exterior Detail:

a. Main entrance door: steel
36″ wide
1 3/4″ thick
white pine for frame

b. Garage overhead doors: steel insulted, 2 9′ × 7′ doors with weather-stripping

c. Garage entrance door: steel
36″ wide
1 3/4″ thick
white pine for frame
Head flashing (aluminum)
Weatherstripping (vinyl)

d. Family room door: double french door
center hinged door with screen

Exterior Millwork: aluminum (vented)
roof louvers

Electric Wiring:

a. General: circuit-breaker
200 amp
20 circuits
all copper cable
washer and dryer hookup in basement
provisions for a sump pump hookup
provisions for a garage door opener
provisions for an overhead fan in the family room
doorbell (front door and side garage door)
2 exterior outlets (front and back)
dimmer switch for dining room, family room and eating area
humidifier hookup on furnace

		freezer outlet in basement (near washer and dryer)
b.	Bathroom #1:	light over vanity
		exhaust fan
c.	Bathroom #2:	lights over vanity
		light/exhaust/heat fan
d.	Bathroom #3:	light over vanity
		light/exhaust fan
e.	Outside Lighting:	1 at side garage door
		1 over large garage door
		1 at front door
		1 at back sliding doors
f.	Inside Lighting:	3 recessed lights over fireplace
		1 in eating area
		1 in kitchen
		1 over kitchen sink
		1 in entrace foyer
		1 inside garage
		2 in second floor hallway
		3 (one in each bedroom)
		1 in dining room
g.	Kitchen:	Exhaust fan and light (hood over range)

Plumbing:

a.	General:	all copper water pipes
		PVC drain pipes
		cast-iron piping below cellar
		washer hookup with sink in basement
		2 exterior faucets (back of house and in garage)
		provisions for a complete bath in basement (sink, shower, toilet)
		glass-lined, gas water heater (50 gal.)
		gas hookup for stove in kitchen
		gas starter in fireplace
		gas line for furnace and gas dryer in basement
		dishwasher hookup in kitchen
		water hookup for humidifier on furnace
b.	Kitchen:	stainless steel double sink
		single-lever faucet with spray
c.	Bathroom #1:	sink and toilet with insuliner (first floor half bath)
		faucet
d.	Bathroom #2:	sink and toilet with insuliner (second floor full bath)
		fiberglass tub-shower combo (5′)
		faucet
e.	Bathroom #3:	sink and toilet with insuliner (master bedroom full bath)
		fiberglass shower stall with door (4′)
		faucet

Heating:

natural gas
fan forced
perimeter system
galvanized steel ducts, supply and return furnace (minimum of 2,400 square feet heating capacity
electric start with pilotless ignition

Porches:

concrete front porch
1 pillar

Garage:

framed as house
4″ concrete floor, with slope

Floor Covering:

carpet to be responsibility of buyer
kitchen and eating area, foyer, baths #1,2,3 inlaid linoleum ($_____per square yard allowance)

Bathroom Accessories:

recessed/chrome
attached/chrome

Cabinets and Interior Detail:

cabinets (oak)
counter tops
medicine cabinets (2)
vanity (all baths)
mirrors (2)

Stairs:

oak handrail
oak balustrade

Fireplace:

ash dump and cleanout
brick facing
firebrick lining
brick hearth
wood mantel (solid)
gas starter

Chimney:

brick facing
block construction
tile flue lining
heater flue size 12″ × 12″
gas furnace vent 5″ with flue damper
water heater vent 3″ with flue damper

Chapter 30

Working With Your Contractor

After selecting the contractor you want, the next step is to draw up an agreement or contract with him. This agreement, along with the final plans, drawings, and specifications will act as guidelines for your relationship with the builder. It has to be signed by all parties before the first shovelful of earth is turned. Most builders will have such an agreement already prepared, with specification blanks to be filled in.

What follows is a sample contract agreement between a general contractor and a party who is arranging the construction of a new house. It's a good example, and covers most of the concerns either party could have, yet it can still be amended, its sections changed, added to, or deleted with the approval of both parties.

The sample contract addresses the scope of work involved, expected time for completion, and lists reasons for legitimate delays. It also states the inability of the builder to assign his responsibilities to others, describes the contract documents, explains insurance particulars, presents procedures for making alterations or extra work, and discusses housekeeping and trash removal from the building site. It covers compliance with ordinances and statutes, arbitration in the event of any disputes, plus what constitutes the acceptance and occupancy by the owners. It further provides a schedule by which the builder is paid, and describes contractor warranties.

When it comes to your contract, you should slowly review each section with the builder and question anything that isn't crystal clear.

CONTRACT AGREEMENT
By and Between

Mr.	Richard C. Jones
Mrs.	Jennifer H. Jones
Address	2014 Warsaw Avenue
	Erie, Pennsylvania

AND

Realto Construction Company
1435 East 12th Street
Erie, Pennsylvania

Building Site   Lot #35, Oakland Hills

THIS AGREEMENT MADE THIS _____ day of _____,
19_____ by and between Richard C. Jones and Jennifer H. Jones of the City of Erie, County
of Erie, and State of Pennsylvania, hereinafter called "Owners"

AND

Realto Construction Company, a corporation with principle offices in the City of Erie, County of
Erie, and State of Pennsylvania, hereinafter called "Contractor."

WITNESSETH   The Owners and Contractor for and in consideration of the mutual covenants of
each other, and for and in consideration of the work to be done by the Contractor and the money
to be paid by the Owners, as hereinafter set forth, it is agreed between the parties as follows:

1. SCOPE OF WORK   The Contractor covenants and agrees to furnish all the labor, perform all
the work that shall be required for the erection of a 2 story frame dwelling which is more fully
set forth in plans, attached hereto and marked Exhibit "A", and specifications attached hereto and
marked Exhibit "B", both of which documents have been initialed by the parties. Said dwelling
house to be built on the property of the Owners, Lot #35 Oakland Hills Subdivision.
   The Contractor covenants and agrees to do and complete all the work set forth in said plans
and specifications for the erection of said dwelling house, in a good and workmanlike manner, and
within a reasonable time after the construction job has been started. The Contractor specifically
covenants and agrees to pursue the work diligently without delay after the construction of said dwelling
house has been started by them. All work shall be new and all workmanship done and performed
under this Contract, by the Contractor, shall be of good quality and shall be performed in a good
and workmanlike manner. The Contractor shall protect all the parts of the work from damage by
cold or other elements. The Contractor shall also be responsible for temporary electrical service.
All the work and materials furnished by the Contractor shall meet or exceed the minimum FHA
requirements. The Contractor shall be responsible for the building permit, gas permit, and sewer
permit and for the expense entailed in obtaining said permits. The Contractor further covenants
and agrees to sign a Release of Mechanic's Lien before any work is started.

2. TIME OF COMPLETION   The work shall be started as soon as possible, weather permitting,
and shall be completed as soon as possible, Acts of God, strikes, material shortages, government
regulations, or catastrophes excepted. The Contractor covenants and agrees to pursue the work of
erecting said dwelling house in a diligent manner after the same has been started.

3. DELAY OF COMPLETION   If after the dwelling house has been substantially completed and
livable, full completion thereof is materially delayed through no fault of the Contractor, the Owners
shall, and without terminating the Contract, make payment for the balance due the Contractor for

that part of the work fully completed and accepted by the Owners.

4. ASSIGNMENTS   The Contractor shall not assign this Contract to others. However, this shall not prohibit the sub-contracting of parts of the work to others by the Contractor.

5. CONTRACT DOCUMENTS   The Contract documents shall consist of the Contract Agreement, the Specifications, and the Plans and they are all as fully a part of the Contract Agreement as if attached hereto and herein repeated. The Parties herewith covenant and agree that upon execution of this agreement they shall, each of them, initial the specifications and the plans.

6. INSURANCE   The Contractor shall insure himself against all claims under Workman's Compensation Acts and all other claims for damage for personal injuries, including death, which may arise from operations under this Contract, whether such operations be by themselves, or by anyone directly or indirectly employed by him. The Contractor shall save the Owners of this protection. The Owners shall maintain fire insurance and vandalism insurance on the structure as soon as the sub-floor is completed, and the Contractor shall be reimbursed from said insurance from any and all loss due to fire.

7. EXTRA WORK OR ALTERATIONS   The Owners shall have the right to make changes or alterations, but any order for change or alterations shall be in writing and signed by the Owners and the Contractor; said amount shall be stated in the written order, and to be paid to the Contractor (or Owner if it shall be a saving) before final payment is made. The extra charges, if any, shall be considered a part of the contract cost.

8. CLEANING UP   The Contractor shall, at all times, keep the premises free from all unnecessary accumulation of waste material or rubbish caused by his employees or the work and at the completion of the work he shall remove all rubbish from and about the building, and all tools, scaffolding, surplus material, and shall leave the work "broom clean."

9. ORDINANCE AND STATUTES COMPLIANCE   The Contractor shall conform in all respects to the provisions and regulations of any general or local building acts or ordinances, or any authority pertinent to the area. The Contractor covenants and agrees that he has examined the land, plans and specifications, and understands any and all difficulties that may arise in the execution of this Contract. The Contractor specifically covenants and agrees that in laying out the house, he shall observe the building line required in the sub-division. The Owners, however, shall be responsible for providing an exact survey of the building site.

10. ARBITRATION CLAUSE   In the event any dispute arises between the parties hereto which cannot be amicably settled between the parties, it is hereby agreed that each party shall appoint an arbitrator within three days after receipt of written request from the other, that the two arbitrators so appointed shall select a third arbitrator within three days after notice of their appointment, and that the arbitrators shall hear the dispute and, by majority decision, make a decision or award. It is agreed that any compensation required by the arbitrators shall be shared equally by the parties thereto regardless of the decision or award made.

11. ACCEPTANCE BY OWNERS AND OCCUPANCY   It is agreed that upon completion, said dwelling shall be inspected by the Owners and the Contractor, and that any repairs or adjustments which are necessary shall be made by the Contractor. It is further agreed that the Owners shall not be permitted to occupy said dwelling until the Contractor is paid the full amount of the Contract. Occupancy of said dwelling by the Owners in violation of the foregoing provisions shall constitute unconditional acceptance of the dwelling house and a waiver of any defects or uncompleted work.

12. TIME OF PAYMENTS
   1st Stage—Platform: 10%
   2nd Stage—Under roof: 35%
   3rd Stage—Plastered: 25%
   4th Stage—Trim completed: 20%
   5th Stage—Completion: 10%
   (Or according to bank regulations that closely resemble the above schedule.)

13. WARRANTY The final payment shall not relieve the Contractor of responsibility for faulty materials or workmanship; and he shall remedy any defects due thereto within a period of one year, material free with minimum service charge. This warranty is only valid when the Contractor is paid contract cost in full.

This contract shall be binding upon parties, their heirs, executors and assigns. And by this agreement, the parties intend to be legally bound in witness whereof, the parties have hereunto set their hands and seals the day and year first written above.

As another example of how the building materials are described, here is a sample specifications form filled out for the same 2-story house: called a Description of Materials list.

VETERANS ADMINISTRATION, U.S.D.A. FARMERS HOME ADMINISTRATION, AND
U.S. DEPARTMENT OF HOUSING AND URBAN DEVELOPMENT
HOUSING — FEDERAL HOUSING COMMISSIONER

Form Approved
OMB No. 2502-0192

(For accurate register of carbon copies, form may be separated along above fold.
Staple completed sheets together in original order.)

☐ Proposed Construction

☐ Under Construction

DESCRIPTION OF MATERIALS

No. _____
(To be inserted by FHA, VA or FmHA)

Property address _____ City _____ State _____

Mortgagor or Sponsor _____
(Name)   (Address)

Contractor or Builder _____
(Name)   (Address)

INSTRUCTIONS

1. For additional information on how this form is to be submitted, number of copies, etc., see the instructions applicable to the HUD Application for Mortgage Insurance, VA Request for Determination of Reasonable Value, or FmHA Property Information and Appraisal Report, as the case may be.

2. Describe all materials and equipment to be used, whether or not shown on the drawings, by marking an X in each appropriate check-box and entering the information called for each space. If space is inadequate, enter "See misc." and describe under Item 27 or on an attached sheet. THE USE OF PAINT CONTAINING MORE THAN THE PERCENTAGE OF LEAD BY WEIGHT PERMITTED BY LAW IS PROHIBITED.

3. Work not specifically described or shown will not be considered unless required, then the minimum acceptable will be assumed. Work exceeding minimum requirements cannot be considered unless specifically described.

4. Include no alternates, "or equal" phrases, or contradictory items. (Consideration of a request for acceptance of substitute materials or equipment is not thereby precluded.)

5. Include signatures required at the end of this form.

6. The construction shall be completed in compliance with the related drawings and specifications, as amended during processing. The specifications include this Description of Materials and the applicable Minimum Property Standards.

1. EXCAVATION:
Bearing soil, type ___Gravel, Clay___

2. FOUNDATIONS:
Footings concrete mix ___1-3-5___ ; strength psi ___2400#___ Reinforcing _____
Foundation wall: material ___10" Concrete Block___ Reinforcing ___Rods & Cement in Core___
Interior foundation wall material _____ Party foundation wall _____
Columns: material and sizes ___4" Steel___ Piers: material and reinforcing _____
Girders: material and sizes ___8" Steel___ Sills: material ___2 x 10___
Basement entrance areaway _____ Window areaways ___Galvanized Steel___
Waterproofing ___Sprayed Asphalt Coating___ Footing drains ___3" Plastic Pipe in Gravel___
Termite protection _____
Basementless space: ground cover _____ ; insulation _____ ; foundation vents _____
Special foundations _____
Additional information: ___Sill Sealer Insulation___

3. CHIMNEYS:
Material ___Stone & Block___ Prefabricated (make and size) _____
Flue lining: material ___Tile___ Heater flue size ___8 x 8___ Fireplace flue size ___12 x 12___
Vents (material and size): gas or oil heater ___5"___ ; water heater ___3"___
Additional information _____

4. FIREPLACES:
Type: ☒ solid fuel; ☐ gas-burning; ☐ circulator (make and size) _____ Ash dump and clean-out ___1 Each___
Fireplace: facing ___Stone___ ; lining ___Firebrick___ ; hearth ___Brick___ ; mantel _____
Additional information: _____

5. EXTERIOR WALLS:
Wood frame: wood grade, and species ___#2 BTR SPF (2x6)___ ☐ Corner bracing. Building paper or felt ___Foil___
Sheathing ___Waferboard___ ; thickness ___7/16___ ; width ___4 x 8___ ; ☒ solid; ☐ spaced _____ " o. c.; ☐ diagonal; _____
Siding ___Aluminum___ ; grade ___A___ ; type ___Bevel___ ; size ___DBL-4___ ; exposure ___8___ "; fastening ___Alum. Nails___

Shingles _____ ; grade _____ ; type _____ ; size _____ ; exposure _____"; fastening _____
Stucco _____ ; thickness _____"; Lath _____ ; weight _____ lb.
Masonry veneer __Stone_____ Sills __Stone____ Lintels _____ Base flashing _____
Masonry: ☐ solid ☐ faced ☐ stuccoed; total wall thickness _____"; facing thickness _____"; facing material _____
Backup material _____ ; thickness _____", bonding _____
Door sills _____ Window sills _____ Lintels _____ Base flashing _____
Interior surfaces dampproofing, _____ coats of _____ ; furring _____
Additional information: _____
Exterior painting: material __Brand X_____ ; number of coats __2__
Gable wall construction ☒ same as main walls, ☐ other construction _____

6. FLOOR FRAMING:
Joists: wood, grade, and species __SPF 2×10__ ; other _____ ; bridging __1 × 3__ ; anchors __½" Bolts__
Concrete slab: ☒ basement floor, ☐ first floor; ☒ ground supported; ☐ self-supporting; mix __1-2-4__ ; thickness __3__";
reinforcing _____ ; insulation _____ ; membrane _____
Fill under slab: material __Gravel_____ ; thickness __4__". Additional information: _____

7. SUBFLOORING: (Describe underflooring for special floors under item 21.)
Material: grade and species ___Plywood_____ ; size __4×8-½__ ; type __CDX__
Laid ☒ first floor; ☒ second floor; ☐ attic _____ sq. ft.; ☐ diagonal; ☒ right angles. Additional information: _____

8. FINISH FLOORING: (Wood only.   Describe other finish flooring under item 21.)

LOCATION	ROOMS	GRADE	SPECIES	THICKNESS	WIDTH	BLDG. PAPER	FINISH
First floor	All Rooms	5/8"	Particle Board			#15 Felt	
Second floor	All Rooms	½"	Particle Board			#15 Felt	
Attic floor		sq. ft.					

Additional information: _____

9. PARTITION FRAMING:
Studs: wood, grade, and species __SPF 2×4-Inside__ 2×6-Outside size and spacing __2×4 - 16" O.C.__ Other _____
Additional information: _____

10. CEILING FRAMING:
Joists: wood, grade, and species __2×10 - 2nd Floor__ Other _____ Bridging __1×3__
Additional information: _____

11. ROOF FRAMING:
Rafters: wood, grade, and species __Yellow Pine__ Roof trusses (see detail): grade and species __24" O.C.__
Additional information: _____

12. ROOFING:
Sheathing: wood, grade, and species __Plywood__ ; ☒ solid; ☐ spaced _____" o.c.
Roofing __Asphalt Shingles__ ; grade __C__ ; size __4×8-½__ ; type __CDX__
Underlay __#15 Felt__ ; weight or thickness _____ ; size _____ ; fastening _____
Built-up roofing _____ ; number of plies _____ ; surfacing material _____
Flashing: material __Tin__ ; gage or weight _____ ; ☐ gravel stops; ☐ snow guards
Additional information: _____

13. GUTTERS AND DOWNSPOUTS:
Gutters: material __Aluminum__ ; gage or weight __.025__ ; size __4"__ ; shape __O.G.__
Downspouts: material __Aluminum__ ; gage or weight __.025__ ; size __3"__ ; shape __Square__ ; number __4__
Downspouts connected to: ☐ Storm sewer; ☐ sanitary sewer; ☐ dry-well. ☐ Splash blocks: material and size _____
Additional information: _____

14. LATH AND PLASTER:
Lath ☒ walls, ☒ ceilings: material _____ ; weight or thickness __½"__ Plaster: coats __1__ ; finish _____
Dry-wall ☐ walls, ☐ ceilings: material _____ ; thickness _____ ; finish _____
Joint treatment _____

15. DECORATING: (Paint, wallpaper, etc.)

ROOMS	WALL FINISH MATERIAL AND APPLICATION	CEILING FINISH MATERIAL AND APPLICATION
Kitchen		
Bath	BY OWNER	
Other		

Additional information: _____

16. INTERIOR DOORS AND TRIM:
Doors: type __Flush__ ; material __Oak__ ; thickness __1-3/8"__
Door trim: type __Modern__ ; material __Oak__ Base: type __Modern__ ; material __Oak__ ; size __3"__
Finish: doors __Stain - Sealer - Varnish__ ; trim __Stain - Sealer - Varnish__
Other trim (item, type and location) _____
Additional information: _____

17. WINDOWS:

Windows: type _DBL-Hung + Cas._ make _____ ; material _Vinyl_ ; sash thickness _1-3/8"_

Glass grade _Insulated_ ; ☐ sash weights ☒ balances, type _____ ; head flashing _Vinyl_

Trim: type _Modern_ ; material _Oak_ Paint _Stain-Sealer-Varnish_ number coats _____

Weatherstripping: type _____ ; material _Vinyl_ Storm sash, number _____

Screens: ☒ full, ☐ half, type _____ ; number _13_ ; screen cloth material _Aluminum_

Basement windows: type _Hopper_ ; material _Steel_ ; screens, number _____ ; Storm sash, number _____

Special windows _____

Additional information _6' Aluminum Glass Sliding Door_

18. ENTRANCES AND EXTERIOR DETAIL:

Main entrance door: material _Steel_ ; width _36"_ ; thickness _1-3/4"_. Frame: material _W. Pine_ , thickness _5/4_"

Other entrance doors: material _Steel_ ; width _32"_ ; thickness _1-3/4"_. Frame: material _W. Pine_ ; thickness _5/4_"

Head flashing _Aluminum_ Weatherstripping: type _Vinyl_ ; saddles _____

Screen doors: thickness _____"; number _____ ; screen cloth material _____ Storm doors: thickness _____"; number _____

Combination storm and screen doors: thickness _____"; number _____ ; screen cloth material _____

Shutters: ☐ hinged, ☒ fixed. Railings _____ , Attic louvers _____

Exterior millwork: grade and species _Aluminum (vented)_ Paint _____ ; number coats _____

Additional information: _Roof Louvers_

19. CABINETS AND INTERIOR DETAIL:

Kitchen cabinets, wall units: material _____ ; lineal feet of shelves _70_ ; shelf width _12_

Base units: material _____ ; counter top _____ ; edging _____

Back and end splash _Formica_ Finish of cabinets _Factory_ ; number coats _____

Medicine cabinets: make _14 x 18 Bath 2_ ; model _Mirror - Bath 1 + Lav 1 + 2_

Other cabinets and built-in furniture _Vanity - Bath 1 + Lav 1 + 2_

Additional information _____

20. STAIRS:

STAIR	TREADS		RISERS		STRINGS		HANDRAIL		BALUSTERS	
	Material	Thickness	Material	Thickness	Material	Size	Material	Size	Material	Size
Basement	SPF	1-1/2	W. Pine	3/4	SPF	2 x 10	W. Pine	2"		
Main	Oak	1-1/8	W. Pine	3/4	W. Pine	1 x 12	W. Iron			
Attic										

Disappearing: make and model number _____

Additional information: _____

21. SPECIAL FLOORS AND WAINSCOT: *(Describe Carpet as listed in Certified Products Directory)*

	LOCATION	MATERIAL, COLOR, BORDER, SIZES, GAGE, ETC.	THRESHOLD MATERIAL	WALL BASE MATERIAL	UNDERFLOOR MATERIAL
FLOORS	Kitchen + Eating Area	Inlaid Lino (# ___ Sq yd Allowance)		Oak	Particle Brd.
	Bath 1 + 2	" " " " "		"	" "
	Lav 1 + 2	" " " " " " "		"	" "
	Foyer	" " " " " " "		"	" "

	LOCATION	MATERIAL, COLOR, BORDER, CAP. SIZES, GAGE, ETC.	HEIGHT	HEIGHT OVER TUB	HEIGHT IN SHOWERS (FROM FLOOR)
WAINSCOT	Bath				

Bathroom accessories: ☒ Recessed; material _Chrome_ ; number _5_ ; ☒ Attached; material _Chrome_ ; number _5_

Additional information: _____

22. PLUMBING:

FIXTURE	NUMBER	LOCATION	MAKE	MFR'S FIXTURE IDENTIFICATION NO.	SIZE	COLOR
Sink	1	Kitchen		or Equal	32 x 21	St. Steel
Lavatory	4	Baths + Lavs		or Equal	18"	White
Water closet	4	" " "	" "	" "	Cadet	White
Bathtub	1	Bath 1	Fiberglass		5'	White
Shower over tub ▲	1	Bath 1				
Stall shower △	1	Bath 2	Fiberglass		4'	White
Laundry trays						

▲☒ Curtain rod △☒ Door ☐ Shower pan: material _Fiberglass_

Water supply: ☒ public; ☐ community system; ☐ individual (private) system. ★

Sewage disposal: ☒ public; ☐ community system; ☐ individual (private) system. ★

★ *Show and describe individual system in complete detail in separate drawings and specifications according to requirements.*

House drain (inside): ☒ cast iron; ☐ tile; ☐ other _____ House sewer (outside): ☒ cast iron; ☐ tile; ☐ other _____

Water piping ☐ galvanized steel, ☒ copper tubing; ☐ other _____ Sill cocks, number **2**

Domestic water heater: type _____ **Gas** _____; make and mode' _____; heating capacity **40**

_____ gph. 100° rise. Storage tank: material **Glass Lined** _____; capacity **40** gallons.

Gas service ☐ utility company, ☐ liq. pet. gas; ☐ other _____ Gas piping ☐ cooking; ☐ house heating.

Footing drains connected to ☐ storm sewer; ☒ sanitary sewer, ☐ dry well. Sump pump; make and model _____

_____ ; capacity _____ ; discharges into _____

23. HEATING:

☐ Hot water ☐ Steam. ☐ Vapor. ☐ One-pipe system. ☐ Two-pipe system.

   ☐ Radiators. ☐ Convectors. ☐ Baseboard radiation. Make and model _____

   Radiant panel: ☐ floor; ☐ wall; ☐ ceiling. Panel coil: material _____

   ☐ Circulator. ☐ Return pump. Make and model _____ ; capacity _____ gpm.

Boiler: make and model _____ Output _____ Btuh.; net rating _____ Btuh.

Additional information _____

Warm air: ☐ Gravity ☒ Forced. Type of system **Per Motor** _____

   Duct material: supply **Galv. steel** ; return **Galv. Steel** Insulation _____ , thickness _____ ☐ Outside air intake.

   Furnace make and model _____ Input **130,000** Btuh.; output _____ Btuh.

   Additional information: _____

☐ Space heater; ☐ floor furnace; ☐ wall heater. Input _____ Btuh.; output _____ Btuh.; number units _____

   Make, model _____ , Additional information: _____

Controls: make and types _____

Additional information: _____

Fuel: ☐ Coal; ☐ oil; ☒ gas; ☐ liq. pet. gas; ☐ electric; ☐ other _____ ; storage capacity _____

   Additional information: _____

Firing equipment furnished separately: ☐ Gas burner, conversion type. ☐ Stoker: hopper feed ☐; bin feed ☐

   Oil burner: ☐ pressure atomizing; ☐ vaporizing _____

   Make and model _____ Control _____

   Additional information: _____

Electric heating system: type _____ Input _____ watts; @ _____ volts; output _____ Btuh.

   Additional information: _____

Ventilating equipment: attic fan, make and model _____ ; capacity _____ cfm.

   kitchen exhaust fan, make and model _____

Other heating, ventilating or cooling equipment **Vent Fan Lav 1 & 2**

24. ELECTRIC WIRING:

Service: ☒ overhead; ☐ underground. Panel: ☐ fuse box; ☒ circuit-breaker; make **GLS** AMP's **150** No. circuits **20**

Wiring ☐ conduit; ☐ armored cable; ☒ nonmetallic cable; ☐ knob and tube; ☐ other _____

Special outlets ☒ range; ☐ water heater; ☐ other **Dryer**

☐ Doorbell. ☒ Chimes. Push-button locations **Front Door** Additional information: _____

25. LIGHTING FIXTURES:

Total number of fixtures **22** Total allowance for fixtures, typical installation, $ _____ **Retail**

Nontypical installation _____

Additional information: _____

26. INSULATION:

LOCATION	THICKNESS	MATERIAL, TYPE, AND METHOD OF INSTALLATION	VAPOR BARRIER
Roof ____			
Ceiling	10"	Fiberglass	Included
Wall	6"	Fiberglass	Included
Floor ____			

27. MISCELLANEOUS: (Describe any main dwelling materials, equipment, or construction items not shown elsewhere; or use to provide additional information where the space provided was inadequate. Always reference by item number to correspond to numbering used on this form.)

Shelf + Rod in closet

HARDWARE: *(make, material, and finish.)* Brass (Quickset or Equal)

SPECIAL EQUIPMENT: *(State material or make, model and quantity. Include only equipment and appliances which are accept-able by local law, custom and applicable FHA standards. Do not include items which, by established custom, are supplied by occupant and removed when he vacates premises or chattles prohibited by law from becoming realty.)*

JKP27 Double Oven

GSD551W Dishwasher

GFC310 Disposal

C221 Range

PORCHES: Concrete Front Porch

TERRACES: By Owner

GARAGES: Framed as house, 4"concrete floor, 2-9'x7' Overhead Doors

WALKS AND DRIVEWAYS: By Owner
Driveway: width _____ ; base material _____ ; thickness _____ "; surfacing material _____ ; thickness _____ "
Front walk: width _____ ; material _____ ; thickness _____ ". Service walk: width _____ ; material _____ ; thickness _____ "
Steps: material _____ ; treads _____ "; risers _____ ". Cheek walls _____

OTHER ONSITE IMPROVEMENTS:
(Specify all exterior onsite improvements not described elsewhere, including items such as unusual grading, drainage structures, retaining walls, fence, railings, and accessory structures.)

Rough Grading Only

LANDSCAPING, PLANTING, AND FINISH GRADING: By Owner
Topsoil _____ " thick: ☐ front yard; ☐ side yards; ☐ rear yard to _____ feet behind main building.
Lawns *(seeded, sodded, or sprigged)*: ☐ front yard _____ ; ☐ side yards _____ ; ☐ rear yard _____
Planting: ☐ as specified and shown on drawings; ☐ as follows:
_____ Shade trees, deciduous. _____ " caliper. | _____ Evergreen trees. _____ ' to _____ ', B & B.
_____ Low flowering trees, deciduous, _____ ' to _____ ' | _____ Evergreen shrubs. _____ ' to _____ ', B & B.
_____ High-growing shrubs, deciduous, _____ ' to _____ ' | _____ Vines, 2-year _____
_____ Medium-growing shrubs, deciduous, _____ ' to _____ ' | _____
_____ Low-growing shrubs, deciduous, _____ ' to _____ ' | _____

IDENTIFICATION.—This exhibit shall be identified by the signature of the builder, or sponsor, and/or the proposed mortgagor if the latter is known at the time of application.

Date_____   Signature_____

Signature_____

DESCRIPTION OF MATERIALS
HUD-92005 (6-79)
VA Form 26-1852, Form FmHA 424-2

INSURANCE AND WARRANTIES

The contractor probably carries general liability and completed operations insurance. Certainly, if you're working with a relatively new contractor, ask to see his insurance certificates to be safe.

Your contractor might also claim that he'll take care of the insurance—all of it—during the early stages of construction. Even if he means well, consider that the contractor's insurance is purchased specifically with the contractor's best interest in mind, not that of the homeowner's.

It's a good idea to obtain your own homeowners

policy before the ground is broken. That will make for an overall combination of the contractor's insurance, and yours. His policy might only cover mistakes that he or his subcontractors might make, plus general liability—but not accidental damage that could occur to the house at any stage of construction. Conversely, your homeowners policy will probably not cover any contractor errors. His insurance will enable him to make repairs if they're needed because his men, or a subcontractor he hires makes errors resulting in a sunken footer, a collapsed basement wall, or a faulty roof. Your homeowners policy will likely cover only hazards and accidents.

For the relatively inexpensive cost of protection, it's best to watch out for your own interests from the very start.

Warranties are another important consideration. Few houses are constructed perfectly. No matter how good or conscientious the builder is, small problems will inevitably occur, such as sticking doors, a defective appliance, a crack in the basement floor or driveway, or even a minor settling of a wall. Many builders will fix practically anything, no questions asked, some within time periods far beyond that outlined in the original warranty. Others will make the owners work (from repeated phone calls, to begging) for the repairs, yet still do them begrudgingly. Some contractors turn around and blame the owners for any problems and refuse to honor the warranty on those grounds, when they can get away with it. Of course, the vanishing builder is nowhere to be found to be made aware of his mistakes.

All new houses carry an "implied warranty of habitability," which forces builders to repair major construction defects such as a caved-in basement. But smaller problems, especially after the standard warranty period of one year, can leave homeowners with no alternative but to make repairs themselves and eat the costs. Although after the first year is up, an owner has little recourse for the small problems, there is an alternative for major defects.

See if your builder can subscribe to any home warranty programs initiated by a Home Builders Association in your area. If so, new home warranties can be extended to up to 10 years of protection. The builder typically pays a single one-time premium that's passed along to the homebuyer in the dwelling's cost. But the major house structure, the potentially dehabilitating plumbing, heating, cooling, and electrical systems, plus the general workmanship and materials are protected.

If for some reasons the builder can't or doesn't follow through on repairs, the home warranty will cover their expenses after a relatively small deductible is paid by the owners. Warranties available from your builder or other individuals and institutions should be thoroughly researched and considered before the contract is completed.

AFTER THE CONTRACT IS SIGNED

After the contract is signed, it's time for the house construction to begin. Between then and the date of completion you'll be locked into a unique and you hope satisfying relationship with the general contractor, or if it's a large construction company, with whichever individual is responsible for supervising the construction.

Naturally, in this relationship, like any, there are things you should do and things you shouldn't. While you want to become involved enough to let the contractor know that you understand the building process, and are aware of the progress he is making, you don't want to become an impedance to that progress. You don't want to irritate him by making unreasonable demands or by asking too much of his time. Certainly, you don't want to make him angry with you.

To keep the relationship on the up and up, here are a few suggestions that have worked for other homebuyers in the past, and are likely to work for you as well.

Keep Communications Honest, Open, and Current

The most critical part of your relationship is to keep the lines of communication honest, open, and current. Yes, the builder should return your calls . . . eventually. To say promptly is probably asking too much. On the other hand, you shouldn't have to trap him to get his attention. Builders, however, are notoriously hard to pin down. They *are* busy, especially during times when construction is booming, and especially if they're any good. In fact, if they call you right back, something's probably amiss. Maybe it's written somewhere, in secret oath, that one irritating characteristic of a master builder is his delinquency in getting back to you. But, and here is how he differs from the vanishing or marginal builder who never calls back, once the good builder does return your call, he gives you his full attention (undoubtedly irritating some other customer who is waiting for a call back) and will usually go way beyond what could be expected as a normal response to your concerns, or he'll explain to you in depth just why it is that you're wrong.

You want the builder to treat you honestly and fairly. If he makes a mistake, he should tell you about it then stand by his work until the error is corrected. And he expects the same from you. Sure, it's kind of a biblical attitude: treat others as you would have them treat you ... but it works. So be honest and above board. Don't sandbag. If something looks wrong, don't wait until the entire house is completed before bringing it to the builder's attention. If you do, more complications will invariably result because the builder didn't repair the error shortly after it occurred. In fact, bring up concerns as soon as you have them. This also shows the builder that you've prepared yourself to intelligently discuss the ongoing construction with him. You don't want to relinquish all control.

Try not to be at the site just anytime. Instead, time your visits to coincide with important steps of the building process that you'd like to watch to make sure that no skimping is done, and so you'll understand fully what was done, for future reference. For example, good times to be there are when the foundation is being poured over the reinforcement rods, or when some of the wall insulation is going up, or when the first layer of roofing is being laid.

If you have a complaint, don't stand on the job site and argue with the builder in front of his crews. Motion him off to the side or call him by phone later and explain why you have misgivings and then give him a reasonable amount of time to correct the situation if he was wrong.

Contractors almost always take care of minor, irritating problems, at the very least because they'd rather have a satisfied customer than a person bad-mouthing them throughout the neighborhood. Satisfied homeowners can be used as references to develop more business.

Don't Try To Get Extras From the Contractor

You should decide exactly what you want *before* the contractor begins construction. But if you must make a change along the way—and a *few* minor changes are often made with little fuss by the contractor—tell him as soon as possible. For some people, the temptation to make one change after another becomes an obsession. Could the builder make the family room a few feet longer? Maybe. Maybe not. Perhaps there isn't enough room at the side of the house to extend any closer to an adjacent lot. Perhaps the house would then require different roof trusses than the ones already delivered. Could the builder install a bay window in the living room,

now that you've just received a surprise income tax refund? Or, since the contract, you have decided not to have any more children and would the builder make the original four bedrooms into three? Can a shower stall be put in the basement? It would only take a few more lengths of pipe and a floor drain . . .

Those kinds of changes, when insisted upon, especially when requested at a middle or last stage of construction, and especially when one change is piled atop one another, have ruined many a healthy relationship between builder and owner. To avoid such conflicts of interest, try to make all of your changes before the construction starts, or at least keep the changes to a bare minimum and make them as early as possible after the construction begins—plus be prepared to reimburse the contractor for any additional costs incurred because of those changes.

Empathize With The Builder

He has his problems: people not showing up for work, the weather, suppliers running out of materials or sending the wrong things, subcontractors making mistakes that the builder takes the blame for. Try to see things from his point of view, too.

Keep a Friendly Eye On the Work

Don't do it in a hypercritical way. While on the site, jot down questions or concerns for later discussion with the builder. The construction of a house is an interesting process, and you should watch a good deal of it simply to understand how it's put together in case you ever want to modify any part of it or effect major repairs. If you know what went into the house, no one can buffalo you later on as to what's there, and where.

Your main reason for keeping tabs on the construction shouldn't be as watch dog—although you can't help appearing a little like one—but as a student, because you've already guaranteed yourself a good builder and good workmanship by virtue or your initial contractor selection.

Be Friendly and Tactful, Yet Firm

You want to foster an amiable relationship so you wouldn't be afraid to use the same contractor to build another house, should you ever want one. To that effect, you should treat him with the same courtesy you'd treat a good friend or relative. However, do remember that you're picking up the tab; you're paying for everything—the supplies, the labor, the contractor's overhead, and a fair profit all come out of your pocket. Be friendly and tactful, yet firm, and the contractor will respect you for it.

These pointers should help you develop a solid working relationship with your builder. But remember, your builder has seen a lot of customers, a lot of problems, and a lot of crazy situations while in the business. As a group, builders know customer psychology, having seen many more customers than you've seen builders. Many contractors who probably follow a list of their *own* pointers (i.e. listen to whatever the customer says, then build it your own way) are experts at convincing you to see things their way.

An example of this occurred when one particularly taciturn builder, a man of few words yet a master at his craft, was called upon by an owner who was considering knocking out a bedroom wall to expand the sleeping space into an adjacent living room. The builder knew from past dealings with the owners that the husband would be continuously wondering if such a modification would be the right thing to do.

After listening to what the couple wanted done, and agreeing it was a smart move to make, the builder slowly got up and sidled over to the plaster wall in question. He leaned against the wall near a framed oil painting and asked the seated owners if they were sure they wanted the job done. The owner looked at his wife and uttered a tenuous yes. Then the builder, in fluid motion, lifted his steel claw hammer from his belt and smashed a hole in the wall the size of his fist, right then and there, in front of the startled owners.

After the shock wore off, the owners realized why he had done it. The builder's action left little room for them to change their minds overnight, and also took some of the worry out of the very decision itself.

They *had* to do something to fix the hole.

Chapter 31

Setting Up Your
Maintenance Program

It's best to set up a maintenance program before the contractors leave and before the final inspection. After all, why wait until all of the contractors are gone? This way the house and its components are still fresh in the builders' and subcontractors' minds, and you can readily get all the information you need to customize a schedule and checklists for your particular house and equipment.

What better time to record the serial numbers, to file away the instruction manuals, purchase receipts, and pertinent information such as the installers' phone numbers and manufacturers' spare parts lists? And after you plan your maintenance program, while in the first part of the final inspection with your builder, pretend it's your first maintenance inspection and draw what feedback you can from him about what you're inspecting for.

Years ago there wasn't the common expectation of maintenance-free houses as there is today. Sure, it has a lot to do with modern materials and improved manufacturing processes, but also, people today want to spend less time being caretakers to their belongings, houses included, and more time pursuing career and leisure activities. People no longer want to spend a month of their summer scraping and painting a house

from foundation to weathervane. They don't want to varnish a porch every year. Or to reseal and caulk windows every fall. Instead, everyone wants trouble-free houses, and for the most part we have them. In fact, compare a drive down a new subdivision street with another ride down a block of older dwellings. There many of the 60- and 70-year-old houses stand, grouped together on narrow lots, with narrow driveways, tiny front yards, and big porches that are chipped and peeling.

Nowdays, aluminum, vinyl, and a host of other weather and time-beating materials are standard fare.

However, even though you go to great lengths to obtain the most advanced, efficient, maintenance-free components available for your new home, it doesn't mean you can just live there trouble-free without caring for the place. While there aren't as many ways for a house to deteriorate as there were in previous years, some maintenance inspections and fine tuning are in order to prevent long-range problems from developing.

Several valid reasons for preventive maintenance exist. Just as people should schedule themselves a physical examination every so often to detect and take care of potential health problems before they can turn into serious conditions with catastrophic results, so should the major components of a house be inspected, to identify and take care of conditions that could lead to serious failures. More specifically, here are the main reasons for preventive maintenance:

1. To prevent failures at inopportune times which could result in safety hazards, such as improper wiring, a malfunctioning furnace or water heater, or a plugged fireplace flue.

2. To allow you to make repairs at your convenience, instead of on an emergency basis.

3. To prevent minor failures which are inexpensive and simple to correct, from becoming complex, expensive repairs.

4. To make sure your house is a safe and comfortable place to live in.

5. To maintain your house's value and ensure your pride of ownership.

Components in a house seldom fail suddenly. There are usually warning signs. If you know what to watch for and act when you see danger, life on the home front will be easier and much less expensive. The smart homeowner is the one who knows his or her house intimately and who routinely follows a checklist of inspections to make sure no problems are developing. To do this you should understand the construction and the operation of all appliances and major pieces of equipment. You should know where things are, even hidden items such as septic tanks, wells, sewer lines and other underground utilities. You should know what the builder and manufacturer guarantees and warranties cover, and what your homeowners insurance is all about.

Preventive maintenance is the idea that you lessen the likelihood of major damages by repairing minor problems. But to repair or even prevent the minor problems, you need to be aware of what they are and where to look for them.

It's a known fact that most breakdowns you're going to encounter will be small ones—many of which you can probably repair or learn to repair yourself. You won't have to worry about replacing washers in a washerless faucet, but you still must contend with items such as plugged drains, cracking asphalt or concrete driveways, sticky windows, and worn-out blower belts on furnaces. Preventive maintenance can't guarantee that breakdowns will never occur, and it won't even prevent surprise failures; but it *will* minimize the time, effort and money you'll spend on major repairs.

Yes, there's something nice about knowing that your house is in top shape. It's a feeling of confidence. The proof is reflected in a house's condition. Find a run-down, shoddy-looking house that's not too old, and chances are it got that way through a lack of maintenance—to a cumulative deterioration that starts out slowly, almost imperceptibly, but gains momentum once it negotiates past the "maintenance-free" characteristics of new materials as the seasons roll by and each tiny problem mushrooms into others.

SAVING MONEY ON SERVICE CALLS

Planning preventive maintenance in advance gives you time to anticipate what's needed to perform the tasks, and lets you make the best schedule possible for your time and pocketbook. Many homeowners will do this hit or miss, counting that their everyday movements throughout the house will reveal anything that's obviously failing or failed, without having to go out of their way to perform specific inspections.

When something does need attention there's always the question "Who should do the work?" The days of old-fashioned handymen who could fix anything for a song are long gone. Instead we have armies of specialists: plumbers for plumbing, roofers for roofs, electricians for electrical repairs, and septic people for sewage systems. Naturally, homeowners must not only pay for their time, but for their training and expertise as well; the same person who comes to your house to fix a leaking pipe can easily be an individual qualified to lay out the plumbing for an entire high rise apartment building. There's no doubt about it: service calls are expensive.

Whenever possible, you can save money by learning how to do the most basic maintenance and repairs. Even if you never want to lay a finger on a furnace or hot water tank, it's worth it just to learn about their operations. Then, when you must call a professional, you can accurately describe the problem so the serviceman can bring the right tools and materials and won't waste any time once he arrives at your house. In fact, a serviceman might even be able to tell you what to do over the phone.

The largest part of the typical service call bill usually consists of labor. And in most cases, the time spent just getting to and from your house makes up the lion's share of that. The trick is to get as much efficiency out of a call as possible. Do that by finding several other jobs that can be done by the same serviceman on the same service call. That's why it's important for you to know what kind of shape your equipment is in, and what can be inspected, adjusted, and serviced. Then, when you call a plumber to fix a leaky pipe, he can also service the sump pump and inspect the hot water tank.

If your preventive maintenance is good enough, you should be able to keep the servicemen away from weekends, late-evening hours, and holidays—all time you must pay premiums.

Line up qualified servicemen *before* you need them, so you'll know who to call without blindly paging through the phone book. Keep away from repairmen who have the same characteristics as vanishing builders.

POTENTIAL TROUBLE SPOTS

For the record, here are most of the items in a typical house that need preventive maintenance inspections and attention:

Heating and Cooling Systems

Poorly maintained, defective heating and cooling units produce less warmth and coolness for the money, and cost substantially more to operate. A more important consideration for the heating unit is that one not working properly can be extremely hazardous, with fire and asphyxiation two possibilities.

Many homeowners take both heating and cooling systems for granted until the systems quit working, the former most likely at the front end of a New Year 3-day holiday weekend, and the latter in the middle of a Death-Valley-like heat wave.

Fuel Burners. Have pilot lights and burners cleaned and adjusted for safety and efficiency by qualified servicemen. This is the stage at which fuel is consumed. A small or yellow flame at the pilot indicates that either the pressure regulator needs adjustment or that the pilot burner is dirty. Oil burners need more frequent attention than gas burners because oil does not burn as cleanly, but even the most modern gas models can still benefit from yearly inspections and adjustments.

Furnaces. Forced air furnaces are reliable workhorses that need to have their filters cleaned or changed at least once per month when in use. Otherwise, a dirty, clogged filter can reduce the efficiency of heat output by as much as 70 percent, and will cause the entire heating unit to overexert itself trying to reach a 68-degree thermostat setting when outside temperatures plummet below freezing. The air that passes through the filter is cool air drawn from the inside of the house. After being filtered clean it gets warmed by the fuel burner flame, then blown back into the house through ducts and registers. Naturally, if you live in a dusty environment or own a long-haired pet, you'll have to clean or replace air filters more frequently than if you lived in a cleaner area and didn't own a pet.

Next comes the V-belt that runs from the electric motor on the furnace to the blower fan unit that forces air through the ductwork. As time passes, the belt will wear and fray at the edges, and it will stretch. Belt stretching can be corrected by taking up whatever slack has developed by loosening the bolts on the blower motor mount and moving the motor slightly farther away from the blower. Do this whenever the belt can be pulled with your fingertips more than 1/4-inch out of line. Replace worn and frayed belts.

Some of the blower motor and fan assemblies have permanently lubricated bearings, while other bearings should be oiled before each heating season begins and periodically throughout the fall and winter months.

Boilers. Boilers are expensive and somewhat temperamental heating units, especially if their operating conditions are allowed to vary from those suggested by the manufacturer. Read the instructions and maintenance literature carefully and have boilers inspected by a professional before each heating season, then do it by yourself at least once per month during the rest of the cold months while the boiler is in use. There should be a low-water emergency cutoff that will prevent the boiler from operating in case not enough water gets into the boiler for it to run safely. The low water cutoff should be purged at least once per month—or some water drained through it—to expel any rust or corrosion that might settle there and plug the line.

If pockets of air form in the pipes, radiators, and baseboards of hot water systems, they must be bled or released so the hot water can freely circulate.

Others. Although electric heating systems have traditionally been some of the most expensive to operate because of the cost of "heating" electricity when compared to the cost of heating with natural gas or oil, they're practically maintenance-free. About all that's needed is to periodically vacuum dust and dirt from the heater and heating coils.

Solar and other alternative heating sources should be inspected and maintained as recommended by the manufacturers. With all equipment, be aware of where emergency shutoff valves and switches are, and how to restart a system if needed. Ask the serviceman what maintenance you should be doing between his service calls, and how you can confirm that everything is working safely and efficiently.

Plumbing

Naturally, the most obvious conditions to inspect plumbing for are leaky pipes and fixtures. You should know quickly if a leak develops by observing a puddle nearby, or water dripping from ceilings, or stained walls. For the most part, plumbing systems are reliable, with only a few points to keep in mind.

If outside temperatures approach freezing, shut off valves to outside connections and faucets. These shutoff valves should be located inside the house, close to where the pipe goes through an outside wall. When the inside shutoffs are closed, then open the outside fixtures to let any standing water drain from those outside pipes and fixtures.

Every few months remove a gallon or so of water from the drain valve at the bottom of your hot water tank. Along with this expelled water will be any rust, sediments, and mineral deposits that settle on the tank bottom and hinder the efficiency of heat transfer. The maximum life of a hot water tank can be realized if the temperature setting is kept between 140 to 150 degrees Farenheit. Higher temperatures are not necessary, and scalding water is both dangerous and will deteriorate the packing in some hot water faucets, causing leaks.

Septic tanks should be examined professionally once a year to determine if pumping and cleaning is needed. If services are in order, you won't want to undertake them yourself. They're for specialists with equipment designed to handle such unpleasant tasks.

Sump pumps and well pumps should also be serviced as noted in their instruction manuals. If they're rarely used, put them into service several times per year to make sure they'll work when needed.

Floor drains in the garage and basement should be tested at least twice per year to make sure they'll work in case needed during an emergency. Just lift their covers and stick a running hose down them. If mud or debris has accumulated beneath the grate cover, shovel it out first.

Fireplaces and Stoves

In this modern age it isn't surprising that most people who use woodburning equipment have had little, if any, practical experience heating with wood. Consequently, a sizable number of fires occur each year due to faulty installation, misuse, and lack of proper maintenance of fireplaces and stoves. Most fires are caused by one or more of: heat radiation from the stove, stovepipe, or chimney igniting adjacent combustible materials; sparks

escaping into the house; sparks from the chimney top; flames catching on creosote accumulation in the chimney flue. (Creosote is a dark, sticky, resin-like material formed by unburned chemicals that are borne by smoke up the chimney flue. They coat the inside of the flue and create the danger of chimney or flue fires.)

Before each heating season make sure that the chimney flue is not partially obstructed with bird or squirrel nests, beehives, a fallen branch, or a child's softball. And see that the dampers are in good working order. Remove any ashes and debris from the ash pit, and if the flue is thick with creosote and ash, have it cleaned.

Central Air-Conditioning

With air-conditioning it's filters, filters, filters. Dirty filters can really block the output of cool air and will greatly increase the effort and length of time needed to cool a house. If air-conditioning filters are too dirty or clogged to see the beam from a flashlight through, it's time to clean or replace them.

It's also necessary to check the air conditioner's outside condenser. Keep condenser vents free of vines, leaves, mud, grass, and any other encroaching materials that hinder air circulation. Keep drainage tubes clear.

When it's time to stop using air-conditioning for the year, remove dust and dirt from each unit with a vacuum cleaner and securely cover the exterior housings until the next cooling season begins.

Electrical Systems

The wiring installed in most modern homes is largely trouble-free. Circuit breaker boxes are included as safety valves. If a circuit breaker trips because too much demand is placed on that part of the wiring from too many appliances plugged in at once, or from operating a defective appliance, just follow the resetting instructions on the panel—usually you flip the circuit breaker switch back on. Of course, the cause of a tripped circuit should be identified and corrected.

Even if your breakers never trip for any reason, it's a good idea to manually trip them a few times per year to make sure they're operable and the contacts are in working order.

Also, whenever power is lost to the entire house, as during a lightning or ice storm, there can be high-voltage surges that burst through the house wiring that can damage plugged-in appliances and equipment such as stereos and personal computers. During violent storms it's best to unplug as many appliances as possible.

Roofs

As mentioned earlier, roofs take a beating—from rain, sleet, hail, wind, sun, and alternate freezing and thawing temperatures.

Even if no inside water marks or leaks are noticed, you should still periodically take a pair of binoculars, and depending on the roof's pitch, inspect the entire surface from either the ground nearby or from the second-story windows of neighbors' homes. Look for bent, ripped, or missing shingles, and for metal flashing that's warped, torn loose by the wind, or rusting and corroding.

If a leak does develop, consider that the actual damage might be a substantial distance away from where the water is dripping. If you decide to take care of your own minor repairs, fine. But be careful. Never venture out on a roof when it's wet, or during winter when the shingles are frozen and ice-laden. If ice dams occur on the roof due to constant melting snow running to the roof edges where it keeps freezing and accumulating, and water eventually backs up under the shingles and seeps through the roof into the house, it's a sign that too much heat is escaping near the roof's peak. The outer surface of the roof, even near the peak, should be kept cold during winter by the use of proper insulation on roof ceilings and by the installation of adequate roof ventilation.

Gutters and Downspouts

Gutters and downspouts absorb much of the same abuse as roofs do, mostly from the elements. Periodically remove collected debris such as twigs, leaves, acorns, and dirt from gutters and downspouts, then flush clean with water. Observe the gutter system while it's raining and check for blockages, leaks, and places where it might be coming loose from the house. Check to see that the water running through the downspouts is flowing away from the house foundation. During winter, don't let large icicles build up and hang on the gutters and downspouts.

Siding and Trim

Few people are aware of just how dirty their siding and trim can get in a year. Weather, dust, soil, and corrosion all conspire to destroy that "new house look" and

insidiously damage outer surfaces, causing these materials to wear out far sooner than expected. It's hard to tell, however, because everything gets dirty at the same rate.

To keep your siding and trim in good shape, choose a nice summer day, hop into your bathing suit, get your garden hose and a scrub brush attached to an extension handle, fill a pail with mild detergent water and go to town. Depending on your house type, you might also need a long ladder.

Doors and Windows

Here, as with many parts of a modern house, maintenance has radically changed for the better from what it was years ago—an unending scenario of scraping, painting, varnishing, de-warping, unsticking wooden doors and window sashes, sills, and frames, and playing with rotton ropes and pulleys from heavy window weights. Today's doors and windows barely need wiping with a damp cloth.

Basement windows are the most notable exceptions. You still have to keep their outside wells or excavations free of leaves, weeds, and rotting grass, and make sure water doesn't collect and rise above the window bottoms. In areas where heavy snowfalls occur, consider using plastic covers designed for protecting basement window wells. Periodically inspect these window frames and sills for rust if metal, and decay if wood.

Wood Porches and Decks

If you used high-quality, pressure-treated wood, you're practically home free. It's a good idea to reseal it once per year, and beyond that, just watch out for and correct cracked boards, or boards that have come loose from walls, foundations, or steps. If you haven't used pressure-treated wood, stock up on paint brushes and lots of paint or varnish. And keep some funds available to replace rotting boards as needed.

Mildew

Mildew is simply a mold, and mold is a fungus of some type. Mold spores are practically everywhere, borne in the air, and need only a cool or warm moist environment to flourish in. Mildew is not only irritating, it's harmful. It can ruin furniture, furnishings, walls, ceilings, and can even present a health problem to the occupants. It's especially persistent in leaky, damp basements, but will readily occur during periods of high humidity, when water vapor present in warm air condenses on the cool surfaces of walls, floors, pipes, and other items that can provide perfect breeding grounds for fungi.

To combat mildew, merely increase the air circulation. Don't place rubber-backed carpets on basement floors. Keep furniture away from walls where air circulation could be blocked, and operate dehumidifiers around the clock during humid summer months. Small fans will help in cramped locations. Mildew in an attic is a sure sign that ventilation there should be improved.

Asphalt

This material might seem tough, and it is . . . kind of. It'll support your car nicely, and even a large boat and trailer, but it can gradually be ruined by a combination of frost damage, oil and gasoline residues, and road salt drippings.

Whenever asphalt loses its original dark black sheen and begins drying out, developing small cracks and fissures, and starts absorbing oil and gasoline that spills onto its surface, it's time to reseal the surface with a waterproof coating. The typical driveway doesn't receive very hard use, and a good sealer applied correctly can last three or four years.

Concrete and Masonry

Concrete and masonry surfaces are substantially more durable than asphalt *if* cracks are repaired when they occur, and not allowed to go untouched through alternate freezing and thawing temperatures.

Mortar joints in brickwork are likewise susceptible to the elements, and they should be inspected closely, especially if any bricks are in direct contact with the ground.

Concrete surfaces will clean up nicely with a stiff scrub brush and a solution of about a half cup of trisodium phosphate mixed with a gallon of warm water. Masonry surfaces can also be cleaned with this handy chemical, but with a reduced ratio of about 1 tablespoon per gallon of water.

A MAINTENANCE CHECKLIST

To give you an idea of how to compile your own maintenance checklists, here's a sample schedule of a year's worth of inspections and duties organized by seasons. It was prepared for someone living in a moderately cold-climate area. You should create your own checklists, based on where you live, the type of house you have, and the equipment that's in it.

Spring

Time to check for winter damage, mostly from ice, water, and dampness. It is also time to keep up with the lawnwork and prepare for the air-conditioning season.

—Inspect roof and flashing
—Clean gutters and downspouts
—Inspect attic for leaks
—Renail or caulk loose siding or trim
—Unstick windows, put up screens
—Vacuum insides of central air-conditioning equipment
—Clear drainage tubes in air-conditioning units
—Drain some water from hot water tank
—Clean out basement window wells
—Check sump pump
—Test floor drains
—Repair any masonry or concrete cracks
—Resurface asphalt driveway, patch holes and cracks
—Check septic system
—Check trees for large broken branches to be removed
—Fertilize lawn and kill weeds
—Trim shrubs away from outside air-conditioning units

Summer

Summer means garden and yard work and some leisurely outdoors cleaning and polishing.

—Check roof and flashing
—Clean siding and trim
—Clean windows and doors
—Clean kitchen exhaust motor and fan assembly
—Inspect crawl spaces for rot and decay
—Clean or replace air-conditioning filters once per month
—Control mildew
—Fertilize lawn
—Inspect house and yard for insects, grubs, moles and other pests

Fall

Fall is the busiest house maintenance season. It's time to prepare for the ravages of winter. Time to get the heating systems ready and to batten down the hatches.

—Inspect roof and flashing

—Clean gutters and downspouts
—Inspect siding, window, door, and trim caulking
—Install storm windows
—Perform any needed exterior paint touchup
—Seal cracks in porches and decks
—Check foundations, walls, and brickwork for cracks
—Service heating unit and thermostat
—Inspect and clean fireplaces and stoves
—Winterize air-conditioning system
—Test circuit breakers
—Drain some water from the water heater
—Check sump pump
—Clean clothes dryer vent
—Clean out basement window wells
—Install plastic covers over basement window wells
—Test floor drains
—Check fire extinguishers and smoke alarms
—Reseal asphalt driveway if needed
—Fertilize lawn and reseed parts if needed
—Protect delicate shrubs and plants by covering before winter

Winter

In winter you can inspect the outside of your house, but in cold-climate areas, there's not much you'll be able to do in harsh winter weather. Make necessary interior repairs, service your appliances, keep up with the heating system, and be on the lookout for energy-wasting drafts and thermal leaks.

—Keep large ice buildups from gutters and downspouts
—Inspect attic for leaks
—Inspect fireplace and chimney flues
—Try circuit breakers and check electrical cords and outlets
—Clean or replace furnace filters once per month
—Inspect doors, windows, and electrical sockets for drafts
—Vacuum refrigerator condensers
—Inspect indoor caulking around sinks, showers, and tubs
—Inspect and clean major appliances
—Make any needed interior touchups
—Prune trees in yard

Remember that a well-maintained house does not come naturally with most of us. It might be a pain, but it's the best defense against the toll otherwise taken by the elements, nature, and time. After you make up your schedule, try to follow it. Concentrate on exterior repairs and maintenance during the nice weather, then go inside during the colder months. Naturally, a new house won't require much attention, especially at first. And you can keep it that way with regular preventive maintenance schedules and by following through with repairs.

Chapter 32

The Final Inspection

A completed house is the complex result of thousands of separate pieces and parts put together by many craftsmen—craftsmen who have their good days and bad. The materials, too, are not always perfect. Consequently, with this or any effort resulting from sundry components and multiple construction steps (an automobile is another example), there are likely to be errors and imperfections in the final product—"bugs" that must be discovered and taken care of. In addition, a house will go through an initial settling and shrinking period, and as it does, plaster can crack, windows can stick, wood steps could start squeaking, or the plumbing could spring a leak.

Before everyone gets together at your closing—when the attorneys, the contractor, and yourself all sign the final papers so you can receive the deed and house keys, and the contractor his remaining payments—it's a good idea that you schedule what will be the first step of the final inspection of the property. The first step is a meeting at the site between yourself and the contractor or foreman in charge of the construction. Schedule the meeting at least a week in advance (and don't forget to *remind* the builder a day or two in advance) and ask him to bring along any warranties for major appliances or materials such as a water heater, refrigerator,

range, or garage door opener. You should bring any warranties you've already received, and the maintenance file you've been putting together.

At this meeting, resolve any questions you have already jotted down on previous visits, and determine as best as you can that what was specified in the contract was actually performed. At the same time, inspect the property for obvious flaws. Ask the contractor what imperfections you can reasonably expect, and get his advice on what to do about them if they actually occur. Who do you report the problems to? How would the most likely problems be solved?

The initial and continuing parts of the final inspection are important because *you* are expected to be the moving force to finding and reporting and making sure steps are taken to correct any imperfections that develop. The lender won't be worried about minor, irritating defects. All he expects is to receive your monthly mortgage check—usually by mail. And even if something major goes wrong, your payments can't be stopped. Rather, they continue, like clockwork. The builder, too, can't fix what he doesn't know about. And you can't tell him about problems you don't recognize.

Fortunately, if you've followed the advice presented in the chapters on selecting and working with your contractor, problems that arise will likely be handled quickly and professionally. This is not so likely if you have elected or have been unsuspectedly duped into going with an inept or vanishing builder.

Before the meeting is over, review the warranties that were supplied at the time the building contract was signed, and check to make sure you have copies of all the other applicable warranties. Be certain you know who all of the subcontractors were and exactly who you should call for maintenance and repairs at a later date. Go through your maintenance file with the contractor. Ask him if there's anything else he would add to it, or if he sees any part of it that doesn't look quite right. Find out if there's anything special you should do to "break in" certain components of the house.

Maybe during the initial inspection you'll discover that the builder has forgotten to include something promised. It happens. Or maybe the builder is planning to finish a few items that were not completed in time because of scheduling or weather delays. You want to make sure you'll receive what you contracted for; nothing less. If anything has not been completed, find out

what the completion schedule is, and for all but the most minor detail, have the attorneys keep a portion of the builder's final payment in escrow—to be released when the remaining work has been completed.

Make sure you know how to operate all of the appliances, including heating and cooling units, sump pumps, and well pumps. Know also where a septic tank is located, or where the sewer connects to the house.

So you take an hour or two to go through the house with the builder. That's the first step of the final inspection. The rest of the steps are inspections that you'll be making alone, especially during the first year that you live in the place, or until the builder's warranty period is up, whether it's a single year or longer.

During that year, keep a sharp eye and ear out for the development of small flaws that could become serious problems if not taken care of. Don't be afraid to contact the builder later on, either. Remember, he wants you to be satisfied. It's in his best interest to have you say nice things about him so you can be counted on as a future business reference.

But really, if you've followed a good portion of the

Fig. 32-1. Enjoying the home.

proven specifications and ideas in this book, there shouldn't be much need for the builder to come back and redo things. You're likely to have a high-quality home, one that will suit your needs, will hold its value, will require a minimum of upkeep, and will provide you with maximum living comfort, convenience, safety, and privacy.

It's more likely that, after the final checkout period of one year, you can spend most of your free time enjoying your home, knowing you've received the most value for your dollars and efforts, and that you're far ahead of the average person who will just take what's available, or what marginal or vanishing builders offer in floorplan, construction quality, and value (Fig. 32-1).

As a parting note, after you home is checked out fully, about a year after the day you move in, send to the builder—as long as he has taken care of you—a brief note, letter, or call, expressing your appreciation, telling him how much you are enjoying the house, if that's the case.

You might even want to send him a case of beer, a fruit basket, a dinner for two at a classy restaurant, or something you know would be appropriate and safe. You'd be surprised at how few people will *ever* call a builder with something nice to say. Instead, builders, even the best builders, hear mostly brickbats and complaints.

Surprise your contractor with a token of appreciation. People in the service industries don't usually get them, no matter how good their services are (especially after the services have been transacted). Such a gesture will keep you in that builder's mind for a long, long time, will likely cement a friendship, and anyway, if he's done a good job for you, it's the right thing to do.

It's just another way for you to lift up the entire housebuilding process from the routine and cap off the entire experience with style.

Appendices

Appendix A

Electrical Usage Chart

Appliance	Average Wattage	Approximate monthly kilowatt-hour use	
air conditioner (window 8000 BTU)	1,067	207	peak season
bed covering	177	37	peak season
broiler	1,430	9	
clock	2	2	
clothes dryer (8 loads/week)	4,856	80	
coffee maker	894	9	
crock pot	60	10	
deep fat fryer	1,448	7	
dehumidifier	460	138	peak season
dishwasher	1,201	30	
electric air cleaner	100	70	
fan (attic)	370	70	peak season
fan (furnace)	292	130	peak season
fan (window)	200	40	peak season
food blender	386	1	
food freezer			
15 cu. ft. standard	341	100	
15 cu. ft. frostless	440	150	

Appliance	Average Wattage	Approximate monthly kilowatt-hour use
20 cu. ft. standard	494	115
20 cu. ft. frostless	575	180
food waste disposer	445	3
frying pan	1,196	16
hair dryer	1,000$T8	
heat lamp or sun lamp	250	1
heating pad	65	1
heat tape	70	35 peak season
(10 foot length)		
hot plate	1,257	8
humidifier	117	20
iron	1,088	10
light bulb (incandescent)		
100 watts (on 5 hrs.)	100	15
light bulb (fluorescent)		
(including ballast)		
20 watts (on 5 hrs.)	25	4
40 watts (on 5 hrs.)	50	8
microwave oven	1,450	25
oil burner	266	100 peak season
portable heater (on 3 hrs.)	1,350	122 peak season
radio	71	7
radio-phonograph	109	9
range	12,200	100
refrigerator		
12 cu. ft. standard	241	60
12 cu. ft. frostless	321	100
refrigerator/freezer		
14 cu. ft. standard	326	100
14 cu. ft. frostless	615	150
16 cu. ft. standard	516	130
16 cu. ft. frostless	638	165
21 cu. ft. frostless	736	185
roaster	1,333	17
television (7 hrs./day)		
B&W tube	160	30
B&W solid state	55	12
color tube	300	60
color solid state	200	40
toaster	1,146	4
toaster/oven	1,425	30
waffle iron	1,160	3
washer, automatic		
(6 loads per week)	512	10
water heater	2,500	400
water pump	460	20
water bed (on 10 hrs./day)	360	108

Appendix B

Burglar-Proofing Your Home

It's true that crimes of violence, though on the rise, occur much less frequently than crimes against property. But it's also true that crimes against property, such as burglary, can easily turn into crimes of violence if an intruder is discovered and confronted.

Most thieves, burglars, and housebreakers are amateurs. Sure, on television shows they sand their fingertips, pick combination locks, scale buildings with grappling hooks, dress in color-coded outfits, and perform near-perfect crimes; but in real life they're not that classy. Instead, most are opportunists searching for unlocked doors, open windows, and unattended pocketbooks.

According to law enforcement experts, these amateur burglars and housebreakers worry about three things:

- Delay in getting into your home.
- Being forced to work where they can be observed.
- Having to make noise in the process.

In other words, they don't want to cause a commotion. They don't want to get caught. That's why they'll pass by a house that has good door and window locks and other characteristics that will delay or announce their entry.

DOOR LOCKS

Older homes and apartments were usually built solid, with doors of thick oak—heavy, sturdy, impossible to kick in. Unfortunately their locks were often flimsy by comparison.

Many were mortised locks that were recessed into a hole or slot cut from the door's outer edge near the knob. Sometimes their locking bolt fit securely into its door frame receptacle, other times not. If the door was "hung loose" or wobbly, often only 1/8 inch of bolt kept the door locked.

Most of the remaining old-fashioned locks had keyholes, the kind you can look through. They could be operated with skeleton keys, the same keys that sell for under a dollar per set in any hardware store. Not too safe.

Because skeleton locks are so easy to defeat (to open through picking or simply using other skeleton keys), many residents installed chain units to supplement their protection. A chain lock is merely a short chain permanently fastened to the door or door frame that, when attached to its receiving fixture, prevents the door from being opened more than a few inches. Again, not safe. The chains afford little safety from a determined intruder. All he has to do is kick or smash the door with enough force to rip loose the screws holding the chain's receptacle to the door or frame.

Since the days of keyhole locks, safer locks have been designed and put into use . . . some of the time. Other, unsafe locks have also come onto the market, and are being frequently used in new construction because they're inexpensive to produce and easy to install.

These relatively new locks are called key-in-knob locks. To open the door from the outside you insert a key directly into a hole in the knob, and turn. To lock it when leaving, you simply push a button on the inside knob, or push the inside knob and turn, then slam the door shut. The locking bolt is spring operated.

A key-in-knob lock is the easiest modern lock to thwart. Unless anti-defeating features are built into the lock, it can be broken into in several ways:

1. By force. Most of these locks are constructed of flimsy materials. By placing a piece of wide-mouth pipe over the outside door handle, the knob can be snapped off. And because its locking bolt is always short and beveled, a crowbar or similar tool can often pry the door open from its frame.

2. By guile. The simplest way to defeat a key-in-knob lock is to loid it. To loid it? Yes, loiding is a new word, invented in honor of the key-in-knob lock. You probably won't find it in your dictionary, but it comes from the word "celluloid." It means to open a lock by inserting a thin plastic strip (like a credit card) between the bolt and jamb so the plastic strip releases a spring-operated catch.

If you are not yet convinced that key-in-knob locks, by themselves, are unsafe, inquire at your local police department.

From the worst, let's jump to the best: a single cylinder deadbolt, operated by a key from outside and a thumb-latch inside. It throws a 1-inch rectangular (not beveled) bolt into its receptacle, and if installed properly, cannot be pried or loided.

There is a hitch, however: if any glass is situated in or near the door someone could break it, reach in, and turn the thumb latch from inside. In this case, two options are available:

1. Replace the glass.
2. Replace the single cylinder lock with a double-cylinder model—a lock operated by key from inside and out.

The first option is the safest. That's what most locksmiths and law enforcement agencies recommend. A lock that requires a key on the inside could be hazardous in case of a fire or other emergency.

Doors already equipped with a key-in-knob lock can be made safer by having good single-cylinder deadbolts installed above the key-in-knob lock.

DOOR HINGES

Door hinge pins should be located on the *inside* of all doors that lead outside your home; otherwise an intruder could gain entrance by prying up the pins with a screwdriver.

DOOR PEEPHOLES

As mentioned in the chapter on doors, peepholes should be installed in doors to the outside that afford no other view of callers. Plastic lenses will fog, scratch, and deteriorate in a short while. Quality glass peepholes, with a viewing field of 180 degrees, will last a lifetime.

SLIDING DOORS

Sliding glass doors are usually situated in a family room or kitchen dining area, facing the rear of the property. That's great for housebreakers. Remember, they like to work in dark places where no one can watch them.

Although most sliding doors consist of two glass panels, usually only one of them is permanently fastened. To safeproof sliders you must—if the window/door manufacturer has not—prevent an intruder from sliding the movable door open by force, and prevent him from prying the movable door up and out of its track.

The sliding glass panel can be secured with a good lock that is key-operated from the outside. A "Charley Bar," a device that prevents a housebreaker from prying the panels off their tracks, may be purchased commercially and should be used.

Probably the most common way to prevent someone from opening the movable panel by force is to place a long wooden dowel or piece of broom stick in the bottom track.

WINDOW LOCKS

If a determined housebreaker can't negotiate your exterior doors, chances are he'll turn his attention to your windows.

The windows most vulnerable in an average house are the basement windows. Secure them by installing decorator wrought-iron bars, or discreet yet sturdy grates. Both are available in many hardware and locksmith stores.

Other windows should be locked or fastened in fixed positions, either closed all the way or opened a few inches for air in warm weather.

Double-Hung Windows

These are probably the most commonly used today. They operate upward and downward, and lock with a simple metal latch. Approach a locked double-hung window from the outside. Try sticking a knife blade up the crack and see if you can jiggle the latch with it. An experienced housebreaker can open all but the most tightly fitting windows in this fashion. Older windows often suffer from dry rot around their latches, and the screws that fasten the latch to the wood can easily be pulled out.

You can secure double-hung windows in this way: When the window is closed, drill a neat hole through the top piece of the bottom window sash, at one side, all the way through and into the bottom part of the top sash. Then insert a metal bolt or pin. Windows fixed this way are impossible to open from the outside. Make sure the pin or bolt protrudes slightly from its hole so you could remove it easily in case of an emergency. In the summer, when you want fresh air, drill a few more holes into the top sash several inches above the first hole. The window can then be opened slightly yet remain securely fastened.

Casement Windows

These windows operate by means of a crank near their bottom. A latch locks the window to a center post. For additional safety you can remove the crank handle.

Sliding Windows

These windows can be secured the same way as sliding glass doors.

Louvered Windows

Any way you look at them, from security standpoints, they aren't very safe. Commercial locking devices are available.

Give extra attention to windows that lead to a fire escape, patio, or deck. And don't forget about second-story windows, especially if either you or your neighbor stores a ladder out behind a garage. Prune large tree limbs away from the house and don't let thick bushes encroach on your windows.

Index

Edited by Cherie R. Blazer

Other Bestsellers From TAB

☐ **MAJOR HOME APPLIANCES: A Common Sense Repair Manual—Rains**

Prolong the life and efficiency of your major appliances . . . save hundreds of dollars in appliance servicing and repair costs . . . eliminate the inconvenience of having an appliance quit just when you need it most *and* the frustration of having to wait days, even weeks, until you can get a serviceman in to repair it! With the help and advice of service professional Darell L. Rains, even the most inexperienced home handyman can easily keep any major appliances working at top efficiency year after year. 160 pp., 387 illus., (7" × 10").

Paper $14.95   **Hard $21.95**
Book No. 2747

☐ **THE COMPLETE BOOK OF BATHROOMS—Ramsey and Self**

Simple redecorating tricks . . . remodeling advice . . . plumbing techniques . . . it's all here. Find literally hundreds of photographs, drawings, and floorplans to help you decide exactly what kind of remodeling project you'd like to undertake; plus, step-by-step directions for accomplishing your remodeling goals. It's all designed to save you time and money on your bathroom renovations! 368 pp., 474 illus. 7" × 10".

Paper $15.95   **Hard $24.95**
Book No. 2708

☐ **ROOFING THE RIGHT WAY—A Step-By-Step Guide for the Homeowner—Bolt**

If you're faced with having to replace your roof because of hidden leaks, torn or missing shingles, or simply worn roofing that makes your whole house look shabby and run down . . . don't assume that you'll have to take out another mortgage to pay for the project. The fact is, *almost anyone can install a new or replacement roof easily and at amazingly low cost compared with professional contractor prices!* All the professional techniques and step-by-step guidance you'll need is here in this complete new roofing manual written by an experienced roofing contractor. 192 pp., 217 illus. (7" × 10").

Paper $11.95   **Hard $19.95**
Book No. 2667

☐ **PLANNING AND BUILDING FENCES AND GATES**

This colorfully illustrated guide gives you all the expert, step-by-step guidelines and instructions you need to plan and build durable, cost-effective fences and gates. You will be able to design and construct just about any kind of fence you can think of—barbed wire, woven wire, cable wire, mesh wire, board fences, electric fences, gates, and much more! 192 pp., 356 illus. 8 1/2" × 11". 2-Color Throughout.

Paper $14.95   **Hard $22.95**
Book No. 2643

☐ **THE BUILDING PLAN BOOK: Complete Plans for 21 Affordable Homes—Ernie Bryant**

Here, in one impressive, well-illustrated volume, are complete building plans for a total of 21 custom-designed homes offering a full range of styles and features—efficiency dwellings, ranches, capes, two-story homes, split-levels, even duplexes. It's a collection of practical, good looking home designs that not only offer comfort, convenience, and charm but can also be built at a reasonable cost. 352 pp., 316 illus., 8 1/2" × 11"

Paper $14.95   **Hard $24.95**
Book No. 2714

☐ **THE GARDENING IDEA BOOK**

Whether you have space for a full-space garden or only a pocket size back yard, this exciting collection of articles from *Farmstead Magazine* shows how you can grow all kinds of delicious, healthful fruits and vegetables. Here's expert advice and guidance that's guaranteed to make your garden more productive, easier to take care of, and less expensive! 208 pp., illustrated.

Paper $10.95   **Hard $15.95**
Book No. 2684

☐ **ALL ABOUT LAMPS; CONSTRUCTION, REPAIR AND RESTORATION—Coggins**

You'll find step-by-step directions for making a wall lamp or a hanging lamp from wood, novelty lamps from PVC plumbing pipe, and designer lamps from acrylic or polyester resins. Shade projects range from needlepoint and fabric models to globes, balls, and tubular forms. There are suggestions for advanced projects, using salvaged and low-cost materials, and more! 192 pp., 196 illus. 7" × 10".

Paper $16.95   **Hard $24.95**
Book No. 2658

☐ **101 PROJECTS, PLANS AND IDEAS FOR THE HIGH-TECH HOUSEHOLD**

If you're looking for decorative effects, you'll be impressed with the number of projects that have been included. And electronics hobbyists will be amazed at the array of projects—all of them with clear building instructions, schematics, and construction drawings. You'll also find exciting ways to use your microcomputer as a key decorative element in your high-tech atmosphere. 352 pp., 176 illus. 7" × 10"

Paper $16.95   **Hard $24.95**
Book No. 2642

Other Bestsellers From TAB

□ **UPHOLSTERY TECHNIQUES ILLUSTRATED—Gheen**

Here's an easy-to-follow, step-by-step guide to modern upholstery techniques that covers everything from stripping off old covers and padding to restoring and installing new foundations, stuffing, cushions, and covers. All the most up-to-date pro techniques are included along with lots of time- and money-saving "tricks-of-the-trade" not usually shared by professional upholsterers. 352 pp., 549 illus., 7" × 10".
Paper $16.95   **Book No. 2602**

□ **CABINETS AND VANITIES—A BUILDER'S HANDBOOK—Godley**

Here in easy-to-follow, step-by-step detail is everything you need to know to design, build, and install your own customized kitchen cabinets and bathroom vanities and cabinets for a fraction of the price charged by professional cabinetmakers or kitchen remodelers . . . and for less than a third of what you'd spend for the most cheaply made, ready-made cabinets and vanities! 142 pp., 126 illus.,
Paper $12.95   **Book No. 1982**

□ **HARDWOOD FLOORS—INSTALLING, MAINTAINING, AND REPAIRING—Ramsey**

This comprehensive guide includes all the guidance you need to install, restore, maintain, or repair all types of hardwood flooring at costs far below those charged by professional builders and maintenance services. From details on how to select the type of wood floors best suited to your home, to time- and money-saving ways to keep your floors in tip-top condition. 160 pp., 230 illus., 4 pages in full color.
Paper $10.95   **Hard $18.95**
Book No. 1928

□ **DO YOUR OWN DRYWALL—AN ILLUSTRATED GUIDE**

Proper installation of interior plaster board or drywall is a must-have skill for successful home building or remodeling. Now, there's a new time- and money-saving alternative: this excellent step-by-step guide to achieving professional-quality drywalling results, the first time and *every* time! Even joint finishing, the drywalling step most dreaded by do-it-yourselfers, is a snap when you know what you're doing. 160 pp., 161 illus.
Paper $10.95   **Book No. 1838**

□ **TILE FLOORS—INSTALLING, MAINTAINING AND REPAIRING—Ramsey**

Now you can easily install resilient or traditional hard tiles on both walls and floors. Find out how to buy quality resilient floor products at reasonable cost . . . and discover the types and sizes of hard tiles available. Get step-by-step instructions for laying out the floor, selecting needed tools and adhesives, cutting tiles, applying adhesives, and more. 192 pp., 200 illus., 4 pages in full color. 7" × 10".
Paper $12.95   **Hard $22.95**
Book No. 1998

□ **BUILDING OUTDOOR PLAYTHINGS FOR KIDS, WITH PROJECT PLANS**

Imagine the delight of your youngsters—children or grandchildren—when you build them their own special backyard play area! Best of all, discover how you can make exciting, custom-designed play equipment at a fraction of the cost of ordinary, ready-made swing sets or sandbox units! It's all here in this step-by-step guide to planning and building safe, sturdy outdoor play equipment! 240 pp., 213 illus. 7" × 10".
Paper $12.95   **Hard $21.95**
Book No. 1971

□ **DO-IT-YOURSELF DESIGNER WINDOWS—Boyle**

If the cost of custom-made draperies puts you in a state of shock . . . if you don't know what to do with a problem window or what type of window decor would look right in your home . . . here's all the advice and information you've been searching for. It's a complete, hands-on guide to selecting, measuring, making, and installing just about any type of window treatment imaginable. 272 pp., 414 illus.7" × 10".
Paper $14.95   **Hard $21.95**
Book No. 1922

□ **PRACTICAL LANDSCAPING AND LAWN CARE—Webb**

Here's absolutely everything you need to successfully plan, plant, and maintain lawn grasses and groundcovers, vines and flowering ornamentals, annual, biennial, and perennial flowers, shade trees, lawn trees, even decorative (and delicious) fruits and berries. Packed with tips from a master gardener who has perfected the art of easy, good looking, low-maintenance landscaping! 240 pp., 84 illus. 7" × 10".
Paper $13.95   **Book No. 1818**

Other Bestsellers From TAB